Metastable Liquids

Metastable Liquids

Concepts and Principles

PABLO G. DEBENEDETTI

PRINCETON UNIVERSITY PRESS · PRINCETON, NEW JERSEY

Library of Congress Cataloging-in-Publication Data
Debenedetti, Pablo G., 1953–
    Metastable liquids: concepts and principles / Pablo G. Debenedetti.
    p.   cm. — (Physical chemistry)
    Includes bibliographical references and index.
    ISBN 0-691-08595-1 (cloth: alk. paper)
    1. Supercooled liquids.   2. Liquids—Thermal properties.   3. Phase
transformations (Statistical physics)   4. Chemistry, Physical and
theoretical.   I. Title.   II. Series: Physical chemistry (Princeton, N.J.)
QC145.48.S9D43   1996
530.4′2—dc20   96-18027

This book has been composed in Times Roman

Princeton University Press books are printed on
acid-free paper, and meet the guidelines for permanence
and durability of the Committee on Production
Guidelines for Book Longevity of the
Council on Library Resources

Printed in the United States of America
by Princeton Academic Press

1   3   5   7   9   10   8   6   4   2

To Silvia

We boil at different degrees.

Ralph Waldo Emerson
Society and Solitude (1870)

The end of our foundation is the
knowledge of causes, and secret motions of
things; and the enlarging of the bounds of
human Empire, to the effecting of all
things possible.

Francis Bacon
New Atlantis (1627)

Contents

3. Kinetics   147

4. Supercooled Liquids   235

5. Outlook   363

Preface and Acknowledgments

Liquids are one of the three principal states of matter. They usually exist and perform their varied functions under conditions of stability with respect to boiling or freezing. Familiar examples include oil, blood, wine, amniotic fluid, the oceans. Liquids can also exist under conditions where the stable state is either a solid, a vapor, or a liquid of different composition. When this occurs, a liquid is said to be metastable with respect to the stable phase; metastability with respect to vapor and crystalline phases defines superheated and supercooled liquids, respectively. Metastable liquids can be found in industrial processing and manufacturing plants, clouds, trees, minerals, insects, and fishes. They can be used to prolong the shelf life of labile biochemicals, but can also cause industrial explosions. Understanding the behavior and properties of metastable liquids, therefore, is important technically and scientifically.

In this book I discuss the basic principles of metastability in liquids. The major topics covered include rigorous and approximate ways of calculating the thermodynamic properties of metastable liquids; the determination of limits of stability; the mechanisms and rates of phase transitions undergone by metastable and unstable liquids; and supercooled liquids and the glass transition. There is a rich literature on each of these topics. However, a comprehensive viewpoint is lacking. The goal of this book is to provide this comprehensive viewpoint, taking into account established topics as well as recent developments.

More than twenty years ago, Skripov's *Metastable Liquids* presented a masterful synthesis of knowledge about superheated liquids. Since then, interest in supercooled liquids in general, and in supercooled water in particular, has grown enormously. Important breakthroughs in the theory of nucleation, such as the use of density functionals and the formulation of fully kinetic theories, have occurred in this period, which has also seen the emergence of computer simulations as an indispensable technique for the study of liquids. These and other modern developments are covered here.

In order to calculate the thermophysical properties of a metastable liquid rigorously, it is necessary to eliminate the stable phase from consideration. The system is thus studied under the action of a constraint that prevents it from sampling more stable configurations. An important theoretical goal is to determine limits of stability beyond which the constrained metastable liquid is unstable to infinitesimal perturbations. In the laboratory, on the other hand, stability is a kinetic concept: a metastable liquid can only be studied over times that are short compared to its lifetime. When the latter becomes so short that measurements can no longer be performed, the liquid has reached a practical limit of stability. Absolute limits of stability are determined by thermodynamics; practical limits of stability are determined by kinetics. The relationship between these two types of limits, and more generally the extent to which kinetics and thermodynamics determine the behavior of metastable liquids, is the basic theme that I address in this book.

I have attempted to illustrate the ubiquity of metastable liquids with examples ranging from ozone depletion to the behavior of proteins at low temperature, and from the prevention of vapor explosions to the development of new technologies for prolonging the shelf life of labile biochemicals. I hope that these examples will give the reader an appreciation for the natural importance and technical potential of metastable liquids.

The sheer breadth of the subject has forced me to be selective in the choice of topics. Numerous examples of experimental techniques are provided, but I have not attempted a systematic discussion. Though many of the concepts presented here apply to polymers too, the book deals exclusively with nonpolymeric liquids. Furthermore, the glass transition is discussed "from the liquid side" only: the physical properties of glasses are not treated except when they are relevant to understanding supercooled liquids.

Each main chapter treats one major topic. The thermodynamics of metastable liquids is treated in Chapter 2, which addresses the calculation of equilibrium properties and limits of stability. These are straightforward problems from the perspective of continuum thermodynamics, but not from a microscopic, statistical mechanical viewpoint. The usefulness and limitations of the continuum and microscopic approaches are also discussed in Chapter 2. The mechanisms by which metastable liquids evolve towards stable equilibrium, and the corresponding rates, are discussed in Chapter 3. Although questions of rates and mechanisms are the proper province of kinetics, thermodynamic considerations play an important role here too. In the first place, few rate theories are truly kinetic. Most require knowledge of the reversible work needed to form a small embryo of the stable phase, a thermodynamic problem. From a purely kinetic perspective, furthermore, there is no clear distinction between unstable and metastable states, and this is incorrect. The reconciliation between the thermodynamic and kinetic viewpoints is also treated in Chapter 3. Supercooled liquids are the subject of Chapter 4. At sufficiently low temperatures, supercooled liquids can fall out of equilibrium and become structurally arrested when the experimental and molecular relaxation times overlap. This process is known as the glass transition; it is a distinguishing feature of low-temperature liquid metastability. Both kinetic and thermodynamic theories of the glass transition are treated in Chapter 4. Also included is a discussion of supercooled and glassy water, with emphasis on the various theoretical interpretations that have been proposed to explain the unusual properties of supercooled water. Detailed proofs and derivations on the thermodynamic theory of stability, and on the thermodynamics of fluid interfaces, are given in Appendixes 1 and 2. Basic statistical mechanical definitions can be found in Appendix 3.

This book could not have been written without the help and support provided by many people. My former advisor and teacher at the Massachusetts Institute of Technology, Robert Reid, gave me an appreciation for the logical structure of thermodynamics, taught me how to think rigorously about the subject, and introduced me to metastable liquids. His knowledge, scholarship, and integrity are a continuing source of inspiration. John Prausnitz suggested to me the idea of writing a book on metastable liquids. With no appreciation for the scope of the task ahead, I accepted. The book that resulted some six years later is very different from the one I envisioned originally. Writing entails much more reading and critical rethinking than I first imagined. I am grateful to John for his confidence in me, and for convincing me to undertake a project from which I have learned so much. I am indebted to Austen Angell, Felix Franks, Howard Reiss, and Robin Speedy for all that I have learned from studying their work. I hope that my writing reflects adequately their influence on me.

Many colleagues read parts of the manuscript and provided valuable criticism and suggestions. I am grateful to all, but especially to Farid Abraham, Berni Alder, Michael Burke, Felix Franks, William Graessley, Gary Grest, Sanat Kumar, David Oxtoby, Thanasis Panagiotopoulos, Peter Poole, John Prausnitz, Robert Reid, Howard Reiss, Lynn Russell, Isaac Sanchez, Srikanth Sastry, Francesco Sciortino, Dendy Sloan, Robin Speedy, and David Young. My colleagues at Princeton have provided an environment of friendship and appreciation for scholarship that I treasure. I especially want to mention Roy Jackson, whose command of all aspects of chemical engineering is unmatched; Dudley Saville, scholar and custodian of high intellectual standards; and William Russel, scholar, and patient and supportive chair during the long gestation of this book.

I thank the National Science Foundation, the Department of Energy, and the Camille and Henry Dreyfus Foundation for supporting my work. A one-year sabbatical in Berkeley as a Guggenheim Fellow in 1991–92 allowed me the luxury of uninterrupted reading, thinking, and writing.

Charlene Hoffner typeset the long manuscript with uncommon skill, and never complained about my endless additions and revisions. I thank her for her professionalism and good humor. I am grateful to Tom Agans, William Allen, and Flavio Robles for doing the illustrations. Marlene Snyder provided valuable assistance with library work.

My father taught me the importance of working hard, having clear goals, and thinking rationally. It is a lesson that guides me to this day. My family, Silvia, Gabriel, and Dina, provided the love, support, and understanding without which this work would not have been completed. I now intend to give them the time and attention that they need and deserve.

Pablo Debenedetti
Princeton
October 1995

Metastable Liquids

1

Introduction: Metastable Liquids
in Nature and Technology

Water at atmospheric pressure can exist as a liquid below 0°C, its normal freezing point. When this happens, we say that water is supercooled. It is also
possible to heat liquid water above 100°C, its normal boiling point. We then
say that water is superheated. Care must be taken to prevent supercooled water
from freezing, or superheated water from boiling. Vibrations, suspended impurities, or irregularities on the walls of a container can trigger the appearance
of a new phase. The more liquid water is cooled below its freezing point, or
the more it is heated above its boiling point, the more the precautions needed
to prevent solidification or vaporization, respectively. Thus supercooled and
superheated water exist in a state of precarious equilibrium. By carefully removing suspended or dissolved impurities, and by minimizing contact with
rough bounding surfaces, the range of temperatures above the boiling point,
or below the freezing point, within which the liquid state can be observed, is
expanded. In the case of water at atmospheric pressure, this range extends
from ca. 280°C to ca. −41°C (Apfel, 1972a; Cwilong, 1947; Mossop, 1955)
although in one experiment (Skripov and Pavlov, 1970), water was apparently
superheated to 302°C. Experiments show, however, that a condition is eventually reached in which boiling or freezing can no longer be prevented, and a new
phase appears suddenly. This is because extraneous impurities merely facilitate
what spontaneous molecular motion can also accomplish, albeit in general more
slowly: to drive the system towards a condition of greater stability through the
formation of a new phase.

The possibility of superheating a liquid above its boiling point, or of supercooling it below its freezing point, is not limited to water. All liquids and their
mixtures can be brought to a condition whereby minor external disturbances
trigger the sudden appearance of one or more new phases, including, in the
case of some liquid mixtures, another liquid phase of different composition.
Such a condition of precarious equilibrium is called metastable equilibrium.
Superheated and supercooled liquids are said to be metastable. In this chapter
we show the ubiquity and importance of metastable liquids through selected
examples taken from nature, industry, and the laboratory. In Section 1.1, we
introduce some basic definitions, and we give two examples of laboratory inves-

tigations of superheated and supercooled water. Sections 1.2 and 1.3 illustrate the natural and technological importance of metastable liquids.

1.1 INTRODUCTION

1.1.1 Definitions

When a substance conforms to the shape of its container without necessarily filling it, it is said to be in the liquid state (Rowlinson and Swinton, 1982). The boundaries between the liquid, solid, and gaseous states[1] of a pure substance are shown schematically in Figure 1.1, in a pressure vs. temperature projection (P, T). Lines ab, bc, and bd are loci of two-phase equilibrium. Line ab, the locus of solid-vapor equilibrium, is called the sublimation curve, and the functionality $P(T)$ along this curve is the solid's vapor pressure. Line bc is the locus of vapor-liquid equilibrium, terminating at the critical point c. The functionality $P(T)$ along bc is the liquid's vapor pressure. Line bd, the locus of solid-liquid equilibrium, is the melting curve. Point b, the triple point, represents the single combination of temperature and pressure at which the solid, liquid, and gaseous states of a pure substance coexist.

Transitions between phases along ab, bc, and bd are accompanied by volume and entropy discontinuities. They are said to be first-order phase transitions, because entropy S and volume V are first-order derivatives of the Gibbs energy G,[2]

$$S = -\left(\frac{\partial G}{\partial T}\right)_{P,N}, \quad V = \left(\frac{\partial G}{\partial P}\right)_{T,N}, \tag{1.1}$$

where T is the temperature, P the pressure, and N the number of molecules in the system. Because of the entropy discontinuity, measurable energy effects accompany first-order transitions, for example, the evolution of heat during freezing or condensation. Figure 1.2 is a schematic pressure vs. volume (P, V) projection of the phase diagram of a pure substance. At the critical point, differences between the gaseous and liquid states disappear. In contrast, experimental evidence (e.g., Cannon, 1974; Godwal et al., 1990; Young, 1991), computer simulations (e.g., Belak et al., 1990), and theoretical arguments (Lifshitz and Pitaevskii, 1980) support the idea that the difference between the solid and liquid phases of matter never disappears. In this book, we are interested in first-order transitions between the liquid and gaseous states, and between the solid and liquid states. There is no ambiguity as to the use of the term liquid in the former case. In the latter situation, we use the term liquid with the understanding that, above the critical pressure, the fluid phase in equilibrium

---

[1] We use the terms vapor and gas interchangeably.

[2] In the classification originally proposed by Ehrenfest (1933), the order of the transition is given by the lowest-order derivative of the Gibbs energy that is discontinuous at the transition.

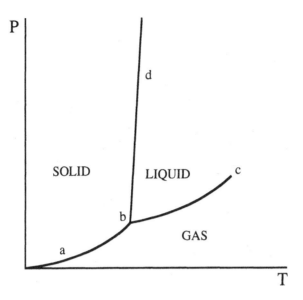

Figure 1.1. Schematic pressure-temperature projection of the phase diagram of a pure substance. *ab*, *bc*, and *bd* are loci of solid-vapor, liquid-vapor, and solid-liquid coexistence, respectively. *c* is the critical point, and *b* is the triple point.

with a solid can be heated at constant pressure indefinitely without undergoing a phase transition.

Returning now to Figure 1.2, if a liquid *e* is heated at constant pressure, it can start to boil at a temperature T_2. At T_2 and P_2, liquid and vapor phases e' and e'' can coexist,[3] and are said to be saturated. However, boiling can be suppressed beyond e', and the liquid can exist as such (j) inside the vapor-liquid coexistence dome $b'cb''$. This dome is the locus of volumes of the vapor and liquid phases that coexist at a given pressure. The suppression of boiling can be done by removing dissolved gases and suspended particles, and by minimizing contact with solid surfaces. Another possible way of penetrating into the coexistence region is by lowering the pressure isothermally along $e'k$. A liquid that is exposed to a pressure lower than its vapor pressure at the given temperature (k), or to a temperature higher than the boiling temperature at the prevailing pressure (j), is called superheated. Just as penetration into the gas-liquid coexistence region defines a superheated liquid, penetration into the liquid-solid coexistence region without solidification defines a supercooled liquid.

As penetration into the coexistence region deepens, ever smaller perturbations can suddenly cause superheated liquids to boil and supercooled liquids to

---

[3]The conditions for phase coexistence are discussed in Chapter 2.

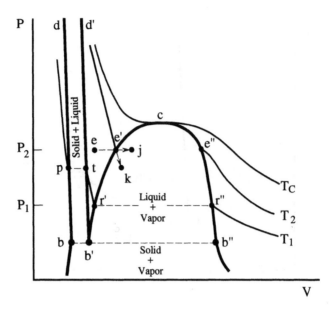

Figure 1.2. Schematic pressure-volume projection of the phase diagram of a pure substance. b, b', and b'' are the coexisting solid, liquid and vapor phases at the triple point. p and t are the coexisting solid and liquid phases, and r' and r'' the coexisting liquid and vapor phases, at a temperature T_1. e is a stable liquid, and e' and e'' the saturated liquid and vapor phases at the same pressure as e, and at a temperature T_2. j and k are metastable, single-phase superheated states obtained by isobaric heating and isothermal decompression, respectively. c is the critical point, and T_c the critical temperature ($T_c > T_2 > T_1$).

freeze. Freezing and boiling can be partial or total, depending upon the external constraints imposed on the system (for example, constant pressure, constant volume). Immediately prior to the phase transition, while the homogeneous liquid still exists inside a coexistence region, it is said to be in metastable equilibrium. Superheated and supercooled liquids are metastable. Metastability is a relative concept: one needs to define with respect to what condition a given system is metastable. This, in turn, requires the specification of external constraints. For example, as shown in Figure 1.3, if a superheated liquid boils, the final state will be a vapor if the temperature and pressure are held constant,[4] or a vapor-liquid mixture if constant volume and temperature are imposed. The rigorous specification of external constraints, as well as the identification of quantitative measures of comparative stability for a given set of constraints, are the province of thermodynamics, and are discussed in Chapter 2.

---

[4]The "constancy" of pressure and temperature applies only to the initial and final states.

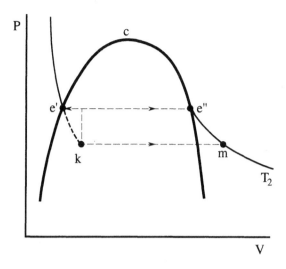

Figure 1.3. Relaxation of a superheated liquid, k (see also Figure 1.2), under different external constraints. If pressure and temperature are imposed, the liquid will boil completely to form a single-phase vapor, m. If total volume and temperature are imposed, the superheated liquid will boil partially to form a mixture of saturated liquid and vapor phases, e' and e''. The region of solid-liquid coexistence is not shown.

1.1.2 Two Experiments: Superheated and Supercooled Water in the Laboratory

Much of what we know about metastable liquids derives from careful experiments on small, high-purity samples. Two such studies are discussed in this section. Figure 1.4 shows a schematic diagram of the apparatus used by Apfel (1972a) to superheat water to 279.5°C at atmospheric pressure. The experiment consisted of isolating carefully purified water droplets from contact with solid surfaces, and heating the droplets as they rose in a denser host liquid. Apfel used benzyl benzoate, whose normal boiling point is 324°C, and in which water is only sparingly soluble, as host liquid. Benzyl benzoate was drawn under vacuum into an inner tube, referred to as the Pyrex tube in the figure. A continuous nitrogen blanket was provided between the inner tube and a concentric external tube (not shown) to prevent oxidation of the host solvent. Water droplets were injected through a hypodermic syringe. A heating wire powered by a variable resistor was wrapped around the Pyrex tube, with the pitch decreasing in the vertical direction. Thus water droplets were exposed to increasing temperatures as they rose. The injected water samples were prepared from distilled, deaerated water, and the loaded syringe was prepressurized to approximately 1,000 bar. This forced residual gas pockets into solution, and insured wetting

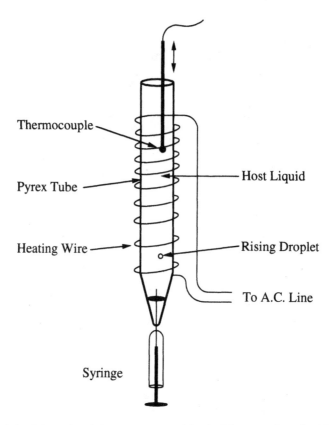

Figure 1.4. Schematic of the apparatus used by Apfel to superheat liquid water to 279.5°C. The liquid under study rises in a heavier, immiscible host liquid and is exposed to vertically increasing temperatures. (Adapted from Apfel, 1972a)

of solid surfaces (Apfel, 1972a). A Chromel-Alumel sliding thermocouple provided accurate temperature measurement inside the host liquid. Apfel observed about one hundred droplets, which ranged in diameter from 0.2 to 0.5 mm. They all rose with velocities of ca. 30 cm sec^{-1}, and boiled explosively when they reached a certain temperature. The highest measured superheat temperature was (279 ± 0.5)°C. Nine droplets exploded between 277 and 279°C, twelve between 270 and 277°C, eleven between 260 and 270°C, and the rest between 240 and 260°C. No correlation between explosion temperature and droplet size was observed.

The technique of droplet superheating in an immiscible host liquid was introduced by Dufour (1861a,b), who heated water droplets in a pool of oil (Avedisian, 1985). The modern implementation of this technique is due to Moore (1956, 1959), and to Wakeshima and Takata (1958). With slight variants, it has been used by several investigators to study many superheated liquids and their mixtures (e.g., Skripov and Kukushkin, 1961; Skripov and Ermakov, 1963, 1964; Ermakov and Skripov, 1967, 1969; Apfel, 1971a,b; Blander et al., 1971; Apfel, 1972a; Eberhart et al., 1975; Porteous and Blander, 1975; Renner et al., 1975; Mori et al., 1976; Holden and Katz, 1978; Patrick-Yeboah and Reid, 1981; Avedisian, 1982; Shepherd and Sturtevant, 1982; Avedisian and Sullivan, 1984). Additional examples are given in Chapters 2 and 3.

Apfel's experiment shows that by eliminating contact with solid surfaces, and by carefully pretreating the sample (deaeration, distillation, pre-pressurization), water can be heated well above its boiling point and still remain a liquid. These precautions serve to eliminate preferential sites for the formation of the new phase. The highly superheated droplets tend to boil explosively, at a temperature that is largely independent of their diameter. This behavior is in fact common to all superheated liquids (e.g., Reid, 1976). Some of the interesting questions prompted by Apfel's experiments include: Can explosive boiling occur in industrial practice? If so, how dangerous is it? Is there an absolute limit to the possible extent of superheating? Why do droplets boil explosively? Can the velocity of this phase transition be meaningfully defined and calculated? How does this velocity depend on the extent of superheating? Why is there no correlation between the temperature at which a droplet explodes and its size? The first two questions are addressed in this chapter; the third in Chapter 2, which deals with the thermodynamics of metastable liquids; the rest in Chapter 3, which deals with the mechanisms and rates of phase transitions.

The idea of facilitating penetration into the coexistence region by isolating droplets from contact with particles and solid surfaces is not unique to investigations of superheated liquids. Surfactant-stabilized emulsions have played an important role in the study of supercooled water (Rasmussen and MacKenzie, 1973; Rasmussen et al., 1973; Kanno et al., 1975; Angell and Kanno, 1976). In this technique, due to Rasmussen and MacKenzie (1972), an emulsion of water in a hydrocarbon is stabilized by a surfactant. Supercooling of the water droplets can then be easily achieved by using a hydrocarbon with a lower freezing temperature than water. In their study of supercooled water under pressure, Kanno et al. (1975) used n-heptane and mixtures of methylcyclohexane and methylcyclopentane as the continuous hydrophobic phase, and sorbitan tristearate as surfactant. Emulsification was achieved by saturating the hydrocarbon with surfactant while heating gently, and then blending water with the hydrophobic phase mechanically. The resulting emulsions had water volume fractions as high as 70%, and droplet diameters in the 1–10 μm range.

Small emulsion samples were pressurized and exposed to liquid nitrogen, leading to a cooling rate of ca. 3°C per minute. The resulting sudden crystallization temperatures of the highly supercooled samples were determined by differential scanning calorimetry, and are shown in Figure 1.5. This important study shows the wide range of temperatures and pressures over which super-

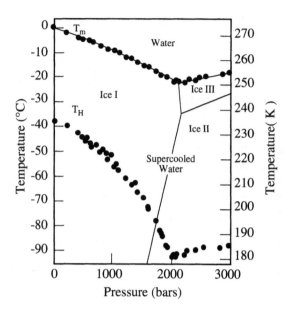

Figure 1.5. Crystallization temperatures of highly supercooled water droplets emulsified in hydrocarbons, as a function of pressure. T_H is the crystallization temperature; T_M is the equilibrium melting temperature. (Adapted from Kanno et al., 1975)

cooled water can exist before crystallizing. The lowest measured crystallization temperature was −92°C, at 2 kbar. Cwilong (1947) observed supercooled water at −41°C in a cloud chamber. Thus water at atmospheric pressure can exist as a liquid over a range that is at least three times wider than the normal liquid range (−41–279 °C vs. 0–100 °C).[5] Much remains to be learned about the properties of liquid water at these extreme conditions. Some of the intriguing questions raised by the study of Kanno et al. (and the chapters where they are addressed) include: Given the widespread occurrence of subfreezing temperatures on earth, is supercooling important for some forms of life (Chapter 1)? Are the crystallization temperatures for highly supercooled water shown in Figure 1.5 absolute limits, or can they be lowered by using emulsions with a smaller droplet size (Chapters 2, 3, and 4)? Why is it important to isolate metastable liquids from foreign particles and solid surfaces in order to achieve substantial penetration into the coexistence region (Chapter 3)?

The experiments of Apfel and of Kanno et al., are based on the possibility of greatly reducing the number of foreign sites that can facilitate the formation of

---

[5]Skripov and Pavlov (1970) reported superheating water up to 302°C using a pulse heating technique. However, the limited number of observations (one) and the comparative difficulty (Avedisian, 1985) in reproducing pulse heating results, or in estimating experimental errors, make this technique less reliable than the droplet superheat method shown in Figure 1.4.

the new phase. The process whereby a new phase is formed on such foreign sites as suspended impurities, or microscopic crevices on the walls of a container, is called heterogeneous nucleation. When the interface between the metastable liquid and the embryonic new phase is formed within the bulk metastable liquid, the process is called homogeneous nucleation. Both topics are discussed in Chapter 3.

The experimental results summarized in Figure 1.5 extend to remarkably low temperatures. Some of the most interesting properties of highly supercooled liquids follow from the fact that molecular motion becomes increasingly sluggish at low temperatures. If crystallization is suppressed, it becomes possible to bring a supercooled liquid to a condition such that spontaneous molecular motion is so slow that different molecular configurations can no longer be sampled on an experimentally accessible time scale. This process of structural arrest is known as vitrification or glass formation.[6] No comparable phenomenon exists for superheated liquids.

1.2 METASTABLE LIQUIDS IN NATURE

1.2.1 Life at Low Temperatures

Liquid water is essential for life. It participates in the four major classes of biological reactions: oxidation, photosynthesis, hydrolysis, and condensation (Franks, 1985). It is the key to the stabilization of biologically significant structures formed by proteins, nucleotides, carbohydrates, and lipids (Franks, 1983). In most animals, blood, an aqueous medium, transports oxygen and nutrients to cells and disposes of carbon dioxide. In plants, a watery fluid, sap, carries water and mineral nutrients upwards from the root system. Cells being roughly 80% water by weight, processes such as protein synthesis occur in aqueous solution.

The temperature at which water freezes at atmospheric pressure, 0°C, occurs widely on earth. Organisms exposed to prolonged subfreezing conditions must develop strategies to deal with the potential freezing of their water inventory. Ice formation inside the cell is injurious and often lethal (Mazur, 1970; Franks, 1985), so the mechanisms of survival at subfreezing conditions fall into two categories: freeze tolerance and freeze avoidance. In the former case, the organism survives extracellular ice formation; in the latter case, freezing is avoided altogether. During partial freezing of extracellular fluids, the concentration of soluble material remaining in solution increases. This creates an osmotic imbalance that causes water to flow out of the cell. As a result, the concentration of solutes in the cell increases. The alteration of the normal levels of solute

---

[6] A glass is a molecularly disordered material that appears not to flow because of its very high viscosity. The formation of glasses from supercooled liquids is discussed in Chapter 4.

concentrations inside the cell is often lethal (Franks, 1985). In addition, water depletion may cause the collapse of the cell membrane (Storey and Storey, 1990). For these reasons, a large number of living systems cope with prolonged exposure to subfreezing temperatures by avoiding ice formation. It is on this type of behavior that we focus our attention.[7]

Freeze avoidance in animals is achieved in two ways: by colligative[8] depression of the freezing point and through the action of antifreeze proteins. Low-molecular-weight polyalcohols and sugars are naturally occurring freezing point depressants, and are commonly referred to as cryoprotectants. Glycerol is the most common freezing point depressant; other examples include sorbitol, ethylene glycol, mannitol, trehalose, fructose, glucose, and ribitol. Colligative cryoprotectants are synthesized by freeze-resistant species in response to cold. Figure 1.6 (Storey and Storey, 1991) shows the seasonal content of glycerol in a freeze-avoiding insect. In many insects exposed to prolonged subfreezing conditions, cryoprotectant synthesis begins in early autumn, and these substances are cleared in early spring (Storey and Storey, 1991). For the case shown in Figure 1.6, the glycerol content reaches 18.7% of the insect's wet weight.

Most freeze-avoiding insects supercool well below their colligatively depressed freezing point. For example, the hemolymph of the beetle *Uloma impressa* melts at $-9.9°$C, but can be supercooled to $-21.5°$C (Franks, 1985). The Antarctic tick *Ixodes uriae* supercools to $-20°$C with only a trace of glycerol (Lee and Baust, 1987). While in many cases insects enter a dormant, quiescent state known as diapause in response to low temperatures, normal activity at very low temperatures is not uncommon: a Himalayan midge, for example, remains active down to $-16°$C (Lee, 1991; Kohshima, 1984). The molecular mechanisms that allow quiescent or active survival with a metastable liquid inventory constantly in danger of freezing are not well understood. In some cases, supercooling is facilitated by a period of starvation prior to exposure to subfreezing conditions: it is believed that food particles provide effective sites for the heterogeneous nucleation of ice (Franks, 1985). In general, though, it is not known whether cryoprotectants inactivate sites for heterogeneous nucleation, or otherwise interfere with the growth of ice nuclei.

Antifreeze proteins are found in cold-water fish and in many insects and small land-based animals. The blood of polar fish, for example, is permanently supercooled: the surface temperature of the Antarctic Ocean is approximately constant at $-1.5°$C (Peixoto and Oort, 1992), one degree below the freezing point of blood. Antifreeze proteins protect Antarctic fish from freezing down to $-2°$C (Franks, 1985; DeVries, 1968). Among land-based animals, antifreeze

---

[7]For detailed discussions of freeze tolerance, see Marchand, 1991; Storey and Storey, 1988; Lee and Denlinger, 1991. For a discussion of cell responses to freezing, see Franks et al., 1983.

[8]The depression of the freezing point of a dilute solution and the increase in its boiling point, with respect to that of the pure solvent, are examples of colligative properties. These are characterized by depending on the number of molecules of solute present, but not on its identity.

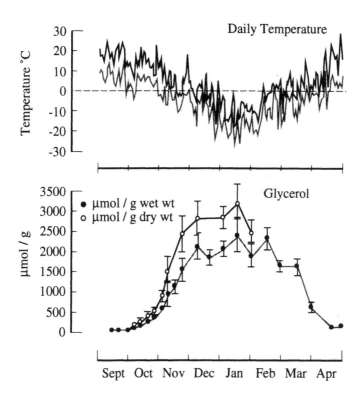

Figure 1.6. Daily maximum and minimum temperatures, and glycerol content in the freeze-avoiding larvae of the goldenrod gall moth. (Adapted from Storey and Storey, 1991)

proteins have been found in beetles, spiders, centipedes, and mites, but not in flies, mosquitoes, wasps, or ants (Duman et al., 1991). Whereas the level of antifreeze proteins is fairly constant in cold-water fish, it varies seasonally in some insects. In general, the amount of supercooling made possible by antifreeze proteins varies between 2 and 10°C.

The colligative freezing point depression due to antifreeze proteins is quite small (Duman, 1982). Furthermore, when the blood of cold-water fish is seeded with ice crystals, the melting point of the seed crystals is considerably higher than the temperature at which ice starts propagating from the seeds (DeVries, 1983). Thus the action of antifreeze proteins is kinetic, not thermodynamic. Chemically, antifreeze proteins can be either glycopeptides or peptides. Glycopeptides were first isolated from the blood of Antarctic fish (DeVries, 1968, 1983), and constitute approximately 3.5% (w/v) of the fish blood. It has been possible to identify a few glycopeptides by electrophoresis. Five consist of an alanyl-alanyl-threonyl repeat unit, with a disaccharide side chain attached via a glycoside linkage to the hydroxyl side chain of each threonine (DeVries et al., 1971). Peptide antifreezes have been isolated from several Arctic fish. Frequently, the repeat units are irregular, roughly 60% of the residues are alanine, and there are no carbohydrate residues (Franks, 1985; Duman, 1982). Several insect antifreeze proteins have been identified and their amino acid composition has been determined (Duman et al., 1991).

A full understanding of the mechanism through with antifreeze proteins inhibit ice formation is not yet available, though significant progress in this direction has been made. The most plausible hypothesis is that inhibition occurs by hydrogen-bond-mediated adsorption of the proteins on growing ice embryos (DeVries, 1983; Zachariasen and Husby, 1982). In vitro studies of solutions of glycopeptide and peptide antifreezes indicate that these substances adsorb on ice (Duman and DeVries, 1972; Raymond and DeVries, 1977; Tomimatsu et al., 1976), and that both the affinity for ice and the antifreeze activity disappear when these molecules are chemically modified (DeVries, 1983). The amino acid sequence of antifreeze proteins isolated from flounder includes the polar cluster threonine-alanine-alanine-aspartate, separated by long sequences of nonpolar alanines. According to DeVries (1983, and references cited therein), these peptides adopt an α-helix secondary structure, in which the polar residues aspartate and threonine are separated by 4.5 Å. This being the same distance that separates adjacent oxygen atoms on the prism faces of ice along the a axes, a plausible mechanism for hydrogen-bond-mediated adsorption is illustrated in Figure 1.7 (DeVries, 1983). In this model, the antifreeze peptide is aligned parallel to an a axis, and hydrogen bonds form between oxygen atoms on the lattice and the polar residues on the peptide. Note that because the polar clusters are separated by nonpolar alanine repeat units, every third row of oxygen does not participate in hydrogen bonding.

Research on the structure and mode of action of antifreeze proteins remains very active (Knight et al., 1993; Chao et al., 1994; Sicheri and Yang, 1995; McDonald et al., 1995). These kinetic inhibitors to ice growth illustrate how metastability arises when constraints that interfere with a system's spontaneous relaxation are present.

The responses of plants to subfreezing temperatures vary widely. Most winter-hardy plants survive by tolerating freezing of extracellular ice. However, extensive supercooling, without ice formation, is also common (George et al., 1982). In woody plants, supercooling is normally restricted to ray parenchima cells (see also Section 1.2.3). These cells function in nutrient storage during winter and lateral transport of nutrients during the summer. Ray cells are killed instantly when they freeze. Differential thermal analysis studies have shown that cell solutions in hickory stem tissue can supercool down to $-40°C$ (Burke et al., 1976). Using the same technique, George et al. (1974) have identified twenty-five species with maximum supercooling ability between -41 and

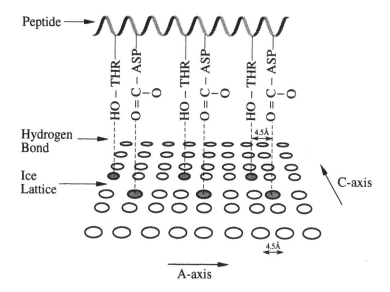

Figure 1.7. Proposed mechanism of antifreeze protein action in flounder. Polar residues in the antifreeze protein (threonine hydroxyl and aspartic acid carbonyl) hydrogen-bond to oxygen atoms in the ice lattice, inhibiting further growth of the ice embryos. (Adapted from DeVries, 1983)

$-47°$C. That such behavior corresponds to supercooling rather than colligative freezing point depression follows from the fact that the equilibrium freezing point of sap[9] is roughly $-2°$C (Franks, 1985). A period of acclimation is required before plants acquire their full capacity for supercooling. The physiological and biochemical changes that occur during acclimation (for example, changes in enzyme activity or in membrane lipid composition) are not fully understood, although it appears that they depend as much on light intensity and duration of exposure to light as on temperature (Franks, 1985). The ability of plants to supercool varies seasonally. Sudden temperature drops during spring or summer (or deliberate chilling of tissue samples) cause nonacclimated plants to freeze at much higher temperatures than during winter.

The vessels through which sap is transported upwards under tension (Section 1.2.3) are empty during winter in many supercooling tree species. In cases where this does not occur, the freezing of sap causes the release of dissolved air because the solubility of air in ice is lower than in liquid water. During spring thawing, bubbles cause loss of hydraulic continuity in xylem[10] chan-

---

[9]Sap, the aqueous fluid that transports nutrients from the root system to the leaves, is conveyed upwards under tension (see also Section 1.2.3).

[10]Xylem is the water-conducting tissue of plants through which sap is transported upwards from

nels, a phenomenon known as embolism[11] in the plant physiology literature (Zimmerman, 1983). Strategies for survival after thawing involve the refilling of bubble-containing vessels by positive root pressure (Zimmerman, 1983).

Partial freezing has been particularly well documented in fruit trees, whose response to low temperature is of great commercial significance. For example, only 5% of the water in peach flower buds supercools to $-30°C$, the bulk of the water freezing at much higher temperatures (Burke et al., 1976). In this case, the fraction that supercools corresponds to living-cell water, while the fraction that freezes corresponds to bark, cortex, and other extracellular water (Franks, 1985). Obviously, some barrier must exist to prevent osmotic water efflux from unfrozen cells to partially frozen extracellular fluid. At present, the nature of this barrier is not known. Strategies for survival after partial or total freezing include the formation of new growth in spring in ring-porous species (e.g., oaks, ashes, elms).

The above discussion of supercooling in plants is purely phenomenological: the actual mechanisms that allow deep supercooling of cell solutions are not well understood.

1.2.2 Proteins at Low Temperatures

The structure of proteins can be described in terms of four hierarchical levels of organization. The primary structure is the sequential order of amino acid residues. Regions of the primary sequence that form locally symmetric structures, such as a helix, define the secondary structure. The three-dimensional arrangement of an indivisible protein unit constitutes its tertiary structure. The non-covalent association of independent tertiary units gives rise to the quaternary structure. The structure of proteins is intimately linked to their biological function, and it is a familiar observation that heat destabilizes proteins by disrupting their secondary and higher levels of organization (denaturation). Potentially of far greater physiological and ecological significance is the apparently general phenomenon of cold denaturation, whereby proteins unfold at sufficiently low temperatures (Franks and Hatley, 1991), thereby losing, albeit reversibly, their biological activity. In this section we discuss the reversible loss of biological activity caused by low temperatures in aqueous, often supercooled, solvents.[12]

Brandts (1964) studied the thermal and pH-induced denaturation of chymotrypsinogen in vitro, spectrophotometrically. Assuming the reversible de-

---

the root system. Xylem structure and sap transport are discussed in greater detail in Section 1.2.3.

[11]The term embolism is used to describe any form of bubble formation in plants, and is not restricted to freeze-induced bubble formation.

[12]This should not be confused with denaturation upon freezing, which is generally irreversible, and is caused by the sharp increase in the concentration of water-soluble substances following ice formation.

naturation process to be a transition between a single native (N, folded) and a single denatured (D, unfolded) state, and using the pH independence of the enthalpy change of denaturation at any given temperature, he was able to construct a single ΔG vs. temperature curve by correcting curves measured under different conditions to the same pH and ionic strength. Here, ΔG is the overall Gibbs energy change of the denaturation process. Brandts' results are shown in Figure 1.8. Note the familiar high-temperature denaturation at 55°C. It can be seen that the ΔG curve is parabolic; this led Brandts to speculate about a possible cold-induced reversion to the D (unfolded) state. This would occur when ΔG vanishes at low temperature. Parabolic Gibbs energy curves such as that shown in Figure 1.8 have been measured for a large number of proteins (Privalov, 1979). In most cases, the extrapolated cold denaturation temperature lies well below 0°C. Thus, to observe cold denaturation, one must add chemical denaturants that narrow the temperature range within which the protein is stable, bringing the cold denaturation temperature into an experimentally accessible range (Chen and Schellman, 1989). Alternatively, the freezing point of the solution can be lowered colligatively (Hatley and Franks, 1989), or the phenomenon can be studied in a supercooled medium (Franks and Hatley, 1985; Hatley and Franks, 1992), using the emulsion technique described in Section 1.1.2. Through the use of these methods, cold denaturation has been directly observed for chymotrypsinogen in an emulsified supercooled aqueous buffer (Franks and Hatley, 1985); for lactate dehydrogenase, using methanol as a colligative freezing point depressant (Hatley and Franks, 1986, 1989); and for myoglobin in a slightly supercooled aqueous buffer (Privalov et al., 1986). Using urea as a destabilizer, cold denaturation has been observed for myoglobin (Privalov, 1990); staphylococcal nuclease (Griko et al., 1988); and lactoglobulin (Privalov, 1990). Chen and Schellman (1989) used guanidium chloride as a destabilizer, and were able to observe cold denaturation of a mutant of the protein phage T4 lysozyme at -3°C. A variety of techniques was used in these experiments to study the protein's stability, including calorimetry, light absorption, circular dichroism, nuclear magnetic resonance, and viscometry.

The reversibility of cold denaturation can be used to refold imperfectly folded proteins. Hatley and Franks (1989) cycled the temperature of colligatively stabilized lactate dehydrogenase around its cold denaturation point and found that unfolding and refolding caused reduction in the number of exposed residues.

The thermodynamic implications of cold denaturation are quite interesting, and follow from writing

$$\left(\frac{\partial \Delta G}{\partial T}\right)_P = -\Delta S, \quad \left(\frac{\partial \Delta G/T}{\partial 1/T}\right)_P = \Delta H. \tag{1.2}$$

Then, from a $\Delta G(T)$ curve such as that shown in Figure 1.8, it follows that high-temperature denaturation must be endothermic ($\Delta H > 0$), and cold denaturation exothermic ($\Delta H < 0$). A simple and elegant theory that explains

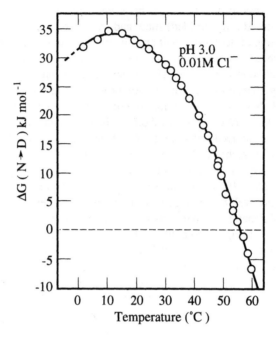

Figure 1.8. Overall Gibbs energy change of denaturation ($\Delta G_{N \to D}$) for chymotrypsino-gen at pH 3 and 0.01 M Cl^-, according to Brandts (1964). (Adapted from Franks, 1985)

the existence of two denaturation temperatures has been proposed by Dill and co-workers (Dill, 1985, 1990; Dill et al., 1989; Chan and Dill, 1991). In this model, two types of segments are considered along the protein's backbone: hydrophobic and hydrophilic. In order to calculate the Gibbs energy of folding (renaturation), a path is constructed from the D to the N state which involves two steps: condensation into a globular structure within which residues are randomly distributed, and reconfiguration of the globular structure into a state in which hydrophilic residues surround a largely hydrophobic core. For each of these steps, expressions are written for the hydrophobic and conformational contributions to the total Gibbs energy change. The former describe the aggregation of hydrophobic residues in the protein's core; the latter, the loss of configurations due to folding. The main result of the theory is the existence of two denaturation temperatures, at each of which a first-order transition occurs between the folded (N) and unfolded (D) states of the protein. At low temperatures, the conformational contribution to the Gibbs energy is almost independent of temperature, whereas the hydrophobic contribution strengthens with increasing temperature. Thus cooling favors unfolding. At high temperatures, the Gibbs energy decreases by the gain in configurations due to unfolding more than it increases by loss of attraction between hydrophobic residues. Thus

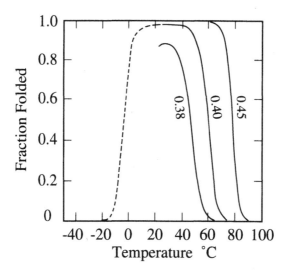

Figure 1.9. Calculated fraction of native protein as a function of temperature, for 100-residue proteins with 38, 40, and 45% hydrophobicity, according to Dill's mean-field model of protein stability. The 40% case shows cold denaturation. (Adapted from Dill et al., 1989)

heating favors unfolding. Figure 1.9 shows denaturation profiles predicted by Dill's model for a one hundred-residue protein, as a function of the fraction of hydrophobic residues. Note the cold denaturation at 40% hydrophobicity.

The entropy, enthalpy, and Gibbs energy changes due to the unfolding process are shown schematically in Figure 1.10. The generality of such parabolic free energy profiles follows from the thermodynamic identity

$$\Delta G\left(T\right) = \Delta H\left(T_h\right)\left(1 - T/T_h\right) - T\int_{T_h}^{T}\frac{\Delta C_p}{T'}\,dT' + \int_{T_h}^{T}\Delta C_p\,dT', \quad (1.3)$$

where T_h is the temperature of thermal denaturation, and ΔC_p, the heat capacity change due to denaturation, is the temperature derivative of ΔH. One way of constructing ΔG profiles experimentally is to measure the enthalpy of thermal denaturation calorimetrically at different pH values (and hence at different temperatures), and then assume that ΔH is insensitive to the pH.[13] Parabolic profiles such as those shown in Figures 1.8 and 1.10 can result from nonzero heat capacity changes upon denaturation. Franks et al. (1988) discuss the effects of various forms of $\Delta C_p(T)$ upon free energy profiles.

---

[13] As pointed out by Franks and Hatley (1991), this is a controversial assumption, and data so obtained must be treated with caution.

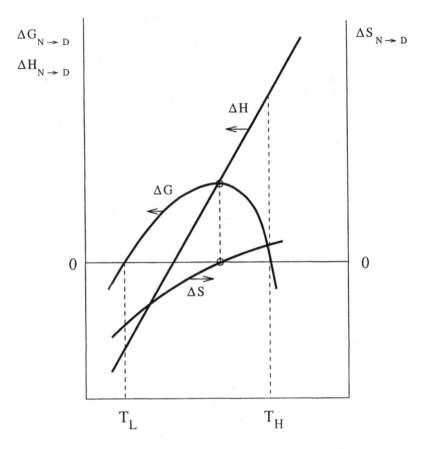

Figure 1.10. Schematic entropy, enthalpy, and Gibbs energy changes due to protein denaturation (N = native; D = denatured). T_H and T_L denote the high and low unfolding temperatures.

Figure 1.11 (Privalov et al., 1986) shows the nuclear magnetic resonance (NMR) spectra of native (N) and heat- and cold-denatured (D) myoglobin. The similarity between the the two D spectra is evident. Privalov et al. point out the similarity between the D spectra and those of free amino acids in solution, and conclude from this that at both high and low temperatures the aromatic and aliphatic groups in the protein change their mobility drastically. Taken together with ultraviolet absorption, circular dichroism, and intrinsic viscosity measurements (Privalov et al., 1986; Hatley and Franks, 1989), these observations strongly suggest that the underlying picture of a common D state for heat and cold denaturation is a very reasonable approximation. Thus denaturation gives rise to comparable disruptions in tertiary structure whether it occurs at

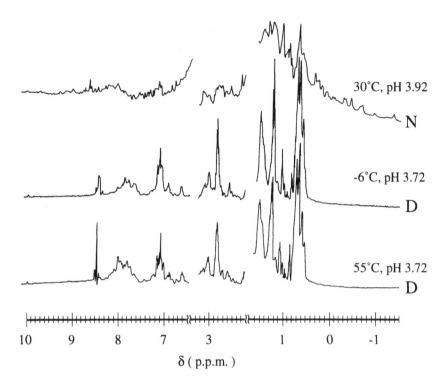

Figure 1.11. Nuclear magnetic resonance (NMR) spectra of native (top, *N*), cold-denatured (middle, *D*), and heat-denatured (bottom, *D*) myoglobin in 10 mM sodium acetate aqueous solution. Note the similarity between the cold- and heat-denatured myoglobin spectra. [Reprinted, with permission, from Privalov et al., *J. Molec. Biol.*, 190: 487 (1986)]

low temperature, where the dominant factor is the temperature dependence of hydrophobic aggregation of residues, or at high temperature, where it is caused by the gain in configurations due to chain unfolding.

Cold denaturation appears to be a general phenomenon. The degree of supercooling required for the laboratory observation of unfolding typically falls within the range of temperatures to which freeze-resistant organisms are either seasonally or permanently exposed. The mechanisms responsible for protecting proteins in vivo from low-temperature unfolding are not well understood. It has been suggested (Franks and Hatley, 1991) that natural cryoprotectants such as low-molecular-weight polyols and sugars (see Section 1.2.1) stabilize proteins and lower their cold denaturation temperature. This remains to be proved: methanol, which is commonly used in the laboratory (but not in nature) as a freezing point depressant actually increases the cold denaturation temperature of lactate dehydrogenase in vitro (Franks and Hatley, 1989). The accumulation of cryoprotectants during cold acclimation (see Figure 1.6) requires seasonal changes in the biosynthetic pathways in many insects. Cryoprotectant accumulation generally results from the massive conversion of glycogen to polyols (Storey, 1990). Franks and Hatley (1991) have suggested the interesting possibility that cold denaturation may play a role in inhibiting glycolytic enzymes, as well as in promoting polyol synthesis from glycolytic intermediates. Cold denaturation of phosphofructokinase may also be an important cause of low-temperature sweetening of potato tubers (Dixon et al., 1981).

To date, cold denaturation has only been observed conclusively in vitro. Caution must be exercised in extrapolating conclusions derived from laboratory studies to in vivo conditions. Experiments and theory suggest, however, that cold denaturation of proteins is an important naturally occurring phenomenon. Supercooled liquids are the obvious medium for the study of this interesting aspect of protein function and stability.

1.2.3 The Ascent of Sap in Plants

Sap is the name given to the aqueous fluid that transports water and mineral nutrients upwards from a plant's roots to its leaves, and photosynthetic products downwards from the leaves. The tissue through which water and minerals ascend is called xylem; the one through which sugars descend is called phloem. As shown in Table 1.1, the composition of xylem and phloem sap is quite different: in particular, the solute content of xylem sap is very low.

More than 90% of the water that ascends in the xylem evaporates from leaf cell walls into the surrounding air. The driving force for this upward motion of sap is the difference in chemical potential between water in the soil and water vapor in the air. The mechanism of sap ascent is a gradient of tension (negative pressure): columns of xylem sap are pulled to the leaves, their tension increasing with height. A liquid under tension is exposed to a pressure that is lower than its vapor pressure at the given temperature; hence it is superheated.[14] Thus trees rely on a metastable liquid for their water and mineral nutrient intake. Experimental proof of the fact that sap ascends under tension dates back to Boehm (1893), who fitted a leafy twig to a water-filled glass tubing from which air had been expelled. This tube was immersed in mercury, and the leaf was allowed

---

[14]At low enough temperatures, sap is simultaneously superheated and supercooled. This condition occurs in general when the temperature is lower than the temperature along the metastable continuation of a substance's melting curve at the given tension (Hayward, 1971). Liquids that are simultaneously superheated and supercooled are called subtriple (D'Antonio, 1989); their pressure is lower than the pressure along the metastable continuation of the boiling curve at the given temperature, and their temperature is lower than the temperature along the metastable continuation of the melting curve at the given pressure. Metastable continuations of melting or boiling curves are examples of phase equilibrium between metastable phases. This topic is discussed in Section 2.2.9.

Table 1.1. Xylem and Phloem Sap Compositions in White Lupine[a]

	Xylem Sap (mg/l)	Phloem Sap (mg/l)
Sucrose	Not detectable	154,000
Amino acids	700	13,000
Potassium	90	1,540
Sodium	60	120
Magnesium	27	85
Calcium	17	21
Iron	1.8	9.8
Manganese	0.6	1.4
Zinc	0.4	5.8
Copper	Trace	0.4
Nitrate	10	Not detectable
pH	6.3	7.9

[a] Adapted from Salisbury and Ross (1992).

to transpire, whereupon the mercury was raised more than 760 mm (Zimmerman and Brown, 1980). Early experiments were also done by Dixon and Joly (1896) and Askenasy (1897). Dixon (1914) first formulated a consistent theory of sap ascent under tension. Since then, numerous experiments, notably by Scholander and collaborators (e.g., Scholander, 1972, and references therein), have established that sap ascends to the leaves of trees under tension. This viewpoint has been questioned occasionally (Zimmermann et al., 1993; Canny, 1995). Two recent experiments, however, have demonstrated clearly the ability of water-filled xylem channels to sustain tensions up to −35 bar (Holbrook et al., 1995; Pockman et al., 1995).

The ascent of sap occurs in conduits formed by dead cells. These conduits are formed by loss of the cytoplasm and the cytoplasmic membrane that separates one xylem cell from another. Cell walls are strengthened by cellulose fibrils and encrusted with lignin (Zimmerman, 1963). This is sometimes referred to as secondary wall formation (Salisbury and Ross, 1992). Thus xylem serves a dual function: water transport and structural rigidity. Figure 1.12 shows different types of xylem conduits. In conifers (pine, spruce), elongated cells called tracheids overlap along their end portions. Hydraulic continuity is provided by orifices called bordered pits that allow the passage of water but trap air bubbles. In birch, flow occurs through partially dissolved end compartments. In oak, there are no end walls, and water flows through long, rigid capillaries

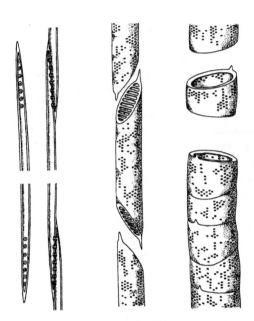

Figure 1.12. Different types of xylem conduits. Spindle-shaped tracheid cells in pine (left); xylem channel in birch, showing partially dissolved end walls (center); rigid oak tubes without end walls (right). [Reprinted from "How Sap Moves in Trees," by M. H. Zimmerman, *Sci. Amer.* 208 (3): 133 (1963). Copyright 1963 by Scientific American, Inc. All rights reserved]

composed of squat segments (Zimmerman, 1963). A schematic section of ash stem, showing the two water-conducting tissues, phloem and xylem, is shown in Figure 1.13. In most trees, the life cycle of xylem moves inwards, with one new growth ring formed annually, while the life cycle of phloem moves outwards to form the innermost layers of bark (Zimmerman, 1963).

Water flows to a tree's leaves during photosynthesis, more than 90% being ultimately transpired to the surrounding air. Thus the flow rate of ascending sap varies during the day in response to photosynthetic activity. Some typical sap peak velocities at midday are listed in Table 1.2, together with the corresponding diameter of a conducting vessel and the calculated pressure (tension) gradient needed to drive the flow.

Xylem vessels are neither rectilinear, circular, nor smooth, as assumed in the last column of Table 1.2. Zimmerman and Brown (1980) report experimental correction factors by which the ideal (Hagen-Poiseuille) pressure drop should be multiplied to reconcile measured (Reidl, 1937; Münch, 1943) pressure gradients, volumetric throughputs, and vessel diameters. These factors range from 1 for vines to 3 for birch and 5 for poplar. Thus one arrives at pressure gradients that are typically in the 0.05–0.2 bar m^{-1} range, including the hydrostatic gradi-

Cork | Phloem | Xylem | Ray

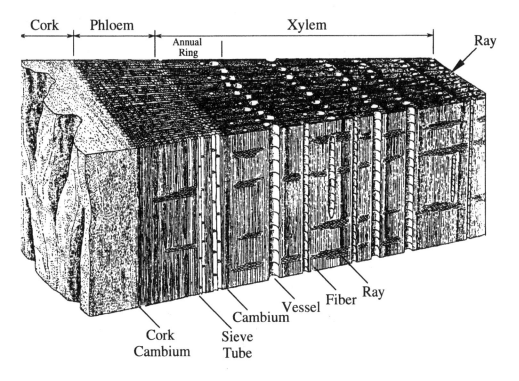

Annual Ring

Cork
Cambium

Sieve
Tube

Cambium

Vessel Fiber Ray

Figure 1.13. Cross section of ash stem showing the two fluid transport systems. Water ascends through the xylem, and photosynthetic products descend through the phloem. [Reprinted from "How Sap Moves in Trees," by M. H. Zimmerman, *Sci. Amer.* 208 (3): 133 (1963). Copyright 1963 by Scientific American, Inc. All rights reserved]

ent, 0.1 bar m^{-1} (Zimmerman and Brown, 1980). This means that the maximum tension at the top of a 30 m tree will typically be between -1.5 and -6 bar.

Relying on metastable liquid columns to convey needed minerals and water poses an interesting optimization problem to trees, namely, balancing safety and efficiency. The latter requires the needed flow rate of sap to be delivered with as small a tension gradient as possible: this calls for large vessels. Since the columns of ascending sap are under tension, hydraulic continuity is constantly in danger of being interrupted by bubble formation. Thus safety requires that bubble formation, should it occur, affect as small a fraction of the sap flow as possible. Evolution has solved this problem in two ways: first, by limiting the diameter of xylem vessels to 500 μm; secondly, by allowing the ascending sap to follow tangential and radial trajectories. Interconnections between xylem channels and radial rays and extensive interchannel pitting between touching conduits allow alternative flow paths. This insures that sap can continue to ascend even when the tree is severely damaged; it is well known that if two saw cuts are made to a tree stem from opposite sides some vertical distance apart,

Table 1.2. Midday Peak Velocities of Xylem Sap in Some Woody Plants, and the Corresponding Vessel Diameters and Calculated Pressure Gradients[a]

Tree	Sap Velocity (m/h)	Vessel Diam. (μm)	Pressure Gradient[b] (bar/m)
Ring porous[c]			
Quercus pedunculata Oak	43.6	200–300	0.043–0.096
Castanea vesca Chestnut	24	300–350	0.017–0.024
Cytius laburnum Broom	3.9	60–250	0.006–0.096
Diffuse porous[c]			
Populus balsamifera Cottonwood	6.25	80–120	0.039–0.087
Alnus glutinosa Alder	2	20–90	0.022–0.444
Aesculus hippocastanum Horse chestnut	0.96	30–60	0.024–0.095
Conifers			
Larix decidua Larch	2.1	Up to 55	0.062
Picea excelsa Spruce	1.2	Up to 45	0.053
Tsuga canadiensis Hemlock	1	Up to 45	0.044

[a] Velocity and diameter data from Zimmerman and Brown (1980).

[b] $\Delta P/L = 8\eta \langle v \rangle / r^2$, where $\Delta P/L$ is the pressure gradient, η the viscosity (= 0.001 kg sec^{-1} m^{-1}; corresponding to water at 20°C), $\langle v \rangle$ the mean velocity in the capillary, and r the capillary radius. Ideal gradient, calculated from the Hagen-Poiseuille equation.

[c] In ring porous trees (e.g., oaks, ashes, elms) the diameter of xylem vessels formed early in the growing season is much larger than that of late-forming vessels. In diffuse porous trees (e.g., poplars, maples, birches) the vessels are of small diameter, and those formed early in the growing season have approximately the same diameter as those formed later (Kramer and Kozlowski, 1979).

sap continues to ascend. This would not be the case if the only possible ascent path were along isolated vertical conduits.

The previous discussion merely outlines some of the basic facts about tree hydraulics and architecture; interested readers should consult the monographs

of Zimmerman (1983) and Zimmerman and Brown (1980) for in-depth discussions. Nevertheless, this brief treatment illustrates the crucial role played
by metastable liquids in the life cycle of trees, and it shows nature's highly
sophisticated strategies for handling liquids under tension.

1.2.4 Mineral Inclusions

Most crystals have grown in a fluid medium, such as a silicate melt or an aqueous
solution. After crystallization, minerals are usually fractured several times, and
these fractures heal in the presence of fluids. During both crystal growth and
fracture healing, small quantities of the surrounding fluid medium are often
trapped as fluid inclusions in the host crystal. Fluid inclusions in minerals
are very common, and their investigation has numerous applications, including
the reconstruction of extraterrestrial processes through the study of meteoritic
samples; the evaluation of the safety of proposed sites for nuclear reactors
or nuclear waste deposits by using inclusions to estimate the time elapsed
since the last movement of faults; guidance of drilling in active geothermal
systems by providing data on deep temperatures and on whether a portion of a
geothermal site is heating or cooling; the direct observation of phase behavior
at extreme conditions; studies of the evolution of the atmosphere and climate,
for example, reconstruction of paleoconcentrations of CO_2 by investigation of
gas inclusions in dated polar ice sheets, or reconstruction of temperature and
climate during the last 350,000 years by investigation of fluid inclusions in cave
deposits; guidance in oil drilling by relating information on fluid inclusions
containing hydrocarbons to the pressure-temperature evolution of petroleum
basins (Roedder, 1984).

Inclusions are small, typically in the 10–100 μm range. The presence of impurities that can trigger the formation of a new phase being highly improbable in
small samples, metastable conditions can persist in inclusions over geological
time scales (Roedder, 1984). Roedder (1967) used naturally occurring water
inclusions in fluorite to study metastable phase equilibrium between liquid water and ice. He observed melting of ice at 6.5°C. Using the Clausius-Clapeyron
equation to extrapolate water's negatively sloped melting curve (in P, T projection) to this temperature, he calculated a tension of 900 bar. This situation
is illustrated in Figure 1.14. Interestingly, ice crystals under conditions such
as are shown in Figure 1.14 are exposed to isotropic tension, and are able to
deform in any direction without influencing surface stresses (Roedder, 1967).
This is not normally possible for crystalline solids.

Of the various types of metastability in mineral inclusions discussed by
Roedder (1984), the presence of ice in metastable equilibrium with liquid water
under tension (see Figure 1.14) is particularly interesting, because it provides
the basis for a well-known method for studying liquid water under very high
(i.e., kilobar range) tensions. Angell and his co-workers (Green et al., 1990;

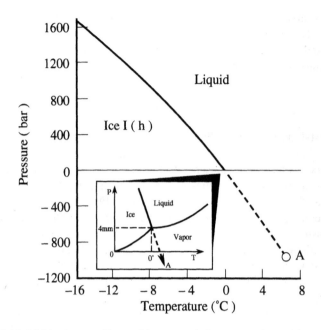

Figure 1.14. Melting curve of ice and its extrapolation to negative pressures. Point *A* corresponds to the maximum temperature observed by Roedder (1967) for inclusions containing ice in metastable equilibrium with liquid water. (Adapted from Roedder, 1967)

Zheng, 1991; Zheng et al., 1991) created water inclusions by fissuring quartz and fluorite crystals and autoclaving them in ultrapure water. Autoclaving caused the fissures to heal, and water was trapped in the interstices. Figure 1.15 shows fluorite crystals with inclusions used by Zheng (1991). The combination of temperature and pressure during autoclaving was chosen so as to achieve the desired density of water in the inclusions. Because quartz and fluorite are virtually incompressible, water inside the inclusion followed a constant-volume (isochoric) path when the temperature was varied. Trapped air bubbles were dissolved by heating. In this approach, then, a liquid sample is subjected to large negative pressures until rupture occurs and bubble formation restores the sample to positive pressure. Zheng et al. attained a maximum tension of approximately 1400 bar at 42°C. This tension was calculated using the Haar-Gallagher-Kell equation of state (Haar et al., 1980), knowing the density of the water inside the inclusion and the temperature. The measurements of Zheng et al. (1991) are summarized in Figure 1.16.

Mineral inclusions are therefore not only a very common and inherently interesting phenomenon, they have also inspired the best-known technique for

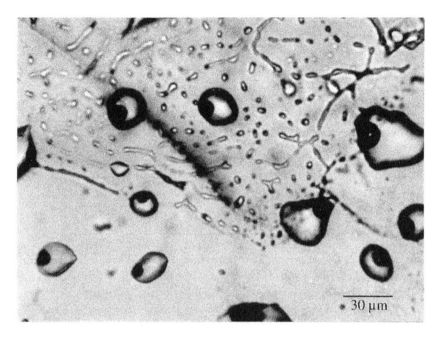

Figure 1.15. Water inclusions in fluorite. [Reprinted, with permission, from Zheng (1991)]

investigating liquids under tension, and may yet become important for studying isotropically stressed crystalline solids.

1.2.5 Clouds

The largest inventory of supercooled water can be found in clouds. Clouds are formed by the ascent of humid air which cools by expansion caused by the hydrostatic decrease of ambient pressure with height. Eventually, the ascending air becomes saturated,[15] and condensation of water vapor into the liquid phase is favored thermodynamically. Were it not for the existence of airborne nuclei that facilitate droplet formation, the water vapor in ascending air would become supercooled. However, the atmosphere contains a variety of particles (condensation nuclei) on which droplet formation can occur, and in practice the relative humidity[16] rarely rises above 101% (Rogers, 1979). Cloud condensation nuclei include sea-salt particles formed by the bursting of air bubbles in the

[15]This happens when the partial pressure of water vapor in the air equals the vapor pressure of water at the given temperature.

[16]If r is the mass of water vapor per unit mass of dry air, the relative humidity is defined as $100(r/r_{sat})$, where the subscript "sat" denotes saturation.

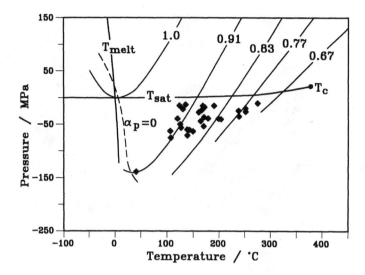

Figure 1.16. Measured temperatures and calculated tensions at which liquid rupture occurred in microscopic inclusions of pure water in quartz crystals. Also shown are the equilibrium boiling and melting curves of water, and several lines of constant density (isochores), in g cm^{-3}. The dashed line is the locus of density maxima. (Adapted from Zheng et al., 1991)

foam of breaking waves (Mason, 1971), ammonium sulfate particles formed by photochemical oxidation of sulfur dioxide and reaction with ammonia in the air, and particulates formed by industrial combustion and forest fires. These particles range in size from 10^{-3} to 10 μm (Rogers, 1979), and in concentration from less than 10 per cm^3 in maritime clouds to more than 10^5 per cm^3 over large cities. Clouds are formed by droplet condensation onto airborne nuclei.

Clouds occurring at altitudes ranging from 5 to 13 km are referred to as high clouds, and include the cirrus, cirrocumulus, and cirrostratus families. Middle clouds occur at heights between 2 and 7 km, and include the altocumulus, altostratus, and nimbostratus families. Clouds found at altitudes below 2 km are called low clouds; they include the stratocumulus, stratus, cumulus, and cumulonimbus families (Mason and Loewe, 1987). The temperature in the troposphere[17] decreases away from the earth's surface at an average rate of approximately 6.5 K km^{-1} (Harte, 1988). This temperature profile is the result

---

[17]Troposphere is the name given to the lowest layer of the atmosphere, in which temperature decreases with height. The troposphere extends up to 8 km at the poles, and 16 km above the equator. The atmospheric layer above the troposphere is called the stratosphere; it extends from about 12 to 50 km above the earth's surface. In the stratosphere, the temperature increases with height, the principal contributing effect being solar radiation (Harte, 1988).

of the interaction between solar radiation, radiation from the earth's surface, and convective cooling of rising air columns. Thus high clouds are colder than middle clouds, which are in turn colder than low clouds. Subfreezing temperatures are the normal state of affairs in medium and high clouds. However, because of the relative scarcity of efficient airborne ice-nucleating particles, it is normal for cloud droplets to be supercooled well below $0°C$.

A cloud is an assembly of droplets having diameters ranging from 5 to 200 μm, and numbering on the order of 100 to 1000 cm^{-3} (Pruppacher and Klett, 1978; Rogers, 1979). The terminal velocity attained by droplets falling through air is the result of a balance between weight, buoyancy, and friction.[18] It ranges from 1 cm sec^{-1} for a 20 μm droplet to 2 m sec^{-1} for a 200 μm one. Growing clouds are sustained by vertical air currents with velocities ranging from a few cm sec^{-1} to several m sec^{-1} (Mason and Loewe, 1987). Thus droplets must grow considerably in order for precipitation to occur: a typical raindrop has a diameter of 2 mm and falls at a velocity of almost 8 m sec^{-1}. In tropical and subtropical regions, where cloud temperatures do not fall below freezing, precipitation occurs as a result of droplet growth and coalescence. Larger droplets fall by gravity, and in so doing collide and coalesce with a fraction of the smaller, slower-moving droplets. Given a sufficiently deep cloud, falling droplets will eventually attain raindrop size, and precipitation will occur.

It is only when the temperature drops below $-15°C$ that ice crystals begin to appear in clouds (Rogers, 1979). This is because efficient nuclei for the freezing of droplets (frequently called ice nuclei) are present in very low concentrations in the atmosphere. There is considerable uncertainty as to the nature of ice nuclei. It appears that 0.1–4 μm particles of silicate minerals of the clay and mica variety, particularly kaolinite, are efficient ice nucleators (Kumai, 1961; Kumai and Francis, 1962; Rogers, 1979). Such materials are common constituents of soil, and are presumably carried to the atmosphere by the wind. Since the vapor pressure of ice is lower than that of supercooled water, ice crystals will grow at the expense of supercooled droplets when both are present in a cloud. Once growing ice particles attain sufficient size, they fall through the cloud. The coalescence of several such falling crystals gives rise to snowflakes. Depending on the temperature and supersaturation[19] at which they are formed,

---

[18]The terminal velocity v of a liquid droplet of radius r in air can be estimated from the equations v (cm/sec) $= 0.0136[r(\mu m)]^2$, if r (μm)$\times v$ (cm/sec) < 700, and v (cm/sec) $= 24.1[r(\mu m)]^{-1/2}$, if r (μm) $\times v$ (cm/sec) > 700. These expressions follow from equating the frictional drag on a falling droplet, $0.5\pi r^2 v^2 \rho C_D$, to its weight $(4\pi/3)r^3 \rho_l g$, with ρ the density of air, ρ_l the density of the droplet, and g the acceleration due to gravity. The drag coefficient C_D is given by $24/Re$ if $Re < 1$, and 0.45 when $Re > 1$. Re, the droplet Reynolds number, is defined as $2rv\rho/\eta$, where η is the viscosity of air. Using $\rho = 1.16$ kg m^{-3}, $\rho_l = 10^3$ kg m^{-3}, and $\eta = 1.6 \times 10^{-5}$ kg m^{-1} sec^{-1} yields the above expressions. These calculations are valid in the continuum regime, where droplet dimensions exceed the molecular mean free path in air. In practice, this means $r > 1\mu$m.

[19]Supersaturation is the ratio of partial pressure of water vapor in the air to the vapor pressure of ice at the given temperature.

snowflakes can adopt one of seven forms: plates, stellars, columns, needles, spatial dendrites, capped columns, or irregular crystals (Mason and Loewe, 1987). If the falling ice crystals collide primarily with supercooled droplets rather than with other crystals, graupel or hail can form. If snowflakes, graupel, or hail melt as they fall through warmer regions, they reach the ground as raindrops.

Until recently, it was believed that cirrus clouds, which exist in the upper troposphere, are composed almost entirely of small ice crystals. However, supercooled droplets in cirrus clouds have been observed at temperatures below $-40°C$ (Sassen et al., 1985; Sassen, 1992). This observation has profound implications for global climate modeling, because the radiative properties of liquid droplets are very different from those of ice particles. Since the distribution of supercooled droplets and ice particles determines the reflective properties of cirrus clouds, metastability affects the transfer of solar and terrestrial radiative fluxes, and hence can influence the temperature distribution in the atmosphere (Sassen et al., 1992).

The idea of stimulating rain and snow by seeding clouds with efficient ice nuclei is one of the basic concepts of artificial climate modification. Using a cloud chamber, Schaeffer (1946) first observed that ice crystals could be nucleated in a supercooled cloud by dropping dry ice (solid carbon dioxide). Subsequently, Vonnegut (1947) discovered that small crystals of silver iodide can act as efficient ice nuclei. The fact that silver iodide crystals have hexagonal symmetry, like ice, is undoubtedly important in explaining this substance's effectiveness as an ice nucleator. Though much work has been done since these early pioneering observations, the basic idea of cloud seeding remains the same, namely, to provide artificial ice nuclei so as to compensate for their relative scarcity, and hence to facilitate the occurrence of precipitation by heterogeneous nucleation.

Subfreezing temperatures and a scarcity of ice nucleation sites make clouds an ideal environment for supercooled water. Metastability, therefore, plays a key role among the many interacting phenomena that determine climate on Earth.

Another atmospheric phenomenon where metastability is important involves polar stratospheric clouds. The surface of particles in polar stratospheric clouds is believed to play an important role in catalyzing the destruction of ozone during late winter and early spring (Farman et al., 1985). These particles have been classified into two types. Type I particles are believed to consist mainly of nitric acid and water, and form between approximately $-78°$ and $-84°C$, the frost point of ice.[20] Type II particles consist of ice, and form below $-84°C$.

---

[20]The frost point is the temperature at which the partial pressure of water vapor in the air equals the vapor pressure of ice. It varies with height and humidity, the quoted value being typical of Arctic winter conditions at heights of ca. 20 km.

Chemical reactions that transform stable inorganic chlorine compounds (e.g., $ClONO_2$, HCl) into photolytically active forms (e.g., Cl_2) are efficiently catalyzed by stratospheric particles,

$$ClONO_2 + HCl \rightarrow Cl_2 + HNO_3. \tag{1.4}$$

This enables the formation of Cl and ClO radicals that participate in ozone destruction (Molina et al., 1993; Beyer et al., 1994),

$$Cl + O_3 \rightarrow ClO + O_2,$$

$$ClO + O \rightarrow Cl + O_2. \tag{1.5}$$

Polar stratospheric clouds are also believed to facilitate the conversion of reactive nitrogen gases such as NO, NO_2, and N_2O_5 to stable nitric acid. Reduced NO_2 levels increase the efficiency of ozone destruction because NO_2 reacts with ClO to form the stable species $ClONO_2$ (Molina et al., 1993; Arnold, 1992).

The composition, mechanism of formation, and state of aggregation of type I polar stratospheric cloud particles are not well understood (Drdla et al., 1994). The stable form of nitric acid particles under stratospheric conditions is nitric acid trihydrate (Molina et al., 1993). One view is that nitric acid and water nucleate heterogeneously on stratospheric sulfate aerosols consisting mostly of frozen sulfuric acid hydrates, thereby forming nitric acid trihydrate (Turco et al., 1989; Iraci et al., 1994). A second proposed mechanism involves the uptake of nitric acid by highly supercooled sulfuric acid/water liquid droplets, followed by crystallization of nitric acid trihydrate and sulfuric acid tetrahydrate. The frozen particles then grow by condensation of nitric acid and water vapors (Molina et al., 1993; Beyer et al., 1994). According to a third proposed mechanism, type I particles are supercooled liquid mixtures of water and sulfuric and nitric acids (Drdla et al., 1994; Tabazadeh et al., 1994).

Some Antarctic measurements are consistent with the mechanism involving deposition of nitric acid trihydrate onto frozen sulfuric acid hydrate particles (Fahey et al., 1989). The Arctic stratosphere is not as cold as the Antarctic one, and mechanisms involving extensive supercooling of sulfuric acid/water and sulfuric acid/nitric acid/water systems may occur.

Both the equilibrium phase behavior and the nucleation kinetics in the water/sulfuric acid/nitric acid ternary system under stratospheric conditions need to be better understood. Extensive supercooling may play a major role in determining the composition, phase, and physical characteristics of polar stratospheric clouds. Little is known about the detailed effects of composition (nitric acid trihydrate vs. sulfuric acid tetrahydrate) and state of aggregation (liquid vs. solid) of type I stratospheric particles on the kinetics of chlorine decomposition, and hence on ozone depletion.

1.3 METASTABLE LIQUIDS IN TECHNOLOGY

1.3.1 Storage of Proteins and Cells by Supercooling

The shelf life of isolated biochemicals is frequently short. The biological activity of most isolated proteins in dilute solution, for example, decays very rapidly (Hatley et al., 1987). This limits their practical usefulness in therapeutic or biochemical applications. Current methods for the in vitro stabilization of proteins include the use of chemical additives, and lyophilization (freeze-drying). In the former case, use is made of the ability of high concentrations of substances such as ammonium sulfate or glycerol to enhance the stability of the native state of proteins (Lee and Timasheff, 1974). Additives, however, must often be removed prior to use, thus rendering the enzyme less stable. Lyophilization is used with proteins that do not denature upon freezing (Briggs and Maxwell, 1975). This technique generally requires that the process be tailored to each protein, and often results in low yields and loss of activity during storage (Montoya and Castell, 1987; Yeo et al., 1993). Thus there is a need for new processes to increase the shelf life of commercially available proteins, and of labile biochemicals in general.

One possibility is to use supercooled water as a storage medium, relying on low temperatures to slow down processes leading to loss of biological activity. The effectiveness of supercooling as a means of prolonging the shelf life of proteins hinges on the availability of reliable techniques for preventing freezing. Franks and co-workers used the emulsion technique developed by Rasmussen and MacKenzie (1972), in which stirring is used to disperse water into small (1–10 μm) droplets in a hydrocarbon phase, the emulsion being stabilized with a surfactant (see also Section 1.1.2). Hatley et al. (1987) measured the biological activity of lactate dehydrogenase (LDH) from rabbit muscle as a function of time, for various storage protocols. The activity was determined from the rate of change of light absorbance at 340 nm in aliquots. Their results are shown in Figure 1.17. Storage in supercooled emulsions at -12 and $-20°$C resulted in complete retention of biological activity after one year, with no noticeable difference between samples stored at these two temperatures. The other storage protocols resulted in loss of activity, rapidly for frozen samples, and gradually for samples stored at 4°C. Figure 1.17 demonstrates the difference between frozen and supercooled storage at the same temperature. Interestingly, LDH-containing emulsions were also found to withstand long-distance transport at refrigerator temperatures prior to storage at subfreezing conditions, with no loss of biological activity (Hatley et al., 1987).

Supercooled, surfactant-stabilized water-in-oil emulsions have also been used for the cryopreservation of cells, and to study the responses of cells to low temperatures. Franks et al. (1983) and Mathias et al. (1985) studied the supercooling of animal, plant, and microbial cells in water-in-oil emulsions.

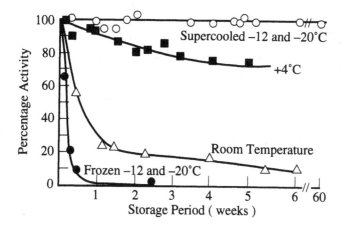

Figure 1.17. Time-dependent activity of lactate dehydrogenase, LDH, for various storage protocols. Note the difference in activity between the supercooled and frozen samples stored at the same temperatures. [Reprinted, with permission, from Hatley et al., *Process Biochem.* Dec.: 169 (1987). Copyright 1987, Elsevier Science Ltd.]

They found that all types of cells survived cooling to $-23°C$, and they were able to cool erythrocytes to $-39°C$. Rewarming of erythrocytes following cooling to $-38°C$ did not lead to hemolysis. Many strains of microbial cultures are used in industry to produce chemicals or food. Cell cultures for these applications are commonly stored frozen, or are freeze-dried. Unfortunately, these storage protocols cause the death of the vast majority of cells. Cultures must therefore be regrown prior to use, or a large excess must be frozen, so that the desired viable concentration is obtained upon recovery. For example, the concentration of lactic-acid-producing organisms needed for cheese production is 10^8 cells ml^{-1}. Typically, up to 10^{12} cells ml^{-1} are frozen, so that the required concentration is obtained upon recovery (Franks, 1988). In these and similar cases, supercooling is a promising alternative to conventional low-temperature cell storage technologies.

It thus appears that low-temperature storage in the form of supercooled water-in-oil emulsions is a viable method for increasing the shelf life of proteins and for preserving cells. The process avoids the use of additives and cryoprotectants, withstands long-distance transport and the accompanying temperature cycling, and can be used to remove folding imperfections by careful cycling around the cold denaturation condition. Practical problems still to be resolved include the scaleup of the oil dispersion process (Franks, 1988), and safeguarding against contamination of the recovered product with oil or surfactant.

An important recent technology for the stabilization of biochemicals involves glasses rather than supercooled liquids. In this approach, sugar-rich, water-soluble glasses are used as physically and chemically stable solid matrices that protect biological molecules from irreversible loss of activity (Franks and Hatley, 1993; Ramanujam et al., 1993; Franks, 1994). Diffusion and chemical reaction are slowed down enormously in the glass matrix, stabilizing trapped biomolecules against chemical reactions that lead to loss of activity. Glasses are formed by vacuum removal of water from sugar solutions. The carbohydrates, in addition to forming the glassy matrix, provide protection from denaturation during water removal when high concentrations of salts and ions occur in solution. This promising technique appears to combine successfully long-term stabilization with storage at ambient temperatures.

1.3.2 Some Uses of Liquids under Tension

A prototype pump that uses liquids under tension was built and tested at the National Engineering Laboratory in Scotland (Hayward, 1970, 1971). It is shown schematically in Figure 1.18. A metal bellows A replaces the piston and cylinder of a conventional reciprocating pump. The bellows is contained inside a pressure vessel B that is in turn connected to a high-pressure oil pump C. The bellows is compressed when oil is pumped into B; it is expanded by the spring E when the oil is released through the valve D. The needle F serves as an indicator of the bellows' level. The suction line dips into a water tank. Slightly above the water level is a nonreturn valve I, above which is located a pressure vessel J, fitted with wire gauze screens to ensure flow uniformity. The pump is primed by filling under vacuum through the discharge line. Vessel B is then pressurized up to several hundred bars while valves H and D are closed. This pressurization is maintained for several minutes, during which gas bubbles in the liquid contained in the bellows, the suction pipe G and pressure vessel J, are forced into solution or into microscopic crevices in the solid walls. This is an effective way of suppressing the formation of bubbles and the consequent loss of tension during operation.[21] The actual pumping cycle consists of three steps: suction, pressurization, and delivery. In suction, valve D is opened, the springs extend the bellows, and water is consequently drawn up under tension. During pressurization, valve D is closed and a pressure of 300 bar is applied to vessel B for a few seconds, the longer pressurization being needed only after the pump is newly primed. Delivery occurs when valve H is opened and oil is pumped into B until the bellows is fully compressed. Valve H is then closed.

Operating at six strokes per minute, with a mechanical efficiency of less than 1%, a delivery rate of 0.4 liters per minute, and with frequent need for repriming due to the formation of bubbles (Hayward, 1970, 1971), the suction pump is far from being of practical use. This illustrates the engineering challenges that must be overcome before negative pressure in liquids can be effectively

---

[21]The suppression of bubble formation by pressurization is discussed in Section 3.4.

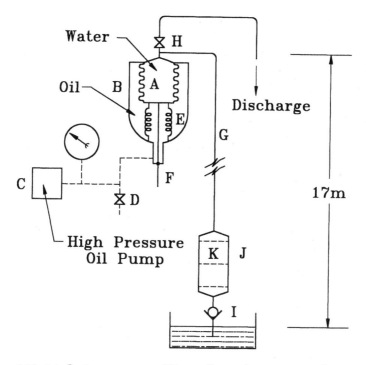

Figure 1.18. Schematic arrangement of Hayward's negative-pressure suction pump. *A*, metal bellows; *B*, pressure vessel; *C*, high-pressure oil pump; *D*, valve; *E*, spring; *F*, bellows position indicator; *G*, suction pipe; *H*, valve; *I*, nonreturn valve; *J*, pressure vessel; *K*, gauze screens. (Adapted from Hayward, 1970)

exploited technologically. Plants provide the ultimate example of reliable, safe, and efficient handling of liquids under tension. In principle, there is no reason why humans cannot also accomplish this. The ability of liquids to withstand appreciable tension could then be applied to the pumping of liquids from deep wells and reservoirs (Hayward, 1971; Apfel, 1972b); or to the development of novel irrigation systems that eliminate evaporation losses (Bulman, 1969; Hayward, 1971). As is always the case with technological progress, other, presently unforeseen applications could result once the major hurdle (handling liquids under tension reliably) is overcome. The laboratory study of liquids under tension has a very long history[22] (e.g., Donny, 1843, 1846; Berthelot, 1850; Reynolds, 1900a,b; Worthington, 1892; Meyer, 1911; Temperley and Chambers, 1946; Temperley, 1946, 1947; Briggs, 1949, 1950, 1951a,b, 1953, 1955; Donoghue et al., 1951; Bull, 1956; Beams, 1959; Hayward, 1964; Couzens and Trevena, 1969, 1974; Apfel, 1971a,b; Winnick and Cho, 1971; Carlson and Henry, 1973;

---

[22]For historical reviews of this topic, see Kell (1983); Henderson (1985); Trevena (1987).

Huang and Winnick, 1974; Carlson, 1975; Carlson and Levine, 1975; Chapman et al., 1975; Wilson et al., 1975; Richards and Trevena, 1976; Sedgewick and Trevena, 1976; Apfel and Smith, 1977; Evans, 1979; Henderson and Speedy, 1980; Overton and Trevena, 1981; Overton et al., 1984; Green et al., 1990; Zheng et al., 1991). It is reasonable to expect that this accumulated knowledge, coupled with technological ingenuity, will eventually result in practical applications.

The tension pump relies on the application of a constant tension as a means of transporting a fluid. In contrast, the cyclic application of tension to a static liquid occurs routinely in ultrasound technology. Sound waves with frequencies higher than 16–18 kHz, the upper frequency range to which the human ear can respond, are called ultrasonic. A longitudinal sound wave causes fluid layers to compress and expand periodically along its direction of propagation. This causes a corresponding periodic variation in pressure,[23] so that the pressure at any given point is the sum of the static and acoustic components. If the amplitude of the acoustic pressure perturbation is sufficiently large, the total pressure can become negative, in which case the liquid is periodically subject to tension. Ultrasonic applications are commonly grouped into two categories: power ultrasound, which involves high-amplitude waves in the 20–100 kHz range, and high-frequency ultrasound, which involves low-amplitude waves in the 1–10 MHz range (Mason and Lorimer, 1988).

The power density I due to a sound wave (that is to say, the amount of energy flowing per unit time per unit area perpendicular to the direction of propagation of sound) is given by

$$I = \frac{P_a^2}{2\rho c}, \tag{1.6}$$

where P_a is the amplitude of the acoustic pressure wave, ρ is the mass density of the liquid, and c is the sound velocity in the liquid. For example, ultrasound propagating in water ($\rho = 1000$ kg m^{-3}; $c = 1500$ m sec^{-1}) at a power density of 10^4 W m^{-2} corresponds to an acoustic pressure amplitude of 1.7 bar. Assuming a frequency of 25 kHz, this means that the pressure varies locally from 2.7 to -0.7 bar, 25,000 times every second.

If a liquid is under tension (or, more generally, if the ambient pressure is lower than the liquid's vapor pressure at the given temperature), small bubbles (cavities) can form. These cavities contain the liquid's own vapor, or previously dissolved gases. In addition to dissolved gases, suspended particles also facilitate bubble formation. This process is called heterogeneous nucleation and is discussed in Section 3.4. In the absence of impurities, bubbles form within the bulk liquid by homogeneous nucleation (Section 3.1). The evolution in time of cavities in a liquid, whether it be small-amplitude oscillations about an equilibrium size or rapid growth and collapse, is commonly known as cavitation. The

---

[23]The displacement and pressure perturbations are out of phase.

numerous applications of ultrasound in liquids involve both bubble formation and collapse, and acoustic stressing in the absence of bubbles.

The first technical application of ultrasound, Langevin's echo method for measuring water depths, dates back to 1917. It involved sending a pulse of ultrasound from an underwater generator/detector; the beam was reflected on the seabed and detected on the same instrument after traversing the original path in reverse. The depth could then be calculated as $0.5ct$, where c is the sound velocity in water and t is the time elapsed between sound emission and detection. This simple idea became the basis for the submarine detection equipment known as ASDIC (for Anti-Submarine Detection Investigation Committee), used widely during the Second World War. Subsequent developments, including the use of arrays of transducers for sector scanning and improved visual display systems, resulted in the modern SONAR (for SOund NAvigation Ranging). Presently, SONAR is used to study the seabed, to search for shipwrecks or for fish shoals, and to inspect offshore drilling installations (Cracknell, 1980).

In biological applications, power ultrasound is used to homogenize cell suspensions and to disrupt cell walls, causing the release of the cells' contents. It is believed that cell wall disruption is caused by small cavitating bubbles. Ultrasonic cleaning involves the scrubbing of surfaces by means of collapsing bubbles. Aqueous solutions of detergents are commonly used as a cleaning medium. Examples include the ultrasonic cleaning of hospital glassware and surgical instruments, photographic lenses, electronic circuit boards, ball bearings, heat exchangers, and automotive engine blocks (Shoh, 1988). In ultrasonic soldering, cavitation of bubbles in the molten solder removes surface oxide layers and thus exposes clean metal surfaces to solder, allowing soldering without flux (Shoh, 1988). Aluminum components such as window frames, cables, and heat exchanger parts are soldered ultrasonically. Ultrasonic emulsification of antibiotic dispersions, lotions, and mineral and essential oils is attractive because it circumvents the use of surfactants (Higgins and Skauen, 1972; Shoh, 1988). Stable suspensions of coal powder in oil can also be stabilized ultrasonically and then burned in existing oil-burning units (Shoh, 1988). Solids can be effectively dispersed in liquids by means of ultrasound. Examples include the dispersion of china clay, mica, and titanium oxide in the rubber and paper industries; the dispersion of pigments and additives in paints, inks, and resins; and the treatment of mineral slurries for flotation and separation processes (Mason and Lorimer, 1988; Shoh, 1988). Low-power 20 kHz ultrasound has been used successfully to enhance filtration rates of contaminated motor oil and coal slurries (Bjørnø et al., 1978; Shoh, 1988). Among other things, ultrasound reduces clogging of the filter element. Home humidifiers and therapeutic nebulizers are two examples of the application of ultrasound technology for the production of fine liquid droplets. In the latter case, ultrasonically generated droplets in the 1–5 μm range, containing a dissolved or suspended therapeutic drug, are inhaled by the patient. This is a very effective means of delivering drugs to the lungs. High-quality metal powders can also be generated by sonicating molten metals (Pohlman et al., 1974). In the metallurgical industry, cooling melts are sonicated to remove dissolved gases, and to reduce the grain size of the resulting crystalline solid. For example, sonication reduces the grain size of carbon steels from 200 to 30 μm, which results in a 20–30 % increase in mechanical strength (Abramov, 1987; Mason and Lorimer, 1988). Ultrasound is also used in the pharmaceutical industry to increase crystallization rates, and for crystal size reduction (Skauen, 1967).

The medical uses of ultrasound fall into two categories: diagnostic and therapeutic. High-frequency, low-power ultrasound, typically in the 3–10 MHz range, is routinely used to scan the human body, one of the best-known examples being fetal imaging (Mason and Lorimer, 1988; Birnholz and Farrell, 1984). In this application, short pulses are scattered as they encounter boundaries between different tissues, or between fluids and tissues. Transducers provide numerous measurements of the resulting changes in sound velocity. The echo patterns of many such pulses generate an image. Ultrasonic fetal imaging has several advantages over x-ray radiography: it can delineate certain soft tissues that are difficult to detect by x rays; it has no harmful health effects if the intensity is sufficiently low (this is particularly important during the early stages of pregnancy); and echoes are obtainable from the fetus several weeks before the skeleton is visible by x rays. The therapeutic applications of ultrasound include treatment of sporting injuries, and the breakup of kidney stones. In all of the above examples, ultrasound waves travel across a complex heterogeneous medium, which includes, but is not limited to, liquids.

Sonochemistry is the use of power ultrasound to affect chemical reactivity. This is one of the most interesting applications of ultrasound technology. It is believed that sonochemical effects originate in the extremely high temperatures and pressures generated during cavitational bubble collapse; theoretical understanding in this field is in its infancy. Among the numerous examples of reactions that are affected by ultrasound we cite the hydrolysis of methyl ethanoate (Fogler and Barnes, 1968), the hydrolysis of 4-nitrophenyl esters of aliphatic carboxylic acids (Kristol et al., 1981), the decomposition of water with production of oxygen and hydrogen peroxide (Makino et al., 1983), the sonolysis of chloroform (Henglein and Fischer, 1984) and alkanes (Suslick et al., 1983), the generation of organic lithium reagents (Grignard reagents) by heterogeneous reaction of organic halides with lithium wires (Luche and Damiano, 1980), the preparation of colloidal alkali metals (Luche et al., 1984), the conversion of aldehydes to alkenes (methylenation) (Yamashita et al., 1984), the disproportionation of benzaldehyde (Fuentes and Sinisterra, 1986), the degradation of polymers such as polystyrene, hydroxyethyl cellulose, and poly(methyl methacrylate) (e.g., Mark, 1945; Schmid and Rommel, 1939), the synthesis of graft and block copolymers (Mason and Lorimer, 1988), and the polymerization of nitrobenzene (Donaldson et al., 1979).

Liquid droplets immersed in a second, immiscible host liquid can be levitated ultrasonically. According to a theory by Yosioka and Kawasima (1955), such droplets can experience a steady force due to acoustic radiation pressure; when this force equals the buoyancy force, droplets are levitated. Apfel and his collaborators have used this concept to develop an acoustic levitation apparatus for the measurement of thermophysical properties of metastable liquids (Apfel, 1976; Trinh and Apfel, 1980). The technique, which has much in common with the droplet superheating apparatus (Apfel, 1972a; see also Figure 1.4), allows the measurement of the density of levitated droplets, as well as of the velocity of sound in the droplets. Although the levitation technique is not limited to metastable liquids, it is particularly useful in this case because of the elimination of sites for heterogeneous nucleation. The combined droplet superheating and levitation technique was also used by Apfel (1971a) and by Apfel and Harbison (1975) to study the tensile strength of liquids. The amplitude of the levitating acoustic wave was suddenly increased, causing the droplets to boil explosively. The difference between the ambient pressure and the amplitude of the acoustic pressure at the time of explosion then gave the attainable tensile strength of the droplet liquid as a function of temperature. In Section 3.1.4 it will be explained why such attainable limits of metastability can effectively be regarded as true properties of the system, provided the phase transition is triggered by homogeneous nucleation. The work of Apfel and his collaborators illustrates very nicely the analytical uses of ultrasound in the study of metastable liquids.

One can conclude that tension, when applied locally and cyclically to a liquid, underlies a number of important industrial, medical, and military technologies. In contrast, the application of steady and uniform tension to a bulk liquid is still a laboratory curiosity, and technologies based on this principle are very much in their infancy.

1.3.3 Vapor Explosions

In several industrial processes, two liquids having different temperatures can come into contact. If the hotter liquid is relatively nonvolatile and the colder liquid relatively volatile, the latter can be heated to a point where it vaporizes suddenly and explosively. Such vapor explosions[24] are industrial hazards that can cause injury to personnel, as well as considerable structural damage. In several cases, the properties of highly superheated liquids provide a plausible explanation of the phenomenon, and hence a rational basis for its avoidance in industrial practice. The excellent reviews by Reid (1976, 1978a,b,c, 1983) are the standard references on this topic.

---

[24] Other names given to the same phenomenon include rapid phase transitions, thermal explosions, explosive boiling, and fuel-coolant interactions. This diverse nomenclature reflects the variety of industries (liquefied natural gas transport and storage, paper and aluminum production, reactive metal processing, nuclear power generation) in which vapor explosions have been reported.

Two common routes to superheating a liquid are rapid (adiabatic) depressurization and isobaric heating. In either case, the extent of superheating is limited by the presence of impurities, such as suspended particles, that provide preferential sites for the formation of the vapor phase by heterogeneous nucleation. The extent of penetration into the metastable region can be increased by reducing the number of foreign particles, and by minimizing contact with irregularities and microcavities on bounding solid surfaces.[25] Eventually, a superheated liquid will boil, even in the absence of foreign particles. When this happens, boiling occurs uniformly throughout the liquid, and it is triggered by homogeneous nucleation. Homogeneous nucleation occurs suddenly, over a very narrow range of experimental conditions, be they temperatures (isobaric heating) or pressures (adiabatic depressurization). For example, most liquids can be superheated at atmospheric pressure up to approximately 90% of their critical temperature T_c. Thus the temperature $0.9T_c$ is commonly referred to as the homogeneous nucleation temperature. Since the volume of the vapor phase greatly exceeds that of the liquid, sudden boiling resembles an explosion, hence the term vapor explosion.

According to the superheated liquid theory (Reid, 1978a,b,c, 1983), vapor explosions occur when a cold, volatile liquid reaches its homogeneous nucleation temperature as a result of being heated by a hot, nonvolatile liquid. For isobaric heating at atmospheric pressure, this would in principle require that

$$T_h > 0.9T_c, \tag{1.7}$$

where T_h is the temperature of the hot liquid, and T_c the critical temperature of the cold liquid. More precisely, the relevant hot temperature should be that of the interface between both liquids. If each liquid is idealized as an infinite slab of uniform temperature, the temperature at the interface, T_i, is given by

$$\frac{T_h - T_i}{T_i - T_q} = \sqrt{\frac{(\lambda \rho c)_q}{(\lambda \rho c)_h}}, \tag{1.8}$$

where the subscript q denotes the cold liquid, and λ, ρ, and c denote the thermal conductivity, density, and heat capacity (Reid, 1983). In practice, there is an upper limit for T_h, above which vapor explosions do not occur. The reasons for this are illustrated in Figure 1.19 (Katz and Sliepcevich, 1971). If the temperature difference between two liquids is small (for example, when the hot liquid is only slightly warmer than the cold liquid's boiling point), heat transfer occurs by convection and conduction, and no bubbles are formed at the interface. Slightly higher values of the temperature difference cause the

---

[25] If the liquid is fragmented into small droplets, for example, by dispersing it in a second liquid with which it is immiscible, the probability of finding an impurity in any given droplet is small. The majority of the droplets can then be extensively superheated, avoiding heterogeneous nucleation. This is the basis of Apfel's droplet superheating technique discussed in Section 1.1.2.

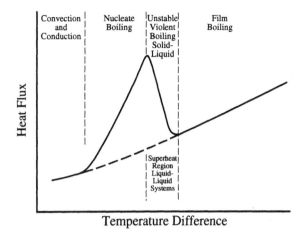

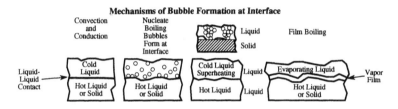

Figure 1.19. Mechanisms of heat transfer and bubble formation at interfaces. (Adapted from Katz and Sliepcevich, 1971)

appearance of bubbles at the interface; this regime is known as nucleate boiling. If the temperature difference is increased further, a point will be reached where bubbles appear in such high numbers that they cause a reduction in the heat flux. Beyond this point, the system enters an unstable transition regime where further increases in the temperature difference cause a reduction in the heat flux. This regime is characterized by vigorous boiling, and it ends when the interface is covered by a continuous vapor film (Leidenfrost point). Thereafter, stable film boiling occurs, with proportionality between flux and driving force reestablished.

According to the superheated liquid theory (Reid, 1978a,b,c, 1983), vapor explosions occur when the temperature difference between the hot, nonvolatile liquid and the cold, volatile liquid is slightly smaller than the Leidenfrost point. The system then no longer supports a vapor blanket at the interface, and high heat fluxes occur. However, because the interface is smooth (that is to say, it does not have the microcavities and heterogeneities that would be present if the cold liquid were in contact with a hot solid surface), the cold liquid is superheated until it reaches its homogeneous nucleation temperature. The smooth interface between the two liquids, in other words, prevents the system from entering the unstable transitional boiling regime. The above-described scenario is better described as a rapid phase transition (RPT), a term coined by Reid (1983), than with the more commonly used term vapor explosion.

The superheated-liquid theory of vapor explosions can explain most of the known facts that occur when liquefied natural gas (LNG; cold, volatile) is spilled on water (hot, nonvolatile). LNG contains typically 70–90 % (molar) methane, 10–15 % ethane, and 1–10 % propane. Natural gas is used as a residential, commercial, and industrial fuel. A convenient method of transporting and shipping natural gas from its geological source to areas of high demand is to liquefy it cryogenically. The product of this operation is LNG. Interest in LNG-water interactions stems from safety concerns about possible spills during shipping and transfer. Boiling of the cryogenic fluid would then result in the formation of a cloud of flammable vapor (Nakanishi and Reid, 1971). In the late 1960s, studies were undertaken by the U.S. Bureau of Mines to investigate spreading and vaporization rates of LNG spills on water (Burgess et al., 1970, 1972). Unexpectedly, vapor explosions occurred during these tests. This finding stimulated research efforts, in both industry and academia, aimed at understanding the phenomenon.[26]

A systematic study of laboratory-scale vapor explosions involving light hydrocarbons on a variety of hot liquid substrates was performed by Porteus and Reid (1976). The experiments consisted of spilling approximately 500 cm^3 of hydrocarbon on a liquid substrate (generally water) housed in a transparent polycarbonate container. Explosions were monitored with a pressure transducer located at the bottom of the container. Their results are summarized in Table 1.3.

It can be seen that vapor explosions occurred over the approximate range $1 < T_h/T_{hn} < 1.1$. The exception, methanol as substrate, can be explained in light of (1.8) once due account is taken of this substance's comparatively small thermal conductivity (Reid, 1983). This causes T_h to be higher (relative to the water case) in order to attain $T_i > T_{hn}$. Taking $T_h/T_{hn} = 1.1$ as the upper limit for vapor explosions to occur, (1.7) can be generalized as

$$0.99 > T_h/T_c > 0.9. \qquad (1.9)$$

In the experiments of Porteous and Reid, the transition from nucleate boiling to vapor explosions was very sharp, whereas that from explosion to film boiling was not. Equation (1.9) allows the construction of diagrams showing the region where vapor explosions can occur if a given hydrocarbon mixture is spilled on

---

[26]Reid (1983) cites several earlier reports of vapor explosions during LNG spills on water. However, it was the U.S. Bureau of Mines study that first attracted widespread attention and generated interest in LNG-water explosions.

Table 1.3. Explosive Ranges for Light Hydrocarbon Spills[a]

Hydrocarbon	Hot Liquid	$T_h{}^b$ (K)	T_h/T_{hn} for Explosion[c]
Ethane	Methanol	306–331	1.14–1.23
	Methanol–water	296–304	1.10–1.13
Propane	Water	326–334	1.00–1.02
Isobutane	Ethylene glycol– water	374–379	1.04–1.05
n-Butane	Water	371	0.98–?
	Ethylene glycol– water	387–398	1.03–1.06
Propylene	Water	317–346	0.97–1.06
Isobutylene	Ethylene glycol	379–408	1.03–1.06
Ethane–propane	Water	293	1.02–1.06
Ethane–n-butane	Water	278	1.00–1.01
		283	1.02–1.04
		293	1.05–1.08
		303	1.06–1.11
Ethane–n-pentane	Water	293	1.04–1.08
Ethylene–n-butane	Water	293	1.05–1.08
Ethylene–n-pentane	Water	293	1.05–1.14
Ethane–propane– n-butane	Water	293	1.02–1.08
Methane–ethane– propane–n-butane	Water	323	1.08–1.13

[a] Adapted from Reid (1983).
[b] Bulk temperature of the hot liquid.
[c] Ratio of hot-liquid bulk temperature to homogeneous nucleation temperature of the cold liquid. For pure hydrocarbons, homogeneous nucleation temperatures were taken from Blander and Katz (1975) and Porteous and Blander (1975). For mixtures, this temperature was calculated as a mole fraction average of component homogeneous nucleation temperatures.

water. One such diagram is Figure 1.20 (Reid, 1983), corresponding to spills of methane-ethane-propane mixtures on water at 25°C. The range of compositions that can give rise to a vapor explosion is bounded by the lines $T_h/T_{hn} = 1$ and $T_h/T_{hn} = 1.1$, and the homogeneous nucleation temperature of a given LNG is estimated as 90% of the mole-fraction-averaged critical temperature.[27]

Large-scale (ca. 40 m³ LNG) spill tests have been conducted (e.g., McRae et al., 1984). Vapor explosions resulting from such large-scale spills are much

[27] This approximation is not valid if the mixture contains dissimilar compounds.

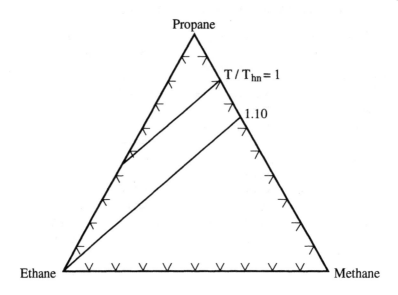

Figure 1.20. Regions where vapor explosions are possible for liquefied methane-ethane-propane spills on water at 25°C. Compositions that can give rise to explosions are bounded by the lines $T / T_{hn} = 1.1$ and 1, where $T = 298$ K and T_{hn} is the homogeneous nucleation temperature of the hydrocarbon mixture. (Adapted from Reid, 1983)

more difficult to predict than their carefully controlled laboratory counterparts. It appears that localized superheating leading to homogeneous nucleation can serve as a trigger, but little is known with certainty about the mechanisms by which such a local phenomenon can propagate into a coherent, large-scale event. Another element of uncertainty is the change in composition of the LNG as a result of boiling, after a spill. Thus, methane-rich LNGs (lower right region of Figure 1.20) will normally be outside the explosive region immediately following a spill. However, as methane is preferentially boiled off, a delayed explosion can occur if locally the mixture enters the explosion regime bounded by the T/T_{hn} lines 1 and 1.1. The explosive damage of large-scale LNG spills has been estimated by measuring air overpressures as a function of distance from the explosion. Air blast equivalent estimates range from a few grams to 5.5 kg of TNT (trinitrotoluene) (McRae et al., 1984).

In the nuclear power generation industry (Cronenberg, 1980), an important safety concern is the analysis of events that would occur should overheating of a reactor's core cause the fuel to melt (hot, nonvolatile liquid), and subsequently to interact with the coolant (cold, volatile liquid). A number of molten-fuel–coolant interactions in test reactors are listed in Table 1.4.

Table 1.4. Some Instances of Molten Fuel–Coolant Interactions in Experimental Nuclear Reactors[a]

Test	Fuel/Coolant	Thermal Interaction Mode	References
SL-1	U/H_2O	Molten uranium fuel into H_2O	Masson (1970) General Electric (1962a,b)
BORAX-1	U-Al/H_2O	Molten U-Al alloy into H_2O	Masson (1970) Goodwin (1954)
SPERT-1	U-Al/H_2O	Molten U-Al alloy into H_2O	Miller et al. (1964) Masson (1970)

[a] Adapted from Cronenberg (1980).

In these tests, vapor explosions occurred. In addition, laboratory tests involving the injection of liquid sodium at 400°C and velocities ranging from 5 to 9.5 m sec^{-1} into molten UO_2 at ca. 3000°C (Anderson and Armstrong, 1972) have resulted in vapor explosions (Fauske, 1973a,b). However, the interpretation of these vapor explosions in terms of the superheated-liquid theory is much less straightforward than in the LNG-water case. The sodium–UO_2 interface temperature, 1100°C, calculated according to (1.8) (Fauske, 1973a) is much lower than the homogeneous nucleation temperature for sodium $(0.9T_c = 2537 \text{ K} = 2264°C)$. Fauske (1973a,b) has proposed a mechanism involving fragmentation of the injected sodium into small droplets, and subsequent heating by the surrounding molten UO_2 to the homogeneous nucleation temperature. In order for this to happen, a time lag must occur during which sodium fractionation and subsequent slow heating take place. Although this is consistent with Anderson and Armstrong's observation of delayed vapor explosions, alternative mechanisms leading to vapor explosions in molten-fuel–coolant systems have been proposed (Cronenberg, 1980). Vapor explosions are an important safety concern in nuclear power generation; however, whether test explosions were caused by homogeneous nucleation of the coolant is not known with certainty.

Vapor explosions are also an important safety concern in the paper, metals-processing, and cryogenic industries (Table 1.5).

Reid (1978a,b,c) discusses alternative theories that have been proposed to explain vapor explosions. While additional factors can play an important role, especially in determining whether a large-scale, coherent vapor explosion occurs in a given situation, the view that vapor explosions occur as a result of

Table 1.5. Components of Vapor Explosions in the Paper, Cryogenic, and Metal Industries

Industry	Hot, Nonvolatile Liquid	Cold, Volatile Liquid	Selected References
Paper	Smelt[a]	Water	Reid (1976, 1978a–c, 1983) Krause et al. (1973) Schick (1980) Schick and Grace (1982)
Cryogenic	Water or oil	Refrigerants	Reid (1976, 1978a–c,1983) Fauske (1973a) Henry et al. (1974) Rausch and Levine (1973)
Aluminum	Molten aluminum	Water	Reid (1976, 1978a–c, 1983) Hess and Brondyke (1969)
Metals processing	Molten metals	Water	Reid (1983)

[a]Molten mixture of sodium carbonate, sodium sulfide, and other sodium or potassium salts.

homogeneous nucleation in a superheated liquid provides the only generally applicable framework for analyzing and interpreting available evidence.[28]

Homogeneous nucleation of superheated liquids could also be responsible for explosions of tank trucks or railroad cars carrying pressurized liquids (Reid, 1979). It has often been observed that when tank trucks or railroad cars are involved in an accident, a violent explosion follows the initial localized failure of the tank metal. A localized fire due to the accident can lead to high internal tank pressure and high metal wall temperature, and thus to localized metal failure. The violent explosion that follows local metal failure often leads to the destruction of the tank, the very rapid release of its contents, and, if the latter are flammable, to very large fires. Reid and co-workers (Reid, 1979; Kim-E and Reid, 1983) have argued plausibly that the rapid depressurization of the liquid inside the tank can lead to extensive superheating of the bulk liquid, and hence to a vapor explosion when the bulk liquid suddenly boils homogeneously. The explosion of tank trucks or railroad cars carrying pressurized

[28]For aluminum-water and smelt-water explosions, interfacial diffusion must be taken into account. When this is done (see, e.g., Schick, 1980; Schick and Grace, 1982; Reid, 1983) experimental observations can be explained in light of the superheated-liquid theory.

liquids is commonly referred to as a boiling-liquid expanding-vapor explosion (BLEVE). While it is probable that other phenomena may also trigger BLEVEs, homogeneous nucleation of a superheated liquid is a likely contributor. Several safety measures suggested by this hypothesis are discussed by Reid (1979) and Kim-E and Reid (1983).

1.3.4 Kinetic Inhibition of Natural Gas Clathrate Hydrates

Natural-gas clathrate hydrates (Holder et al., 1988; Sloan, 1990; Englezos, 1993) are crystalline solids in which water molecules form assemblies of hydrogen-bonded polyhedral cavities, each of which contains one guest molecule, such as methane. The plugging of pipelines by gas hydrates is a serious problem during transport of wet natural gas from offshore fields.[29] There is a strong economic incentive for developing efficient methods for preventing hydrate formation during exploration, production, transmission, and processing of natural gas (Sloan et al., 1994). A promising approach to this problem is the kinetic inhibition of hydrate formation, which has much in common with the action of antifreeze proteins in cold-water fish (Section 1.2.1).

There are three clathrate hydrate structures for which numerous guests have been identified. Structure I (sI) is a body-centered cubic structure consisting of small pentagonal dodecahedral cavities (abbreviated 5^{12}: twelve pentagonal faces) and large tetrakaidecahedral cavities ($5^{12}6^2$: twelve pentagonal and two hexagonal faces). Structure II (sII) is a diamond lattice composed of small 5^{12} and large hexakaidecahedral ($5^{12}6^4$) cavities. Structure H (sH) contains three types of cavities: 5^{12} and $4^3 5^6 6^3$ (small), and $5^{12}6^8$ (large) (Ripmeester et al., 1994). Until recently, sH had only been found in the laboratory (Ripmeester et al., 1987);[30] a hexagonal structure has been proposed, but definitive structural determination is still lacking. sH can accommodate large (7.5–8.5 Å) guest molecules, such as adamantane and methylcyclohexane, but requires the simultaneous presence of smaller guests such as Xe or H_2S to stabilize the lattice (Lederhos et al., 1992). Laboratory studies suggest that sH hydrate may be important in petroleum processing (Mehta and Sloan, 1993, 1994). The following discussion pertains to sI and sII, both of which are known to occur both naturally and in man-made natural gas production and transportation facilities.

Table 1.6 gives geometric characteristics of sI and sII hydrate architecture, and Figure 1.21 shows the relationship between the size of common guest molecules, the cavities they stabilize, and the hydrate structures that result.

The (P, T) locus where a pure gas coexists with gas-saturated liquid water and with a hydrate solid phase defines a line in the pressure vs. temperature

---

[29]Three phases, gas, oil, and water, usually flow out of a well when hydrocarbons are extracted from the earth (Erickson and Brown, 1994).

[30]The first observation of sH outside of the laboratory has recently been reported (Sassen and MacDonald, 1994).

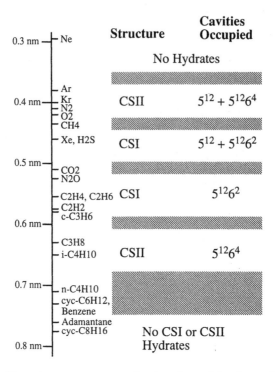

Figure 1.21. Size of common guest molecules, the cavities they stabilize, and the clathrate hydrate structures that result. [Reprinted, with permission, from Christiansen and Sloan, in *Ann. N.Y. Acad. Sci.* 715: 283 (1994)]

Table 1.6. Geometry of sI and sII Hydrate Crystal Structures[a]

	sI	sII
Symmetry	Cubic	Cubic
Small Cavity (Diameter, Å)	5^{12} (7.88)	5^{12} (7.82)
Water Molecules per Cavity	20	20
Number of Cavities per Cell	2	16
Large Cavity (Diameter, Å)	$5^{12}6^2$ (8.60)	$5^{12}6^4$ (9.46)
Water Molecules per Cavity	24	28
Number of Cavities per Cell	6	8
Water Molecules per Cell	46	136

[a] Adapted from Holder et al. (1994) and Long et al. (1994).

plane.[31] This line has positive slope. At fixed temperature, it gives the limiting pressure below which hydrates do not form; at fixed pressure, it gives the limiting temperature above which hydrates do not form. For binary gas mixtures, there is one curve for each gas-phase composition. Hydrate formation is always favored by lowering the temperature (it is exothermic) and increasing the pressure (it is accompanied by a volume contraction).

Conventional means of hydrate inhibition include water removal, heating, depressurization, and addition of thermodynamic inhibitors. Alcohols, such as methanol and glycols, are common thermodynamic inhibitors. They act by displacing the equilibrium hydrate formation locus towards higher pressures and lower temperatures, thereby enlarging the range of stability of the liquid phase with respect to hydrate formation. Large amounts of thermodynamic inhibitors are needed: 50% by weight methanol loadings in the aqueous phase are typical. The costs associated with this mode of hydrate prevention are substantial: injection operating costs can be 1% of annual revenue, and processing capital costs for hydrate prevention range between 5% and 8% of total capital costs. Thus there is a considerable economic incentive to developing better ways of preventing hydrate blockage of transmission lines (Long et al., 1994).

An interesting idea that is currently being studied in the laboratory is to find suitable kinetic inhibitors to hydrate formation. Kinetic inhibitors retard the formation of a solid phase, but they do not provide thermodynamic stability. A thermodynamic inhibitor eliminates the supersaturation with respect to hydrate formation at the given temperature and pressure; a kinetic inhibitor does not, hence the liquid phase on which it acts is metastable. The advantage of kinetic inhibitors over thermodynamic ones is that much smaller amounts are required, typically 0.5% by weight. Two polymers, polyvinylpirrolidone (PVP) and hydroxyethylcellulose (HEC), have recently been found to be promising kinetic inhibitor candidates in laboratory tests. Addition of 0.5% PVP (molecular weight = 360, 000) retarded the initial appearance of cloudiness due to hydrate formation at 0°C in a 17:1 (mol) water-tetrahydrofuran[32] mixture by almost ninety minutes, and retarded complete solidification by more than six hours (Long et al., 1994).

As with antifreeze proteins, the mechanism of action of kinetic inhibitors is not well understood. Some possibilities include adsorption on growth planes of crystalline nuclei, crystal habit modification (Gryte, 1994), and interference with the growth of subcritical nuclei (Sloan and Fleyfel, 1991). Improved knowledge of the mechanisms of kinetic inhibition will provide a rational basis

---

[31] This follows from the phase rule, $F + P = C + 2$, where F is the number of degrees of freedom, P the number of coexisting phases, and C the number of components. For a single-component gas and water, three-phase coexistence has one degree of freedom: fixing the temperature determines the coexistence pressure and vice versa. This defines a line on the (P, T) plane.

[32] Tetrahydrofuran forms hydrates in water. Experimentally, it has the advantage with respect to gas hydrate formers that it is miscible with water.

for the search for better inhibitors. Nuclear magnetic resonance (Ripmeester et al., 1994; Fleyfel et al., 1994) and molecular simulation methods (Hwang et al., 1993; Long and Sloan, 1993; Báez and Clancy, 1994) hold considerable promise for the microscopic study of kinetics and mechanisms of hydrate formation, and hence will be important in future studies of kinetic inhibition.

The promising concept of retarding the formation of hydrates inside the metastable region through the addition of small amounts of kinetic inhibitors is currently undergoing pilot-plant and field testing (Long et al., 1994).

References

Abramov, O.V. 1987. "Action of High Intensity Ultrasound on Solidifying Metal." *Ultrasonics*. 25: 73.

Anderson, R.P., and D.R. Armstrong. 1972. "Laboratory Tests of Molten-Fuel-Coolant Interactions." *Trans. Amer. Nucl. Soc.* 15: 313.

Angell, C.A., and H. Kanno. 1976. "Density Maxima in High-Pressure Supercooled Water and Liquid Silicon Dioxide." *Science*. 193: 1121.

Apfel, R.E. 1971a. "A Novel Technique for Measuring the Strength of Liquids." *J. Acoust. Soc. Amer.* 49: 145.

Apfel, R.E. 1971b. "Tensile Strength of Superheated n-Hexane Droplets." *Nature Phys. Sci.* 233: 119.

Apfel, R.E. 1972a. "Water Superheated to 279.5°C at Atmospheric Pressure." *Nature Phys. Sci.* 238: 63.

Apfel, R.E. 1972b. "The Tensile Strength of Liquids." *Sci. Amer.* 227: 58.

Apfel, R.E. 1976. "Technique for Measuring the Adiabatic Compressibility, Density, and Sound Speed of Submicrometer Liquid Samples." *J. Acoust. Soc. Amer.* 59: 339.

Apfel, R.E., and J.P. Harbison. 1975. "Acoustically Induced Explosions of Superheated Droplets." *J. Acoust. Soc. Amer.* 57: 1371.

Apfel, R.E., and M.P. Smith. 1977. "The Tensile Strength of Di-Ethyl Ether Using Briggs' Method." *J. Appl. Phys.* 48: 2077.

Arnold, F. 1992. "Stratospheric Aerosol Increases and Ozone Destruction: Implications from Mass Spectrometer Measurements." *Ber. Bunsenges. Phys. Chem.* 96: 339.

Askenasy, E. 1897. "Beiträge zur Erklärung des Saftsteigens." *Naturhist.-Med. Verein, Heidelberg*. 5: 429.

Avedisian, C.T. 1982. "Effect of Pressure on Bubble Growth within Liquid Droplets at the Superheat Limit." *J. Heat Transf.* 104: 750.

Avedisian, C.T. 1985. "The Homogeneous Nucleation Limits of Liquids." *J. Phys. Chem. Ref. Data*. 14: 695.

Avedisian, C.T., and J.R. Sullivan. 1984. "A Generalized Corresponding States Method for Predicting the Limits of Superheat of Mixtures. Application to the Normal Alcohols." *Chem. Eng. Sci.* 39: 1033.

Báez, L.A., and P. Clancy. 1994. "Computer Simulation of the Crystal Growth and Dissolution of Natural Gas Hydrates." In *International Conference on Natural Gas Hydrates*, E.D. Sloan, J. Happel, and M.A. Hnatow, eds. *Ann. N.Y. Acad. Sci.* 715: 177.

Beams, J.W. 1959. "Tensile Strength of Liquid Argon, Helium, Nitrogen, and Oxygen." *Phys. Fluids*. 2: 1.

Belak, J., R. LeSar, and R.D. Etters. 1990. "Calculated Thermodynamic Properties and Phase Transitions of Solid N_2 at Temperatures $0 \leq T \leq 300$ K and Pressures $0 \leq P \leq 100$ GPa." *J. Chem. Phys.* 92: 5430.

Berthelot, M. 1850. "Sur Quelques Phénomènes de Dilation Forcée des Liquides." *Ann. Chim. Phys.* 30: 232.

Beyer, K.D., S.W. Seago, H.Y. Chang, and M.J. Molina. 1994. "Composition and Freezing of Aqueous H_2SO_4/HNO_3 Solutions under Polar Stratospheric Conditions." *Geophys. Res. Lett.* 21: 871.

Birnholz, J.C., and E.E. Farrell. 1984. "Ultrasound Images of Human Fetal Development." *Amer. Sci.* 72: 608.

Bjørnø, L., S. Gram, and P.R. Steenstrup. 1978. "Some Studies of Ultrasound Assisted Filtration Rates." *Ultrasonics.* 16: 103.

Blander, M., and J.L. Katz. 1975. "Bubble Nucleation in Liquids." *Amer. Inst. Chem. Eng. J.* 21: 833.

Blander, M., D. Hengstenberg, and J.L. Katz. 1971. "Bubble Nucleation in n-Pentane, n-Hexane, n-Pentane + Hexadecane Mixtures, and Water." *J. Phys. Chem.* 75: 3613.

Boehm, J. 1893. "Capillarität und Saftsteigen." *Ber. Deutsch. Bot. Ges.* 11: 203.

Brandts, J.F. 1964. "Thermodynamics of Protein Denaturation. I. The Denaturation of Chymotrypsinogen." *J. Amer. Chem. Soc.* 86: 4291.

Briggs, A., and T. Maxwell. 1975. "Method of Preparation of Lyophilized Biological Products." U.S. Patent No. 3, 928, 566.

Briggs, L.J. 1949. "A New Method for Measuring the Limiting Negative Pressure in Liquids." *Science.* 109: 440.

Briggs, L.J. 1950. "Limiting Negative Pressure of Water." *J. Appl. Phys.* 21: 721.

Briggs, L.J. 1951a. "The Limiting Negative Pressure of Acetic Acid, Aniline, Carbon Tetrachloride, and Chloroform." *J. Chem. Phys.* 19: 970.

Briggs, L.J. 1951b. "The Limiting Negative Pressure of Five Organic Liquids and the 2-Phase System, Water-Ice." *Science.* 113: 483.

Briggs, L.J. 1953. "Limiting Negative Pressure of Mercury in Pyrex Glass." *J. Appl. Phys.* 24: 488.

Briggs, L.J. 1955. "Maximum Superheating of Water as a Measure of Negative Pressure." *J. Appl. Phys.* 26: 1001.

Bull, T.H. 1956. "The Tensile Strength of Viscous Liquids under Dynamic Loading." *Brit. J. Appl. Phys.* 7: 416.

Bulman, R.B. 1969. *Effective Rainfall.* Bushby Bros.: Carlisle.

Burgess, D.S., J.N. Murphy, and M.G. Zabetakis. 1970. *Hazards Associated with the Spillage of Liquefied Natural Gas on Water.* U.S. Bureau of Mines Rep. Invest. No. 7448.

Burgess, D.S., J. Biordi, and J. Murphy. 1972. *Hazards of Spillage of LNG into Water.* U.S. Bureau of Mines. PMSRC Report No. 4177.

Burke, M.J., L.V. Gusta, H.A. Quamme, C.J. Weiser, and P.H. Li. 1976. "Freezing and Injury in Plants." *Annu. Rev. Plant Physiol.* 27: 507.

Cannon, J.F. 1974. "Behavior of the Elements at High Pressure." *J. Phys. Chem. Ref. Data.* 3: 781.

Canny, M.J. 1995. "A New Theory for the Ascent of Sap—Cohesion Supported by Tissue Pressure." *Ann. Bot.* 75: 343.

Carlson, G.A. 1975. "Dynamic Tensile Strength of Mercury." *J. Appl. Phys.* 46: 4069.

Carlson, G.A., and K.W. Henry. 1973. "Technique for Studying Dynamic Tensile Failure in Liquids: Application to Glycerol." *J. Appl. Phys.* 44: 2201.

Carlson, G.A., and H.S. Levine. 1975. "Dynamic Tensile Strength of Glycerol." *J. Appl. Phys.* 46: 1594.

Chan, H.S., and K.A. Dill. 1991. "Polymer Principles in Protein Structure and Stability." *Annu. Rev. Biophys. Biophys. Chem.* 20: 447.

Chao, H., F.D. Sönnichsen, C.I. DeLuca, B.D. Skyes, and P.L. Davies. 1994. "Structure-Function Relationship in the Globular Type III Antifreeze Protein: Identification of a Cluster of Surface Residues Required for Binding to Ice." *Protein Sci.* 3: 1760.

Chapman, P.J., B.E. Richards, and D.H. Trevena. 1975. "Monitoring the Growth of Tension in a Liquid Contained in a Berthelot Tube." *J. Phys. E. Sci. Instrum.* 8: 731.

Chen, B.-L., and J.A. Schellman. 1989. "Low-Temperature Unfolding of a Mutant of Phage T4 Lysozyme. 1. Equilibrium Studies." *Biochem.* 28: 685.

Christiansen, R.L., and E.D. Sloan, Jr. 1994. "Mechanisms and Kinetics of Hydrate Formation." In *International Conference on Natural Gas Hydrates*, E.D. Sloan, J. Happel, and M.A. Hnatow, eds. *Ann. N.Y. Acad. Sci.* 715: 283.

Couzens, D.C.F., and D.H. Trevena. 1969. "Critical Tension in a Liquid Under Dynamic Conditions of Stressing." *Nature.* 222: 473.

Couzens, D.C.F., and D.H. Trevena. 1974. "Tensile Fracture of Liquids Under Dynamic Stressing." *J. Phys. D. Appl. Phys.* 7: 2277.

Cracknell, A.P. 1980. *Ultrasonics.* Wykeham: London.

Cronenberg, A.W. 1980. "Recent Developments in the Understanding of Energetic Molten Fuel–Coolant Interactions." *Nucl. Safety.* 21: 319.

Cwilong, B.M. 1947. "Sublimation in a Wilson Chamber." *Proc. Roy. Soc. A.* 190: 137.

D'Antonio, M.C. 1989. "A Thermodynamic Investigation of Tensile Instabilities and Sub-Triple Liquids." Ph.D. thesis. Princeton University, Princeton, N.J.

DeVries, A.L. 1968. "Freezing Resistance in Some Antarctic Fishes." Ph.D. diss., Stanford University, Stanford, Calif.

DeVries, A.L. 1983. "Antifreeze Peptides and Glycopeptides in Cold-Water Fishes." *Annu. Rev. Physiol.* 45: 245.

DeVries, A.L., J. Vandenheede, and R.E. Fenney. 1971. "Primary Structure of Freezing Point Depressing Proteins." *J. Biol. Chem.* 246: 305.

Dill, K.A. 1985. "Theory of Folding and Stability of Globular Proteins." *Biochem.* 24: 1501.

Dill, K.A. 1990. "Dominant Forces in Protein Folding." *Biochem.* 29: 7133.

Dill, K.A., D.O.V. Alonso, and K. Hutchinson. 1989. "Thermal Stabilities of Globular Proteins." *Biochem.* 28: 5439.

Dixon, H.H. 1914. *Transpiration and the Ascent of Sap in Plants.* Macmillan: London.

Dixon, H.H., and J. Joly. 1896. "On the Ascent of Sap." *Phil. Trans. Roy. Soc. London B.* 186: 563.

Dixon, W.L., F. Franks, and T. ap Rees. 1981. "Cold-Lability of Phosphofructokinase from Potato Tubers." *Phytochem.* 20: 969.

Donaldson, D.J., M.D. Farrington, and P. Kruus. 1979. "Cavitation-Induced Polymerization of Nitrobenzene." *J. Phys. Chem.* 83: 3130.

Donny, F. 1843. "Mémorie sur la Cohésion des Liquides et sur leur Adhérence aux Corps Solides." *Acad. Bruxelles.* 17: 1.

Donny, F. 1846. "Mémorie sur la Cohésion des Liquides et sur leur Adhérence aux Corps Solides." *Ann. Chim. Phys.* 16: 167.

Donoghue, J.J., R.E. Vollrath, and E. Gerjuoy. 1951. "The Tensile Strength of Benzene." *J. Chem. Phys.* 19: 55.

Drdla, K., A. Tabazadeh, R.P. Turco, M.Z. Jacobson, J.E. Dye, C. Twohy, and D. Baum-gardner. 1994. "Analysis of the Physical State of One Arctic Polar Stratospheric Cloud Based on Observations." *Geophys. Res. Lett.* 21: 2475.

Dufour, M.L. 1861a. "Sur l'Ebullition des Liquides." *C. R. Acad. Sci.* 52: 986.

Dufour, M.L. 1861b. "Sur l'Ebullition des Liquides." *C. R. Acad. Sci.* 53: 846.

Duman, J. 1982. "Insect Antifreezes and Ice-Nucleating Agents." *Cryobiol.* 19: 613.

Duman, J.G., and A.L. DeVries. 1972. "Freezing Behavior of Aqueous Solutions of Glycoproteins from the Blood of Antarctic Fish." *Cryobiol.* 9: 469.

Duman, J.G., L. Xu, L. G. Neven, D. Tursman, and D.W. Wu. 1991. "Haemolymph Proteins Involved in Insect Subzero-Temperature Tolerance: Ice Nucleators and An-tifreeze Proteins." In *Insects at Low Temperature*, R.E. Lee and D.L. Denlinger, eds., chap. 5. Chapman and Hall: New York.

Eberhart, J.G., W. Kremsner, and M. Blander. 1975. "Metastability Limits of Super-heated Liquids: Bubble Nucleation Temperatures of Hydrocarbons and their Mix-tures." *J. Colloid Interf. Sci.* 50: 369.

Ehrenfest, P. 1933. "Phasenumwandlungen im Ueblichen und Erweiterten Sinn, Clas-sifiziert nach den Entsprechenden Singularitaeten des Thermodynamischen Poten-tiales." *Comm. Kamerlingh Onnes Lab. Leiden.* Suppl. 75b: 628.

Englezos, P. 1993. "Clathrate Hydrates." *Ind. Eng. Chem. Res.* 32: 1251.

Erickson, D., and T. Brown. 1994. "Hydrate Occurrence in Multiphase Flowlines." In *International Conference on Natural Gas Hydrates*, E.D. Sloan, J. Happel, and M.A. Hnatow, eds. *Ann. N.Y. Acad. Sci.* 715: 40.

Ermakov, G.V., and V.P. Skripov. 1967. "Saturation Line, Critical Parameters, and the Maximum Degree of Superheating of Perfluoro-Paraffins." *Russ. J. Phys. Chem.* 41: 39.

Ermakov, G.V., and V.P. Skripov. 1969. "Experimental Test of the Theory of Homoge-neous Nucleus Formation in Superheated Liquids." *Russ. J. Phys. Chem.* 43: 1242.

Evans, A. 1979. "A Transparent Recording Berthelot Tensiometer." *J. Phys. E. Sci. Instrum.* 12: 276.

Fahey, D.W., K.K. Kelly, G.V. Ferry, L.R. Poole, J.C. Wilson, D.M. Murphy, M. Loewen-stein, and K.R. Chan. 1989. "In Situ Measurements of Total Reactive Nitrogen, Total Water, and Aerosol in a Polar Stratospheric Cloud in the Antarctic." *J. Geophys. Res.* 94: 11299.

Farman, J.C., B.G. Gardiner, and J.D. Shanklin. 1985. "Large Losses of Total Ozone in Antarctica Reveal Seasonal ClO_x/NO_x Interaction." *Nature.* 315: 207.

Fauske, H.K. 1973a. "On the Mechanism of Uranium Dioxide-Sodium Explosive Inter-actions." *Nucl. Sci. Eng.* 51: 95.

Fauske, H.K. 1973b. "Nucleation of Liquid Sodium in Fast Reactors." *React. Technol.* 15: 278.

Fleyfel, F., K.Y. Song, A. Kook, R. Martin, and R. Kobayashi. 1994. "^{13}C NMR of Hydrate Precursors in Metastable Regions." In *International Conference on National Gas Hydrates*, E.D. Sloan, J. Happel, and M.A. Hnatow, eds. *Ann. N.Y. Acad. Sci.* 751: 212.

Fogler, S., and D. Barnes. 1968. "Shift in the Optimal Power Input in an Ultrasonic Reaction." *Ind. Eng. Chem. Fundam.* 7: 222.

Franks, F. 1983. *Water*. Chap 1. The Royal Society of Chemistry: London.

Franks, F. 1985. *Biochemistry and Biophysics at Low Temperatures*. Cambridge University Press: Cambridge.

Franks, F. 1988. "Storage in the Undercooled State." In *Low Temperature Biotechnology,* J.J. McGrath and K.R. Dilles, eds., p. 107. American Society of Mechanical Engineers: New York.

Franks, F. 1994. "Long-Term Stabilization of Biologicals." *Biotech.* 12: 253.

Franks, F., and R.H.M. Hatley. 1985. "Low Temperature Unfolding of Chymotrypsinogen." *Cryo-Lett.* 6: 171.

Franks, F., and R.H.M. Hatley. 1991. "Stability of Proteins at Subzero Temperatures: Thermodynamics and Some Ecological Consequences." *Pure Appl. Chem.* 63: 1367.

Franks, F., and R.H.M. Hatley. 1993. "Stable Enzymes by Water Removal." In *Stability and Stabilization of Enzymes*, W.J.J. von der Tweel, A. Harder, and R.M. Buitelaar, eds., p. 45. Elsevier: Amsterdam.

Franks, F., S.F. Mathias, P. Galfre, S.D. Webster, and D. Brown. 1983. "Ice Nucleation and Freezing in Undercooled Cells." *Cryobiol.* 20: 298.

Franks, F., R.H.M. Hatley, and H.L. Friedman. 1988. "The Thermodynamics of Protein Stability. Cold Destabilization as a General Phenomenon." *Biophys. Chem.* 31: 307.

Fuentes, A., and J.V. Sinisterra. 1986. "Single Electron Transfer Mechanism of the Cannizzaro Reaction in Heterogeneous Phase under Ultrasonic Conditions." *Tetrahed. Lett.* 27: 2967.

General Electric Company, Idaho Test Station. 1962a. *Additional Analysis of the SL-1 Excursion: Final Report of Progress, July–October 1962*. U.S. Atomic Energy Commission Report No. IDO-19313.

General Electric Company, Idaho Test Station. 1962b. *Final Report of SL-1 Recovery Operation*. U.S. Atomic Energy Commission Report No. IDO-19311.

George, M.F., M.J. Burke, H.M. Pellett, and A.G. Johnson. 1974. "Low Temperature Exotherms and Woody Plant Distribution." *Hort. Sci.* 9: 519.

George, M.F., M.R. Beckwar, and M.J. Burke. 1982. "Freezing Avoidance by Deep Undercooling of Tissue Water in Winter-Hardy Plants." *Cryobiol.* 19: 628.

Godwal, B.K., C. Meade, R. Jeanloz, A. Garcia, A.Y. Liu, and M.L. Cohen. 1990. "Ultrahigh-Pressure Melting of Lead: A Multidisciplinary Study." *Science.* 248: 462.

Goodwin, J.G. 1954. *Evaluation of a Modified Zircaloy-2 Ingot F-1071 Melted at WAPD*. U.S. Atomic Energy Commission Report No. AECD-3688.

Green, J.L., D.J. Durben, G.H. Wolf, and C.A. Angell. 1990. "Water and Solutions at Negative Pressure: Raman Spectroscopic Study to −80 Megapascals." *Science.* 249: 649.

Griko, Yu.V., P.L. Privalov, J.M. Sturtevant, and S.Yu. Venyaminov. 1988. "Cold Denaturation of Staphylococcal Nuclease." *Proc. Natl. Acad. Sci. U.S.A.* 85: 3343.

Gryte, C.C. 1994. "On the Control of Nucleation and Growth by Inhibition of Gas Hydrate Formation." In *International Conference on Natural Gas Hydrates*, E.D. Sloan, J. Happel, and M.A. Hnatow, eds. *Ann. N.Y. Acad. Sci.* 751: 323.

Haar, L., J.S. Gallagher, and G.S. Kell. 1980. "Thermodynamic Properties for Fluid Water." In *Water and Steam. Their Properties and Current Industrial Applications. Proceedings of the 9th International Conference on the Properties of Steam*, J. Straub and K. Scheffler, eds., p. 69. Pergamon Press: Oxford.

Harte, J. 1988. *Consider a Spherical Cow. A Course in Environmental Problem Solving.*

University Science Books: Mill Valley, California.

Hatley, R.H.M., and F. Franks. 1986. "Denaturation of Lactate Dehydrogenase at Subzero Temperatures." *Cryo-Lett.* 7: 226.

Hatley, R.H.M., and F. Franks. 1989. "The Effect of Aqueous Methanol Cryosolvents on the Heat- and Cold-Induced Denaturation of Lactate Dehydrogenase."*Eur. J. Biochem.* 184: 237.

Hatley, R.H.M., and F. Franks. 1992. "Cold Destabilization of Enzymes." *Faraday Discuss.* 93: 249.

Hatley, R.H.M., F. Franks, and S.F. Mathias. 1987. "The Stabilization of Labile Biochemicals by Undercooling." *Process Biochem.* 12/87: 169.

Hayward, A.T.J. 1964. "Measuring the Extensibility of Liquids." *Nature.* 202: 481.

Hayward, A.T.J. 1970. "Mechanical Pump with a Suction Lift of 17 Metres." *Nature.* 225: 376.

Hayward, A.T.J. 1971. "Negative Pressure in Liquids: Can It Be Harnessed to Serve Man?" *Amer. Sci.* 59: 434.

Henderson, S.J. 1985. "Water at Negative Pressure." Ph.D. thesis, Victoria University, Wellington, New Zealand.

Henderson, S.J., and R.J. Speedy. 1980. "A Berthelot-Bourdon Tube Method for Studying Water Under Tension." *J. Phys. E. Sci. Instrum.* 13: 778.

Henglein, A., and C.H. Fischer. 1984. "Sonolysis of Chloroform." *Ber. Bunsenges. Phys. Chem.* 88: 1196.

Henry, R.E., J.D. Gabor, I.O. Winsch, E.A. Spleha, D.J. Quinn, E.G. Erickson, J.J. Heiberger, and G.T. Goldfuss. 1974. "Large Scale Vapor Explosions." In *Proceedings of the Fast Reactor Safety Meeting, Argonne, Illinois*, p. 922. Conference Report No. 740 401-P2.

Hess, P.D., and K.J. Brondyke. 1969. "Molten Aluminum-Water Explosions." *Met. Prog.* 95: 93.

Higgins, D.M., and D.M. Skauen. 1972. "Influence of Power on Quality of Emulsions Prepared by Ultrasound." *J. Pharm. Sci.* 61: 1567.

Holbrook, N.M., M.J. Burns, and C.B. Field. 1995. "Negative Xylem Pressures in Plants: A Test of the Balancing Pressure Technique." *Science.* 270: 1193.

Holden, B.S., and J.L. Katz. 1978. "The Homogeneous Nucleation of Bubbles in Superheated Binary Liquid Mixtures." *Amer. Inst. Chem. Eng. J.* 24: 260.

Holder, G.D., S.P. Zetts, and N. Pradhan. 1988. "Phase Behavior in Systems Containing Clathrate Hydrates. A Review." *Rev. Chem. Eng.* 5: 1.

Holder, G.D., S. Zele, R. Enick, and C. LeBlond. 1994. "Modeling Thermodynamics and Kinetics of Hydrate Formation." In *International Conference on Natural Gas Hydrates,* E.D. Sloan, J. Happel, and M.A. Hnatow, eds. *Ann. N.Y. Acad. Sci.* 715: 344.

Huang, H.-S., and J. Winnick. 1974. "P, V, T Behavior of Liquid n-Dodecane at Negative Pressures." *J. Chem. Soc. Faraday Trans. I.* 70: 1944.

Hwang, M.-J., G.D. Holder, and S.R. Zele. 1993. "Lattice Distortion by Guest Molecules in Gas-Hydrates." *Fluid Phase Equil.* 83: 437.

Iraci, L.T., A.M. Middlebrook, M.A. Wilson, and M.A. Tolbert. 1994. "Growth of Nitric Acid Hydrates on Thin Sulfuric Acid Films." *Geophys. Res. Lett.* 21: 867.

Kanno, H., R.J. Speedy, and C.A. Angell. 1975. "Supercooling of Water to $-92°C$." *Science.* 189: 881.

Katz, D.L., and C.M. Sliepcevich. 1971. "LNG/Water Explosions: Cause & Effect." *Hydrocarb. Process.* 50: 240.

Kell, G.S. 1983. "Early Observations of Negative Pressures in Liquids." *Amer. J. Phys.* 51: 1038.

Kim-E., M., and R.C. Reid. 1983. "The Rapid Depressurization of Hot, High Pressure Liquids or Supercritical Fluids." In *Chemical Engineering at Supercritical Fluid Conditions,* M.E. Paulaitis, J.M.L. Penninger, R.D. Gray, Jr., and P. Davidson, eds., chap. 3. Ann Arbor Science: Ann Arbor, Mich.

Knight, C.A., E. Driggers, and A.L. DeVries. 1993. "Adsorption to Ice of Fish Antifreeze Glycopeptides 7 and 8." *Biophys. J.* 64: 252.

Kohshima, S. 1984. "A Novel Cold-Tolerant Insect Found in a Himalayan Glacier." *Nature.* 310: 225.

Kramer, P.J., and T.K. Kozlowski. 1979. *Physiology of Woody Plants.* Chap. 2. Academic Press: Orlando.

Krause, H.H., R. Simon, and A. Levy. 1973. *Smelt-Water Explosions.* Final Report to Fourdrinier Kraft Board Institute, Inc., Battelle Laboratories, Columbus, Ohio.

Kristol, D.S., H. Klotz, and R.C. Parker. 1981. "The Effect of Ultrasound on the Alkaline Hydrolysis of Nitrophenyl Esters." *Tetrahed. Lett.* 22: 907.

Kumai, M. 1961. "Snow Crystals and the Identification of the Nuclei in the Northern United States of America." *J. Meteorol.* 18: 139.

Kumai, M., and K.E. Francis. 1962. "Nuclei in Snow and Ice Crystals on the Greenland Ice Cap under Naturally and Artificially Stimulated Conditions." *J. Atmosph. Sci.* 19: 474.

Lederhos, J.P., A.P. Mehta, G.B. Nyberg, K.J. Warn, and E.D. Sloan. 1992. "Structure *H* Clathrate Hydrate Equilibria of Methane and Adamantane." *Amer. Inst. Chem. Eng. J.* 38: 1045.

Lee, J.C., and S.N. Timasheff. 1974. "Partial Specific Volumes and Interactions with Solvent Components of Proteins in Guanidine Hydrochloride." *Biochem.* 13: 257.

Lee, R.E. 1991. "Principles of Insect Low Temperature Tolerance." In *Insects at Low Temperature,* R.E. Lee and D.L. Denlinger, eds., chap. 2. Chapman and Hall: New York.

Lee, R.E., and J.G. Baust. 1987. "Cold-Hardiness in the Antarctic Tick, *Ixodes uriae.*" *Physiol. Zool.* 60: 499.

Lee, R.E., and D.L. Denlinger, eds. 1991. *Insects at Low Temperature.* Chapman and Hall: New York.

Lifshitz, E.M., and L.P. Pitaevskii. 1980. *Statistical Physics, Part 1.* Volume 5 of Course of Theoretical Physics, by L.D. Landau and E.M. Lifshitz. 3rd ed., chap. 14. Pergamon Press: Oxford.

Long, J., and E.D. Sloan. 1993. "Quantized Water Clusters around Apolar Molecules." *Molec. Simulation.* 11: 145.

Long, J., J. Lederhos, A. Sum, R. Christiansen, and E.D. Sloan. 1994. "Kinetic Inhibitors of Natural Gas Hydrates." 73rd Annual Convention of Gas Processors Association, March 7–9, New Orleans, La.

Luche, J.-L., and J.-C. Damiano. 1980. "Ultrasound in Organic Syntheses. 1. Effects on the Formation of Lithium Organometallic Reagents." *J. Amer. Chem. Soc.* 102: 7926.

Luche, J.-L., C. Petrier, and C. Dupuy. 1984. "Ultrasound in Organic Synthesis 5. Preparation and Some Reactions of Colloidal Potassium." *Tetrahed. Lett.* 25: 753.

Makino, K., M.M. Mossoba, and P. Riesz. 1983. "Chemical Effects of Ultrasound on Aqueous Solutions. Formation of Hydroxyl Radicals and Hydrogen Atoms." *J. Phys. Chem.* 87: 1369.

Marchand, P.J. 1991. *Life in the Cold: An Introduction to Winter Ecology*, 2nd ed. University Press of New England: Hanover, N.H.

Mark, H.F. 1945. "Some Applications of Ultrasonics in High-Polymer Research." *J. Acoust. Soc. Amer.* 16: 183.

Mason, B.J. 1971. *The Physics of Clouds.* 2nd ed. Clarendon Press: Oxford.

Mason, B.J., and F.P. Loewe. 1987. "Atmospheric Humidity and Precipitation." In "Climate and Weather." *Encyclopaedia Britannica*, 15th ed., vol. 16. University of Chicago Press: Chicago.

Mason, T.J., and J.P. Lorimer. 1988. *Sonochemistry: Theory, Applications and Uses of Ultrasound in Chemistry.* Ellis Horwood: Chichester.

Masson, L.S. 1970. *Power Burst Facility In-Pile Tube System Design Basis Report.* Idaho National Engineering Laboratory Report No. TR-150.

Mathias, S.F., F.Franks, and R.H.M. Hatley. 1985. "Preservation of Viable Cells in the Undercooled State." *Cryobiol.* 22: 537.

Mazur, P. 1970. "Cryobiology: The Freezing of Biological Systems." *Science.* 168: 939.

McDonald, S.M., A. White, P. Clancy, and J.W. Brandy. 1995. "Binding of an Antifreeze Polypeptide to an Ice/Water Interface via Computer Simulation." *Amer. Inst. Chem. Eng. J.* 41: 959.

McRae, T.G., H.C. Goldwire, W.J. Hogan, and D.L. Morgan. 1984. *The Effects of Large-Scale LNG/Water RPT Explosions.* Lawrence Livermore National Laboratory Report No. UCRL-90502.

Mehta, A.P., and E.D. Sloan. 1993. "Structure *H* Hydrate Phase Equilibria of Methane + Liquid Hydrocarbon Mixtures." *J. Chem. Eng. Data.* 38: 580.

Mehta, A.P., and E.D. Sloan. 1994. "Structure *H* Hydrate Phase Equilibria of Paraffins, Naphthenes, and Olefins with Methane." *J. Chem. Eng. Data.* 39: 887.

Meyer, J. 1911. "Zur Kenntnis des Negativen Druckes in Flüssigkeiten." *Abh. Bunsenges.* 3: 1.

Miller, R.W., A. Sola, and R.K. McCardell. 1964. *Report of the SPERT 1 Destructive Test Program on an Aluminum, Plate-Type, Water-Moderated Reactor.* U.S. Atomic Energy Commission Report No. IDO-16883.

Molina, M.J., R. Zhang, P.J. Wooldridge, J.R. McMahon, J.E. Kim, H.Y. Chang, and K.D. Beyer. 1993. "Physical Chemistry of the $H_2SO_4/HNO_3/H_2O$ System: Implications for Polar Stratospheric Clouds." *Science.* 261: 1418.

Montoya, A., and J.V. Castell. 1987. "Long-Term Storage of Peroxidase-Labelled Immunoglobins for Use in Enzyme Immunoassay." *Immunol. Meth.* 99: 13.

Mossop, S.C. 1955. "The Freezing of Supercooled Water." *Proc. Phys. Soc. B.* 68: 193.

Moore, G.R. 1956. "Vaporization of Superheated Drops in Liquids." Ph.D. thesis. University of Wisconsin, Madison, Wis.

Moore, G.R. 1959. "Vaporization of Superheated Drops in Liquids." *Amer. Inst. Chem. Eng. J.* 5: 458.

Mori, Y., K. Hijikata, and T. Nagatani. 1976. "Effect of Dissolved Gas on Bubble Nucleation." *Int. J. Heat Mass Transf.* 19: 1153.

Münch, E. 1943. "Durchlässigkeit der Siebröhren für Druckströmungen." *Flora.* 136: 223.

Nakanishi, E., and R.C. Reid. 1971. "Liquid Natural Gas–Water Reactions." *Chem. Eng. Prog.* 67: 36.

Overton, G.D.N., and D.H. Trevena. 1981. "Cavitation Phenomena and the Occurrence of Pressure-Tension Cycles under Dynamic Stressing." *J. Phys. D. Appl. Phys.* 14: 241.

Overton, G.D.N., P.R. Williams, and D.H. Trevena. 1984. "The Influence of Cavitation History and Entrained Gas on Liquid Tensile Strength." *J. Phys. D. Appl. Phys.* 17: 979.

Patrick-Yeboah, J.R., and R.C. Reid. 1981. "Superheat-Limit Temperatures of Polar Liquids." *Ind. Eng. Chem. Fundam.* 20: 315.

Peixoto, J.P., and A.H. Oort. 1992. *Physics of Climate*. Chap. 8. American Institute of Physics: New York.

Pockman, W.T., J.S. Sperry, and J.W. O'Leary. 1995. "Sustained and Significant Negative Water Pressure in Xylem." *Nature.* 378: 715.

Pohlman, R., K. Heisler, and M. Cichos. 1974. "Powdering Aluminium and Aluminium Alloys by Ultrasound." *Ultrasonics.* 12: 11.

Porteous, W.M., and M. Blander. 1975. "Limits of Superheat and Explosive Boiling of Light Hydrocarbons, Hydrocarbons, and Hydrdocarbon Mixtures." *Amer. Inst. Chem. Eng. J.* 21: 560.

Porteous, W.M., and R.C. Reid. 1976. "Light Hydrocarbon Vapor Explosions." *Chem. Eng. Prog.* 72: 83.

Privalov, P.L. 1979. "Stability of Proteins." *Adv. Protein Chem.* 33: 167.

Privalov, P.L. 1990. "Cold Denaturation of Proteins." *Crit. Rev. Biochem. Molec. Biol.* 25: 281.

Privalov, P.L., Yu.V. Griko, S.Yu. Venyaminov, and V.P. Kutyshenko. 1986. "Cold Denaturation of Myoglobin." *J. Molec. Biol.* 190: 487.

Pruppacher, H.R., and J.D. Klett. 1978. *Microphysics of Clouds and Precipitation*. D. Reidel: Dordrecht.

Ramanujam, P., J. Heaster, C. Huang, J. Jolly, C. Lively, E. Ogutu, E. Ting, S. Treml, B. Aldous, R. Hatley, S. Mathias, F. Franks, and B. Burdick. 1993. "Ambient-Temperature-Stable Molecular Biology Reagents." *Biotechniques.* 14: 470.

Rasmussen, D.H., and A.P. MacKenzie. 1972. "Effect of Solute on Ice-Solution Interfacial Free Energy; Calculation from Measured Homogeneous Nucleation Temperatures." In *Water Structure and the Water-Polymer Interface*, H.H.G. Jellinek, ed., p. 126. Plenum: New York.

Rasmussen, D.H., and A.P. MacKenzie. 1973. "Clustering in Supercooled Water." *J. Chem. Phys.* 59: 1003.

Rasmussen, D.H., A.P. MacKenzie, C.A. Angell, and J.C. Tucker. 1973. "Anomalous Heat Capacities of Supercooled Water and Heavy Water." *Science.* 181: 342.

Rausch, A.H., and A.D. Levine. 1973. "Rapid Phase Transformations Caused by Thermodynamic Instability in Cryogens." *Cryogenics.* 13: 224.

Raymond, J.A., and A.L. DeVries. 1977. "Adsorption Inhibition as a Mechanism of Freezing Resistance in Polar Fishes." *Proc. Natl. Acad. Sci. U.S.A.* 74: 2589.

Reid, R.C. 1976. "Superheated Liquids." *Amer. Sci.* 64: 146.

Reid, R.C. 1978a. "Superheated Liquids: A Laboratory Curiosity and, Possibly, an Industrial Curse. Part 1: Laboratory Studies and Theory." *Chem. Eng. Ed.* 12: 60.

Reid, R.C. 1978b. "Superheated Liquids: A Laboratory Curiosity and, Possibly, an

Industrial Curse. Part 2: Industrial and Vapor Explosions." *Chem. Eng. Ed.* 12: 108.

Reid, R.C. 1978c. "Superheated Liquids: A Laboratory Curiosity and, Possibly, an Industrial Curse. Part 3: Discussion and Conclusions." *Chem. Eng. Ed.* 12: 194.

Reid, R.C. 1979. "Possible Mechanism for Pressurized-Liquid Tank Explosions or BLEVE's." *Science.* 203: 1263.

Reid, R.C. 1983. "Rapid Phase Transitions from Liquid to Vapor." *Adv. Chem. Eng.* 12: 105.

Reidl, H. 1937. "Bau und Leistungen des Wurzelholzes." *Jahrb. f. Wiss. Botanik* 85: 1.

Renner, T.A., G.H. Kucera, and M.Blander. 1975. "Explosive Boiling in Light Hydrocarbons and their Mixtures." *J. Colloid Interf. Sci.* 52: 319.

Reynolds, O. 1900a. "On the Internal Cohesion of Liquids and the Suspension of a Column of Mercury to a Height More Than Double that of the Barometer." In *Papers on Mechanical and Physical Subjects*, vol. I, p. 231. Cambridge University Press: Cambridge.

Reynolds, O. 1900b. "Some Further Experiments on the Cohesion of Water and Mercury." In *Papers on Mechanical and Physical Subjects*, vol. I, p. 394. Cambridge University Press: Cambridge.

Richards, B.E., and D.H. Trevena. 1976. "The Measurement of Positive and Negative Pressures in a Liquid Confined in a Berthelot Tube." *J. Phys. D. Appl. Phys.* 9: 122.

Ripmeester, J.A., J.S. Tse, C.I. Ratcliffe, and B.M. Powell. 1987. "A New Clathrate Hydrate Structure." *Nature.* 325: 135.

Ripmeester, J.A., C.I. Radcliffe, D.D. Klug, and J.S. Tse. 1994. "Molecular Perspectives on Structure and Dynamics in Clathrate Hydrates." In *International Conference on Natural Gas Hydrates*, E.D. Sloan, J. Happel, and M.A. Hnatow, eds. *Ann. N.Y. Acad. Sci.* 715: 161.

Roedder, E. 1967. "Metastable Superheated Ice in Liquid-Water Inclusions under High Negative Pressure." *Science.* 155: 1413.

Roedder, E. 1984. *Fluid Inclusions.* Reviews in Mineralogy, vol. 12. U.S. Geological Survey: Reston, Virginia.

Rogers, R.R. 1979. *A Short Course in Cloud Physics.* 2nd ed. Pergamon Press: Oxford.

Rowlinson, J.S., and F.L. Swinton. 1982. *Liquids and Liquid Mixtures.* 3rd ed. Butterworths: London.

Salisbury, F.B., and C.W. Ross. 1992. *Plant Physiology.* 4th ed. Chaps. 5 and 8. Wadsworth: Belmont.

Sassen, K. 1992. Evidence for Liquid-Phase Cirrus Cloud Formation from Volcanic Aerosols: Climatic Implications." *Science.* 257: 516.

Sassen, K., K.N. Liou, S. Kinne, and M. Griffin. 1985. "Highly Supercooled Cirrus Cloud Water: Confirmation and Climatic Implications." *Science.* 227: 411.

Sassen, R., and I.R. MacDonald. 1994. "Evidence of Structure *H* Hydrate, Gulf of Mexico Continental Slope." *Organic Geochem.* 22: 1029.

Schaeffer, V.J. 1946. "The Production of Ice Crystals in a Cloud of Supercooled Water Droplets." *Science.* 104: 457.

Schick, P.E. 1980. "Concentration-Gradient Trigger Mechanism for Smelt-Water Explosions." American Paper Institute Annual Recovery Boiler Committee Meeting, Chicago, Ill.

Schick, P.E., and T.M. Grace. 1982. *Review of Smelt-Water Explosions.* Project No. 3473-2. Institute of Paper Chemistry: Appleton, Wisconsin.

Schmid, G., and O. Rommel. 1939. "Zerreissen von Makromolekülen mit Ultraschall." *Z. Phys. Chem.* A185: 97.

Scholander, P.F. 1972. "Tensile Water." *Amer. Sci.* 60: 584.

Sedgewick, S.A., and D.H. Trevena. 1976. "Limiting Negative Pressure of Water Under Dynamic Stressing." *J. Phys. D. Appl. Phys.* 9: 1983.

Shepherd, J.E., and B. Sturtevant. 1982. "Rapid Evaporation at the Superheat Limit." *J. Fluid Mech.* 121: 379.

Shoh, A. 1988. "Industrial Applications of Ultrasound." In *Ultrasound. Its Chemical, Physical, and Biological Effects*, K.S. Suslick, ed., Chap. 3. VCH: Weinheim, Germany.

Sicheri, F., and D.S.C. Yang. 1995. "Ice-Binding Structure and Mechanism of an Antifreeze Protein from Winter Flounder." *Nature.* 375: 427.

Skauen, D.M. 1967. "Some Pharmaceutical Applications of Ultrasonics." *J. Pharm. Sci.* 56: 1373.

Skripov, V.P., and G.V. Ermakov. 1963. "The Limit of Superheating of Liquids." *Russ. J. Phys. Chem.* 37: 1047.

Skripov, V.P., and G.V. Ermakov. 1964. "Pressure Dependence of the Limiting Superheating of a Liquid." *Russ. J. Phys. Chem.* 38: 208.

Skripov, V.P., and V.I. Kukushkin. 1961. "Apparatus for Measuring the Superheating Limits of Liquids." *Russ. J. Phys. Chem.* 35: 1393.

Skripov, V.P., and P.A. Pavlov. 1970. "Explosive Boiling of Liquids and Fluctuation Nucleus Formation." *High Temp.* 8: 782.

Sloan, E.D. 1990. *Clathrate Hydrates of Natural Gases.* Marcel Dekker: New York.

Sloan, E.D., and F. Fleyfel. 1991. "A Molecular Mechanism for Gas Hydrate Nucleation from Ice." *Am. Inst. Chem. Eng. J.* 37: 1281.

Sloan, E.D., J. Happel, and M.A. Hnatow, eds. 1994. *International Conference on Natural Gas Hydrates. Ann. N.Y. Acad. Sci.* 715.

Storey, K.B. 1990. "Biochemical Adaptation for Cold Hardiness in Insects." *Phil. Trans. Roy. Soc. London B.* 326: 635.

Storey, K.B., and J.M. Storey. 1988. "Freeze Tolerance in Animals." *Physiol. Rev.* 68: 27.

Storey, K.B., and J.M. Storey. 1990. "Frozen and Alive." *Sci. Amer.* 263: 92.

Storey, K.B., and J.M. Storey. 1991. "Biochemistry of Cryoprotectants." In *Insects at Low Temperature*, R.E. Lee and D.L. Denlinger, eds., chap. 4. Chapman and Hall: New York.

Suslick, K.S., J.J. Gawienowski, F.P. Schubert, and H.H. Wang. 1983. "Alkane Sonochemistry." *J. Phys. Chem.* 87: 2229.

Tabazadeh, A., R.P. Turco, K. Drdla, M.Z. Jacobson, and O.B. Toon. 1994. "A Study of Type I Polar Stratospheric Cloud Formation." *Geophys. Res. Lett.* 21: 1619.

Temperley, H.N.V. 1946. "The Behaviour of Water under Hydrostatic Tension: II." *Proc. Phys. Soc.* 58: 436.

Temperley, H.N.V. 1947. "The Behaviour of Water under Hydrostatic Tension: III." *Proc. Phys. Soc.* 59: 199.

Temperley, H.N.V., and L.G. Chambers. 1946. "The Behaviour of Water under Hydrostatic Tension: I." *Proc. Phys. Soc.* 58: 420.

Tomimatsu, Y., J. Scherer, Y. Yeh, and R.E. Fenney. 1976. "Raman Spectra of a Solid

Antifreeze Glycoprotein and its Liquid and Frozen Aqueous Solutions." *J. Biol. Chem.* 251: 2290.

Trevena, D.H. 1987. *Cavitation and Tension in Liquids*. Adam Hilger: Bristol.

Trinh, E., and R.E. Apfel. 1980. "Sound Velocity of Supercooled Water Down to $-33°C$ Using Acoustic Levitation." *J. Chem. Phys.* 72: 6731.

Turco, R.P., O.B. Toon, and P. Hamill. 1989. "Heterogeneous Physicochemistry of the Polar Ozone Hole." *J. Geophys. Res.* 94: 16493.

Vonnegut, B. 1947. "The Nucleation of Ice Formation by Silver Iodide." *J. Appl. Phys.* 18: 593.

Wakeshima, H., and K. Takata. 1958. "On the Limit of Superheat." *J. Phys. Soc. Jpn.* 13: 1398.

Wilson, D.A., J.W. Hoyt, and J.W. McKune. 1975. "Measurement of Tensile Strength of Liquids by an Explosion Technique." *Nature*. 253: 723.

Winnick, J., and S.J. Cho. 1971. "*PVT* Behavior of Water at Negative Pressures." *J. Chem. Phys.* 55: 2092.

Worthington, A.M. 1892. "On the Mechanical Stretching of Liquids: An Experimental Determination of the Volume-Extensibility of Ethyl-Alcohol." *Phil. Trans. Roy. Soc. London A*. 183: 355.

Yamashita, J., Y. Inoue, T. Kondo, and H. Hashimoto. 1984. "Ultrasounds in Synthetic Reactions. II. Convenient Methylenation of Carbonyl Compounds with Zinc-Diiodomethane." *Bull. Chem. Soc. Jpn.* 57: 2335.

Yeo, S.-D., G.-B. Lim, P.G. Debenedetti, and H. Bernstein. 1993. "Formation of Microparticulate Protein Powders Using a Supercritical Fluid Antisolvent." *Biotech. Bioeng.* 41: 341.

Yosioka, K., and Y. Kawasima. 1955. "Acoustic Radiation Pressure on a Compressible Sphere." *Acoustica*. 5: 167.

Young, D.A. 1991. *Phase Diagrams of the Elements*. University of California Press: Berkeley.

Zachariasen, K.E., and J.A. Husby. 1982. "Antifreeze Effect of Thermal Hysteresis Agents Protects Highly Supercooled Insects." *Nature*. 298: 865.

Zheng, Q. 1991. "Liquids under Tension and Glasses under Stress." Ph.D. thesis. Purdue University, West Lafayette, Ind.

Zheng, Q., D.J. Durben, G.H. Wolf, and C.A. Angell. 1991. "Liquids at Large Negative Pressures: Water at the Homogeneous Nucleation Limit." *Science*. 254: 829.

Zimmerman, M.H. 1963. "How Sap Moves in Trees." *Sci. Amer.* 208 (3): 133.

Zimmerman, M.H. 1983. *Xylem Structure and the Ascent of Sap*. Springer-Verlag: Berlin.

Zimmerman, M.H., and C.L. Brown. 1980. *Trees. Structure and Function*. Chap. 4. Springer-Verlag: New York.

Zimmermann, U., A. Haase, D. Langbein, and F. Meinzer. 1993. "Mechanisms of Long-Distance Water Transport in Plants: A Re-Examination of Some Paradigms in the Light of New Evidence." *Phil. Trans. Roy. Soc. London B*. 341: 19.

2

Thermodynamics

In order to study a metastable system, its lifetime τ must be longer than the observation time τ_{obs}. In order to measure a given property, we require in addition that its characteristic molecular relaxation time be much shorter than the system's lifetime ($\tau_{rel} \ll \tau$).[1] Thus, for a property with relaxation time τ_{rel} to be defined under metastable conditions there must be an experimentally accessible time scale that is intermediate between the system's lifetime and the property's internal relaxation time. Provided that $\tau > \tau_{obs} \gg \tau_{rel}$, a given property of a metastable liquid is measurable and reproducible (see, for example, Skripov, 1974, and Skripov et al., 1988, for extensive tabulations). Under these conditions, metastable liquids are amenable to a thermodynamic description, insofar as it makes sense to ask what are their thermophysical properties, and how they can be calculated. In this chapter we address this question. The calculation of the lifetime of a metastable system is the proper province of kinetics; it is discussed in Chapter 3.

There are two approaches to the thermodynamic investigation of metastability; we shall refer to them as phenomenological and microscopic, respectively. In the phenomenological approach, we apply continuum thermodynamic arguments to derive criteria for the stability of macroscopic systems. Stability criteria take the form of inequalities, the violation of which defines a sharp limit-of-stability locus: the spinodal curve. From this perspective, there is an absolute boundary beyond which a given phase cannot exist. Also associated with the phenomenological viewpoint are the actual calculation of spinodal curves, and the estimation of thermodynamic properties of metastable phases, using equations of state. Both types of calculations, estimation of thermodynamic properties and stability limits, are used in a variety of practical applications, hence the importance of the phenomenological approach. It is discussed in Sections 2.1 (derivation of stability criteria), 2.2 (stability of pure fluids), and 2.3 (stability of fluid mixtures).

The microscopic approach discussed in Section 2.4 aims at arriving at a description of metastability based on statistical mechanics. Its starting point is the recognition of basic inconsistencies between the phenomenological and

---

[1] For example, if we are measuring the isobaric heat capacity we require that the relaxation time for enthalpy fluctuations be much shorter than the system's lifetime.

statistical viewpoints. In particular, in a rigorous microscopic theory the transition from metastability to instability is not sharp, and the spinodal becomes a useful approximation rather than a physical locus. Important as this critique of the phenomenological approach is, however, the development of rigorous, microscopically based tools for the calculation of thermophysical properties of metastable fluids is still an open problem. Few results that can be applied to real systems have been derived to date.

Section 2.5 addresses briefly the statistical mechanical treatment of the stability of liquids with respect to the formation of ordered crystalline phases.

Even if it existed, a rigorous and general theory based on statistical mechanics would still give an incomplete picture of metastability. A metastable system will eventually evolve towards a condition of greater stability, and its properties can only be measured if its lifetime is sufficiently long. Hence the very definition of a metastable state involves the notion of time, and calls also for a kinetic description (Binder, 1987): this is the topic of Chapter 3. Naturally, if we base the study of a given system on asking how fast it tends to leave its present condition, the question of stability becomes a relative one, and the notion of a sharply defined stability boundary loses significance.

2.1 PHENOMENOLOGICAL APPROACH: STABILITY CRITERIA

When a macroscopic body cannot interact with its surroundings, it is said to be isolated. For the systems of interest here, isolation, that is to say the impossibility of exchanging mass and of undergoing heat or work interactions with the surroundings, implies fixed energy, volume, and mass (and fixed composition for a nonreacting mixture). If all changes away from a given state that are consistent with the isolation constraints lead to a decrease in a system's entropy (or to no change in entropy), the system under consideration is in equilibrium (Gibbs, 1875–76, 1877–78, 1961).[2] In symbols,

$$[\Delta S]_{U,V,N} \leq 0, \qquad (2.1)$$

where S, U, V, and N denote the system's entropy, energy, volume, and mass (number of moles). The above inequality can also be written in an entirely equivalent way

$$[\Delta U]_{S,V,N} \geq 0 \qquad (2.2)$$

(see Appendix 1; Gibbs, 1875–76, 1877–78, 1961). The above inequalities are necessary and sufficient conditions for equilibrium (Gibbs, 1875–76, 1877–78,

---

[2]More precisely, the system is in equilibrium if all changes away from a given condition that involve no heat interaction or flow of matter across its boundaries (adiabatic interactions of a closed system) lead to an entropy decrease. The isolated system is a particular case of a closed system undergoing adiabatic interactions.

1961). Since entropy and energy are not easily controlled experimental variables, alternative criteria are needed to describe the equilibrium of macroscopic systems subject to different constraints. These read

$$[\Delta A]_{T,V,N} \geq 0, \tag{2.3}$$

$$[\Delta G]_{T,P,N} \geq 0 \tag{2.4}$$

(Appendix 1), where the Helmholtz energy A is defined as

$$A = U - TS \tag{2.5}$$

and the Gibbs energy G as

$$G = U - TS + PV = A + PV. \tag{2.6}$$

Thus a system that cannot exchange mass across its boundaries (a closed system) is in equilibrium when it has the minimum Helmholtz energy consistent with its volume and temperature (V, T), or the minimum Gibbs energy consistent with its pressure and temperature (P, T). Inequalities (2.1)–(2.4) are entirely equivalent (Appendix 1; see also Modell and Reid, 1983): they are all statements of the fact that any spontaneous process undergone by a closed system that does not involve exchange of heat with the surroundings is accompanied by an entropy increase.

Consider now Equation (2.2). Expanding the energy variation,

$$[\delta U + \tfrac{1}{2}\delta^2 U + \cdots]_{S,V,N} \geq 0. \tag{2.7}$$

The vanishing of the linear term yields the equilibrium criterion

$$\delta U|_{S,V,N} = 0. \tag{2.8}$$

For the equilibrium state to be stable, we require that its energy be minimum for all variations subject to constant S, V, N. Therefore,

$$\delta^2 U|_{S,V,N} > 0. \tag{2.9}$$

When, starting from a stable equilibrium state, Equation (2.9) is first violated, that is to say, when

$$\delta^2 U|_{S,V,N} = 0, \tag{2.10}$$

the system has reached a limit of stability. To express the conditions (2.9) and (2.10) in a form that is useful for calculations involving fluids, we write

$$dU = TdS - PdV + \mu dN, \tag{2.11}$$

which relates changes in energy, entropy, volume, and number of molecules of a pure fluid along reversible or quasistatic paths (μ is the chemical potential per molecule).[3] For an n-component mixture, Equation (2.11) reads

$$dU = TdS - PdV + \sum_{j=1}^{n} \mu_j dN_j. \tag{2.12}$$

Equations (2.11) and (2.12) can be written symbolically

$$dU = \sum_{j=1}^{n+2} \xi_j dX_j \tag{2.13}$$

(Modell and Reid, 1983), where X represents the natural independent variables for the energy (entropy, volume, number of molecules), and ξ the conjugate intensive variables (temperature, minus pressure, chemical potential). Thus $[X, \xi] = [S, T; V, -P; N_i, \mu_i \ (i = 1, \ldots, n)]$. The conjugate intensive variables are obtained by partial differentiation,

$$\xi_j = \left(\frac{\partial U}{\partial X_j}\right)_{X_1, X_2, \ldots, X_{j-1}, X_{j+1}, \ldots, X_{n+2}}. \tag{2.14}$$

As shown in Appendix 1, Equation (2.9) can be written as

$$\left(\frac{\partial \xi_{n+1}}{\partial X_{n+1}}\right)_{\xi_1 \cdot \xi_2 \cdots \xi_n \cdot X_{n+2}} > 0. \tag{2.15}$$

When a macroscopic system satisfies Equation (2.15), it is in stable (or metastable) equilibrium. At a limit of stability, we have

$$\left(\frac{\partial \xi_{n+1}}{\partial X_{n+1}}\right)_{\xi_1 \cdot \xi_2 \cdots \xi_n \cdot X_{n+2}} = 0 \tag{2.16}$$

(Beegle et al., 1974a,b). Equations (2.15) and (2.16) are valid for verifying the stability, or for locating the limit of stability, of a homogeneous fluid system with respect to the appearance of another homogeneous fluid phase.

2.2 PHENOMENOLOGICAL APPROACH: STABILITY OF PURE FLUIDS

For a pure substance, $n = 1$, and the general stability criterion [Equation (2.15)] reads

$$\left(\frac{\partial \xi_2}{\partial X_2}\right)_{\xi_1, X_3} > 0 \tag{2.17}$$

---

[3] A process is reversible if the system and all of its surroundings can be restored to their respective initial states. A path is quasistatic if all of its intermediate states are equilibrium states (Modell and Reid, 1983).

(Beegle et al., 1974a,b). Now, ξ_i $(i = 1, 2, 3)$ can be either T, $-P$, or μ. Correspondingly, X_i $(i = 1, 2, 3)$ can be either S, V, or N. Since three quantities can be ordered in six ways, we obtain the six criteria

$$(V, S, N) \Rightarrow \left(\frac{\partial T}{\partial S}\right)_{P,N} > 0, \tag{2.18}$$

$$(N, S, V) \Rightarrow \left(\frac{\partial T}{\partial S}\right)_{\mu,V} > 0, \tag{2.19}$$

$$(N, V, S) \Rightarrow -\left(\frac{\partial P}{\partial V}\right)_{\mu,S} > 0, \tag{2.20}$$

$$(S, V, N) \Rightarrow -\left(\frac{\partial P}{\partial V}\right)_{T,N} > 0, \tag{2.21}$$

$$(V, N, S) \Rightarrow \left(\frac{\partial \mu}{\partial N}\right)_{P,S} > 0, \tag{2.22}$$

$$(S, N, V) \Rightarrow \left(\frac{\partial \mu}{\partial N}\right)_{T,V} > 0. \tag{2.23}$$

The above inequalities are equivalent, since they differ only in the ordering of the extensive variables S, V, N. They are therefore all satisfied or violated simultaneously. The partial derivatives in Equations (2.18)–(2.23) are called stability coefficients.

Equations (2.18) and (2.21) are particularly useful for calculations, since they can be expressed in terms of measurable quantities. In particular, from Equation (2.18) we obtain

$$\left(\frac{\partial T}{\partial s}\right)_P = \frac{T}{c_p} > 0 \tag{2.24}$$

and, from Equation (2.21),

$$\left(-\frac{\partial P}{\partial v}\right)_T = \frac{1}{v K_T} > 0. \tag{2.25}$$

The necessary and sufficient condition for the stability of a pure fluid is therefore that its isobaric heat capacity (c_p) or its isothermal compressibility (K_T) be positive. In the above equations, v and s are specific quantities (e.g., entropy per mole or entropy per molecule). Equation (2.24) implies that if heat is added to a stable or metastable substance at constant pressure, its temperature will rise. Equation (2.25) implies that if a substance is compressed isothermally its pressure will increase. Because Equation (2.25) involves mechanical quantities, the stability criterion for pure substances is often described as a condition of mechanical stability. Note, however, that the thermal criterion (2.24) is always satisfied or violated simultaneously with the mechanical criterion (2.25).

2.2.1 Superheated Liquids

The essential features of the phenomenological description of metastability are illustrated in Figure 2.1. This approach is based on the assumption that the relationship between the Helmholtz energy (A) and the volume (V) of a fixed mass of a fluid can be represented by a continuous, smooth curve such as the one shown in Figure 2.1(a), corresponding to a subcritical temperature T_1. The same situation is shown in Figure 2.1(b) in pressure-volume (P, V) coordinates. The (A, V) and (P, V) diagrams are related by virtue of the fact that

$$P = -\left(\frac{\partial A}{\partial V}\right)_{T,N} \quad \Leftrightarrow \quad A(T, V_2, N) - A(T, V_1, N) = -\int_{V_1}^{V_2} P\,dV. \quad (2.26)$$

Points intermediate between b and e do not have the lowest Helmholtz energy consistent with their volume and temperature. The (A, V) coordinates of all possible mixtures of saturated liquid (b) and vapor (b') are linear combinations of the specific Helmholtz energy and volume corresponding to the saturated phases, and they define line bb'. Therefore any combination of saturated liquid and vapor will have a lower Helmholtz energy than the superheated liquid at the same temperature and total volume. In the (A, V) diagram, points b and b' have the same temperature, pressure [parallel tangents, see Equation (2.26)], and chemical potential $(y$ intercept$)$, and are therefore in equilibrium. The relationship between the y intercept and the chemical potential follows from Equation (2.6) and the fact that $G = \mu N$.

Combining Equations (2.6) and (2.26), the equilibrium condition (equality of chemical potential and pressure) can be expressed as

$$P[V(b') - V(b)] = \int_{V(b)}^{V(b')} P\,dV, \quad (2.27)$$

where the integration is along isotherm T_1. This is the Maxwell equal-area construction: equality of chemical potentials between the vapor and liquid phases occurs at a pressure (P_1) such that areas $b'fd$ and deb are equal [Figure 2.1(b)]. Curve bcb', the equilibrium curve (also called the binodal or coexistence curve), is the locus of points satisfying a Maxwell construction. Points along the isotherm be correspond to a superheated condition: at the given temperature, the liquid has a pressure that is lower than its vapor pressure.

At point e, the superheated liquid has reached a limit of stability: inequalities (2.18)–(2.23) are all violated simultaneously. In particular, it follows from Equation (2.25) that the isothermal compressibility diverges at point e [i.e., $K_T \to \infty$ or, equivalently, $(\partial P/\partial v)_T = 0$]. In (A, V) coordinates, a limit of stability is an inflection point along an isotherm,

$$\left(\frac{\partial^2 A}{\partial V^2}\right)_{T,N} = 0. \quad (2.28)$$

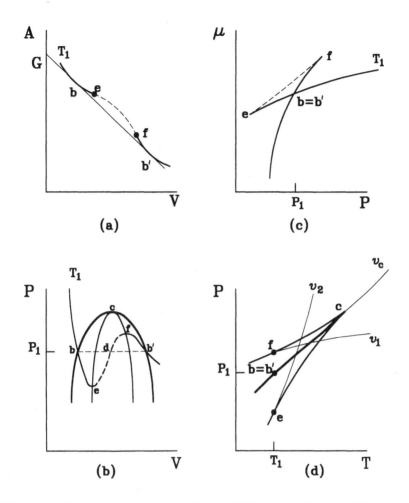

Figure 2.1. Phenomenological picture of metastability in vapor-liquid equilibrium. b and b' are equilibrium states on the binodal. e and f are limits of stability on the spinodal. Unstable states are shown by a dashed curve. (b): bcb' is the binodal, and ecf is the spinodal. (d): v_1, v_2, v_c are isochores ($v_1 > v_c > v_2$), b ($\equiv b'$) c is the binodal. fc is the supercooled vapor spinodal, and ec is the superheated liquid spinodal.

At a limit of stability the second-order energy variation vanishes $[\delta^2 U|_{S,V,N} = 0$; see (2.10)]. To test the stability of states where $\delta^2 U = 0$, we must look at higher-order terms in the energy expansion [Equation (2.7)]. The resulting criterion given below is called the criticality condition, because critical points are stable limits of stability. As shown in Appendix 1, it is necessary and sufficient in order for a limit of stability to be stable that the following conditions be satisfied:

$$\left(\frac{\partial^2 \xi_2}{\partial X_2^2}\right)_{\xi_1, X_3} = 0 \tag{2.29}$$

and

$$\left(\frac{\partial^3 \xi_2}{\partial X_2^3}\right)_{\xi_1, X_3} > 0, \tag{2.30}$$

or, equivalently, for the extensive-variable ordering (S, V, N)

$$\left(\frac{\partial^2 P}{\partial v^2}\right)_T = 0 \tag{2.31}$$

and

$$\left(\frac{\partial^3 P}{\partial v^3}\right)_T < 0 \tag{2.32}$$

(Beegle et al., 1974a,b; Reid and Beegle, 1977). If the third-order derivative [Equation (2.30)] vanishes, the necessary and sufficient conditions for the stability of a limit of stability are that the fourth-order derivative also vanish and the fifth-order derivative be positive (see Appendix 1). In general, then, stability requires that the lowest-order nonvanishing odd derivative $(\partial^{2k+1} \xi_2 / \partial X_2^{2k+1})$ be positive, and all lower-order derivatives be zero. Equations (2.28), (2.31), and (2.32) are satisfied at point c (Figure 2.1), the critical point. Only (2.28) is satisfied at point e. Point c is a stable limit of stability; point e is an unstable limit of stability.

A state such as e is unstable to arbitrarily small perturbations. In macroscopic systems, such perturbations are always present in the form of spontaneous molecular fluctuations, and thus it is impossible for a superheated liquid to exist at a limit of stability. The locus of limits of stability [line ecf in Figures 2.1(b) and 2.1(d)] is called the spinodal curve. Point c, the critical point, is the only stable point on a pure fluid's spinodal.

States between b (saturated liquid) and e (limit of stability) correspond to the superheated liquid. States between b' (saturated vapor) and f (limit of stability) correspond to the supercooled vapor. Unstable states are shown by a dashed line. Schematic (P, T) projections of the spinodal branches corresponding to a the superheated liquid and supercooled vapor conditions are shown in Figure 2.1(d). The corresponding (μ, P) projection of isotherm T_1 is shown in Figure 2.1(c). In this representation, the fluid's stability is related to the curvature of the isotherms through the equation

$$\left(\frac{\partial^2 \mu}{\partial P^2}\right)_T = -v K_T. \tag{2.33}$$

2.2.2 The Spinodal Envelope

The (P, T) projection of a spinodal curve has an important property (Skripov, 1966). It plays a key role in theories of supercooled water (see Sections 2.2.8 and 4.6). To derive this property, we first write

$$dP = \left(\frac{\partial P}{\partial T}\right)_\xi dT + \left(\frac{\partial P}{\partial \xi}\right)_T d\xi, \qquad (2.34)$$

where ξ is a generic intensive property. Therefore, along the spinodal,

$$\left(\frac{dP}{dT}\right)_{sp} = \left(\frac{\partial P}{\partial T}\right)_\xi + \left(\frac{\partial P}{\partial v}\right)_T \left(\frac{\partial v}{\partial \xi}\right)_T \left(\frac{d\xi}{dT}\right)_{sp}. \qquad (2.35)$$

Equation (2.35) signifies that the (P, T) projection of a spinodal curve is an envelope of constant-ξ lines as long as $(\partial v/\partial \xi)_T$ and the (ξ, T) projection of the spinodal both have finite slopes. The former condition is satisfied when ξ is v (specific volume), h (specific enthalpy, $H = Nh = U + PV$), or s (specific entropy). Since a fluid's volume, entropy, and enthalpy vary smoothly along a spinodal, it follows that the (P, T) projection of a spinodal curve is an envelope of isochores (lines of constant specific volume), isentropes (constant s), and isenthalpics (constant h). This is illustrated in Figure 2.1(d), which shows vapor, critical, and liquid isochores ($v_1 > v_c > v_2$, respectively). When v_1 and v_2 are extrapolated to the limit of stability, they become tangent to their respective branches of the spinodal. This geometric property is useful because it places constraints on the possible shape of the spinodal from knowledge of (P, v, T) data for mildly metastable states (see Sections 2.2.8 and 4.6).

2.2.3 The van der Waals Fluid

The calculation of the spinodal locus for a pure substance requires a (P, v, T) equation of state or equivalent experimental information. For superheated liquids, we require an equation of state capable of describing vapor-liquid equilibrium. More than a century has passed since van der Waals proposed his celebrated equation (van der Waals, 1873; Rowlinson, 1988). Its success originated from the fact that this equation succeeded in providing a picture of vapor-liquid equilibrium in qualitative agreement with the observations of Andrews (1869) on the continuity of the fluid states of matter. Though quantitatively inaccurate, this equation (and various modifications thereof) is still widely used, since it provides a simple picture of vapor-liquid equilibrium. We invoke it to illustrate phenomenological calculations of limits of stability. The van der Waals equation reads

$$P = \frac{\rho k T}{1 - \rho b} - \rho^2 a, \qquad (2.36)$$

where ρ is the number density (molecules per unit volume, $\rho = 1/v$), k is Boltzmann's constant, and the constants a and b are obtained from the criticality conditions

$$\left(\frac{\partial P}{\partial v}\right)_T = 0, \quad \left(\frac{\partial^2 P}{\partial v^2}\right)_T = 0, \tag{2.37}$$

which yield

$$\rho_c = \frac{1}{3b}, \quad kT_c = \frac{8a}{27b}, \quad P_c = \frac{a}{27b^2}, \tag{2.38}$$

where the subscript c denotes the critical point. Along the spinodal, both the equation of state and the stability limit condition [i.e., the first relation in Equation (2.37)] must be satisfied. Solving for the condition $(\partial P/\partial v)_T = 0$ simultaneously with the equation of state (2.36), an explicit expression for the spinodal curve of the van der Waals fluid is obtained. In terms of pressure and density, it reads

$$P_r = (3 - 2\rho_r)\rho_r^2 \tag{2.39}$$

and, in terms of temperature and density,

$$4T_r = \rho_r(3 - \rho_r)^2, \tag{2.40}$$

where the subscript r denotes a reduced quantity (that is to say, the value of a quantity divided by its value at the critical point). Figures 2.2 and 2.3 show the limits of stability and the coexistence region for the van der Waals fluid. Curve vcl is the spinodal, branch lc corresponding to the superheated liquid. The binodal, or phase equilibrium locus, is curve bcb [or line bc in Figure 2.2(b)]. Also shown in Figure 2.2(b) are isochores corresponding to reduced densities of 1/2, 1, and 3/2. Note that they become tangent to the spinodal, as discussed in Section 2.2.2.

Below $T_r = 27/32 = 0.8438\ldots$, the van der Waals fluid is capable of withstanding tension: at this temperature, the pressure along the superheated liquid spinodal branch vanishes, and it is negative for $T_r < 27/32$. In order to withstand tension, a van der Waals liquid must have a minimum density of $\rho_r = 3/2$. Figure 2.4 shows the behavior of a van der Waals fluid's Helmholtz energy (referred to that of an ideal gas at the fluid's critical density) and chemical potential (referred to that of an ideal gas at the fluid's critical pressure) at $T_r = 0.8$. The isotherm shown in Figure 2.4 corresponds to a temperature lower than 27/32, and therefore the superheat limit (e) occurs at a negative pressure: note the positive slope in the (a, v) diagram. States between e and f are unstable ($K_T < 0$).

2.2.4 Pseudocritical and Critical Exponents

The metastable portions of a van der Waals fluid's isotherms represent a system constrained to having a uniform density inside the coexistence region. The

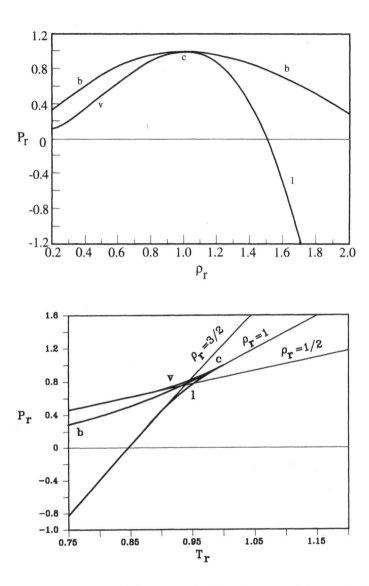

Figure 2.2. (a, top): Binodal (*bcb*) and spinodal (*vcl*) curves of the van der Waals fluid. *c* is the critical point. *v* and *l* denote the supercooled vapor and superheated liquid branches of the spinodal. Reduced pressure and density projection. (b, bottom): The same information in reduced pressure and temperature projection. *b* is the binodal. Also shown are three isochores.

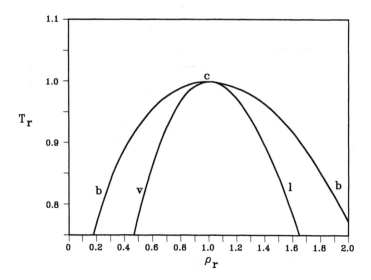

Figure 2.3. Binodal (*bcb*) and spinodal curves of the van der Waals fluid. *c* is the critical point. *v* and *l* denote the supercooled vapor and superheated liquid branches of the spinodal. Reduced temperature and density projection.

microscopic treatment of metastability discussed in Section 2.4 aims at introducing such a constraint rigorously. Approximate theories in which this uniformity constraint results from mathematical simplifications are commonly called mean-field approaches because either they are explicitly based on a picture of molecules immersed in a continuum having the fluid's average properties, or they can be derived by making such an assumption. The van der Waals theory and most other approximate statistical mechanical models are mean-field theories. They all assume that the Helmholtz energy of a fluid is an analytic function of volume and temperature everywhere in the phase diagram. Consequently, they all yield the same values for the exponents that describe the divergence of thermodynamic quantities near a limit of stability. These exponents are called classical, to distinguish them from the true numbers that characterize the behavior of thermodynamic quantities near the critical point; the latter are called nonclassical. The classical exponents that apply near unstable limits of stability are called pseudocritical (Compagner, 1974). In this section we first derive the pseudocritical exponents, and then we discuss nonclassical critical behavior.

The term mean field requires clarification. In the study of critical phenomena, theories that assume the Helmholtz energy to be analytical at the critical point (that is to say, that the Helmholtz energy can be written as a truncated Taylor expansion in terms of temperature and volume around the critical point) are called mean field theories. They all yield the classical critical exponents.

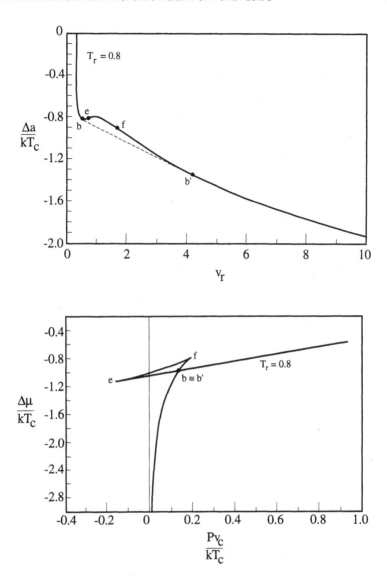

Figure 2.4. (a, top): Relationship between the specific Helmholtz energy and the reduced volume of the van der Waals fluid at a reduced temperature of 0.8. Δa is the difference between the specific Helmholtz energy of the van der Waals fluid and that of an ideal gas at the same temperature and at the van der Waals fluid's critical density. (b, bottom): Relationship between the van der Waals fluid's chemical potential and pressure at a reduced temperature of 0.8. $\Delta \mu$ is the difference between the chemical potential of the van der Waals fluid and that of an ideal gas at the same temperature and at the van der Waals fluid's critical pressure.

In the study of lattice theories of fluids and magnets, the term mean field denotes the simplest solution technique, also know as the zero-order, Bragg-Williams, or molecular-field approximation, in which no account is taken of short-range order (Bragg and Williams, 1934). Higher-order approximations, such as the first-order or Bethe-Guggenheim approximation (e.g., Guggenheim and McGlashan, 1951; Debenedetti et al., 1991); or Kikuchi's cluster variation technique (e.g., Kikuchi, 1951; Kurata et al., 1953; Borick and Debenedetti, 1993), take into account short-range order, and give more realistic descriptions of some aspects of fluid (or magnetic) behavior. All of these approximate techniques, however, yield classical critical behavior. In this section, we use the term mean field in the former sense.

In order to derive the pseudocritical exponents, we expand the Helmholtz energy per unit mass in terms of changes in volume and temperature away from a limit of stability:

$$a = a_{sp} + a_v \delta v + a_T \delta T + \frac{1}{2}[a_{TT}(\delta T)^2 + 2a_{Tv}\delta v \delta T] + \frac{1}{6}a_{vvv}(\delta v)^3, \quad (2.41)$$

where a is the specific Helmholtz energy, a_{sp} the value of this quantity at a point along the spinodal (v_{sp}, T_{sp}), $\delta v = v - v_{sp}$, and $\delta T = T - T_{sp}$. The coefficients in the expansion are partial derivatives evaluated at the spinodal, for example,

$$a_{Tv} = \left(\frac{\partial^2 a}{\partial T \partial v}\right)_{sp} \quad (2.42)$$

and we have used the fact that a_{vv} is identically zero at the spinodal. Successive differentiations with respect to v yield the equation of state and the isothermal compressibility. [$P = -(\partial a/\partial v)_T$; $K_T^{-1} = v(\partial^2 a/\partial v^2)_T$]. Using the thermodynamic relationships

$$c_p = T\left(\frac{\partial s}{\partial T}\right)_P = c_v + TvK_T\left(\frac{\partial P}{\partial T}\right)_v^2, \quad (2.43)$$

$$\alpha_p = \frac{1}{v}\left(\frac{\partial v}{\partial T}\right)_P = K_T\left(\frac{\partial P}{\partial T}\right)_v, \quad (2.44)$$

where α_p is the thermal expansion coefficient, we obtain, along an isotherm,

$$K_T \propto c_p \propto \alpha_p \propto [v_{sp}(T) - v]^{-1}. \quad (2.45)$$

Equation (2.45) assumes that isochores have finite slope, i.e., $(\partial P/\partial T)_v$ is finite. Furthermore, we have used the fact that c_v is finite at the spinodal. Formal proof of this is given in Section 2.2.7.

Since the spinodal is a smooth function in (v, T) coordinates (see Figure 2.3), it can be linearized locally (Speedy, 1982b), and we can write

$$\frac{T - T_{sp}(v)}{v_{sp}(T) - v} = d, \quad (2.46)$$

where d is a smoothly varying function of T. Then, using Equations (2.45) and (2.46), we obtain, along an isochore,

$$K_T \propto c_p \propto \alpha_p \propto [T - T_{\mathrm{sp}}(v)]^{-1}. \tag{2.47}$$

The relationship between volume and pressure along an isotherm in the vicinity of the spinodal follows from differentiating (2.41) to obtain the equation of state $P(v, T)$. One then obtains

$$v_{\mathrm{sp}}(T) - v = \sqrt{\frac{-2}{a_{vvv}}} \left[P - P_{\mathrm{sp}}(T) \right]^{1/2}, \tag{2.48}$$

which, combined with Equation (2.45) yields, along an isotherm,

$$K_T \propto c_p \propto \alpha_p \propto [P - P_{\mathrm{sp}}(T)]^{-1/2}. \tag{2.49}$$

Since the spinodal is a smooth function in (P, T) coordinates, it can be linearized locally (Speedy, 1982b), and we can write

$$f = \frac{P - P_{\mathrm{sp}}(T)}{T_{\mathrm{sp}}(P) - T}, \tag{2.50}$$

where f is a smoothly varying function of T. Then, using Equations (2.49) and (2.50), we obtain, along an isobar,

$$K_T \propto c_p \propto \alpha_p \propto [T_{\mathrm{sp}}(P) - T]^{-1/2}. \tag{2.51}$$

Equations (2.45), (2.47), (2.49), and (2.51) give the pseudocritical exponents corresponding to the divergence of the isothermal compressibility, isobaric heat capacity, and thermal expansion coefficient as the spinodal is approached along isochoric, isothermal, and isobaric paths (Speedy, 1982b). They are common to any theory based upon an analytic Helmholtz energy, such as is shown in Figures 2.1 and 2.4.

Even assuming an analytic Helmholtz energy, the pseudocritical exponents are not valid arbitrarily close to the critical point. To see this, we write the pressure expansions about a point on the spinodal and about the critical point (or, equivalently, we differentiate the corresponding Helmholtz energy expansions):

$$P = P_{\mathrm{sp}} + \frac{1}{2} P_{vv}(\delta v)^2 + P_T \delta T + P_{Tv} \delta T \delta v, \tag{2.52}$$

$$P = P_c + \frac{1}{6} P_{vvv}(\delta v)^3 + P_T \delta T + P_{Tv} \delta T \delta v, \tag{2.53}$$

where the partial derivatives in Equations (2.52) and (2.53) are evaluated at the spinodal and at the critical point, respectively. In Equation (2.52) we have not taken into account explicitly that P_{vv} must vanish if we use the critical point as the reference point for the spinodal expansion. To this end we write, in the spirit of the Landau theory of phase transitions (Landau, 1965; Lifshitz and Pitaevskii, 1980; Boiko et al., 1984),

$$\frac{1}{2}P_{vv} \equiv B\Delta v_c \equiv B(v_c - v_{sp}), \quad B > 0, \tag{2.54}$$

and the expansion (2.52) now reads

$$P = P_{sp} + B\Delta v_c(\delta v)^2 + C(\delta v)^3 + D\delta T + E\delta T\delta v, \tag{2.55}$$

where

$$6C = P_{vvv}, \quad C < 0, \tag{2.56}$$

$$D = P_T, \quad D > 0, \tag{2.57}$$

$$E = P_{Tv}, \quad E < 0. \tag{2.58}$$

The sign of C follows from Equation (2.32); that of E from the fact that starting from the spinodal (superheated liquid branch), P_v decreases as T is increased isochorically.

From Equation (2.55), which is formally valid for any choice of reference condition along the spinodal, including the critical point, we obtain

$$
\begin{aligned}
K_T &= -\frac{1}{v_{sp} + \delta v} \cdot \frac{1}{2B\Delta v_c\delta v + 3C(\delta v)^2 + E\delta T} \\
&= -\frac{1}{v_{sp}} \cdot \frac{1}{2B\Delta v_c\delta v + 3C(\delta v)^2 + E\delta T}.
\end{aligned}
\tag{2.59}
$$

Then, invoking Equation (2.55) to derive the relationship between temperature and volume changes along the critical and subcritical isobars, and using the notation

$$K_T \propto |\delta T|^{-\gamma} \text{ (isobar)}, \quad K_T \propto (\delta T)^{-\gamma} \text{ (isochore)},$$

$$K_T \propto |\delta v|^{-\gamma'}, \quad K_T \propto (\delta P)^{-\gamma''}, \tag{2.60}$$

we obtain the exponents listed in Table 2.1, where the lower values describe the approach to the spinodal and the upper values the approach to the critical point.

The exponents listed in Table 2.1 follow directly from the assumed analyticity of the Helmholtz energy. The change from spinodal to critical behavior, on the other hand, is independent of the analyticity assumption. This change is a consequence of the different conditions that define limits of stability and critical points, regardless of whether the latter can be defined simply by Equations (2.31) and (2.32) in addition to the limit of stability condition.

Table 2.1. Transition from Spinodal to Critical Behavior: Mean-Field Results

Path	γ	γ'	γ''
Isotherm ($\delta T = 0$)	—	$1 \leq \gamma' \leq 2$	$1/2 \leq \gamma'' \leq 2/3$
Isochore ($\delta v = 0$)	1	—	1
Isobar ($\delta P = 0$)	$1/2 \leq \gamma \leq 2/3$	$1 \leq \gamma' \leq 2$	—

Table 2.2. Critical Exponents of Fluids: Definitions and Theoretical Values

Definition[a]	Path	Nonclassical	Classical
$K_T \propto \varepsilon^{-\gamma}$	Critical isochore $(\rho = \rho_c)$	1.239 ± 0.002	1
$c_v \propto \varepsilon^{-\alpha}$	Critical isochore $(\rho = \rho_c)$	0.110 ± 0.003	0
$\rho_l - \rho_v \propto (-\varepsilon)^{\beta}$	Coexistence curve	0.326 ± 0.002	1/2
$P - P_c \propto [(\rho - \rho_c)/\rho_c]^{\delta}$	Critical isotherm $(T = T_c)$	4.8 ± 0.02	3
$G(r) \propto r^{-(1+\eta)}$	Critical point; large r	0.031 ± 0.004	0
$\xi \propto \varepsilon^{-\nu}$	Critical isochore $(\rho = \rho_c)$	0.63 ± 0.001	1/2

[a] $\varepsilon = (T - T_c)/T_c$.

The critical behavior of real fluids differs from that shown in Table 2.1. The best theoretical estimates of the critical exponents of fluids are given in Table 2.2 (Sengers and Levelt Sengers, 1986; Levelt Sengers, 1991).

In this table, $G(r)$ is the correlation function

$$G(r) = \langle [\rho(r) - \langle \rho(r) \rangle][\rho(0) - \langle \rho(0) \rangle] \rangle = \langle \rho(r) \rho(0) \rangle - \rho^2 \quad (2.61)$$

(Stanley, 1971) where r denotes distance, angle brackets denote thermodynamic averaging, and ρ is the number of molecules per unit volume [$\langle \rho(r) \rangle = \langle \rho(0) \rangle = \langle \rho \rangle = \rho$ for a fluid in the absence of external fields]. The quantity ξ is the correlation length, defined such that

$$G(r) \propto \frac{1}{r} e^{-r/\xi}. \quad (2.62)$$

Equation (2.62) applies away from the critical point and at large separations. Physically, ξ gives the distance over which density fluctuations are correlated.

According to the modern theory of critical phenomena (e.g., Binney et al., 1992), systems having the same spatial and order parameter dimensionality belong to the same universality class and have the same critical exponents. The order parameter is a conveniently chosen quantity that vanishes at the critical point, is nonzero on one side of the transition (generally below the critical temperature), and whose fluctuations diverge at the critical point. For pure fluids, the order parameter is $(\rho - \rho_c)$. Below the critical temperature, the order parameter is nonzero, since the saturated densities differ from the critical density. Above the critical temperature and along the critical isochore, the order parameter is zero. Thus, for liquids, the order parameter is a scalar; hence its dimensionality is 1. Magnets are analogous to fluids: below the Curie temperature the spontaneous magnetization is nonzero; at and above the Curie temperature, it is zero. Thus magnets also have a scalar order parameter. Fluids and magnets, then, belong to the same universality class. The critical exponents of liquids and magnets are indeed equal, to within experimental error.[4] The remarkable achievement of the theory of critical phenomena (Wilson and Kogut, 1974) is precisely that it explains the paradox whereby intermolecular forces are responsible for causing a phase transition, yet their detailed microphysics plays no role in determining the behavior near the critical point (Binney et al., 1992).

Because critical exponents depend only on a system's universality class, the most reliable values come from theoretical calculations (e.g., Wilson, 1971; Wilson and Fisher, 1972; Le Guillou and Zinn-Justin, 1985). These calculated values are more accurate than those that can be obtained experimentally for fluids (Levelt Sengers, 1991). The numbers in Table 2.2 are taken from a critical review of theoretical calculations for the three-dimensional Ising universality class[5] (Sengers and Levelt Sengers, 1986).

The six critical exponents are not independent. The following relations apply among them:

$$\alpha + 2\beta + \gamma \geq 2, \tag{2.63}$$

$$\alpha + \beta (1 + \delta) \geq 2, \tag{2.64}$$

$$(2 - \eta) \nu \geq \gamma, \tag{2.65}$$

$$d\nu \geq 2 - \alpha, \tag{2.66}$$

where d is the system's dimensionality. Equations (2.63)–(2.66) are known as the Rushbrooke, Griffiths, Fisher, and Josephson inequalities, respectively. These authors first showed that thermodynamics imposes the above constraints on the critical exponents (Rushbrooke, 1963; Griffiths, 1965; Fisher, 1969; Josephson, 1967a,b). Scaling arguments suggest (Widom, 1965; Kadanoff,

[4]In order to compare the critical exponents of fluids and magnets, the proper correspondence between fluid variables (density, pressure, compressibility) and magnetic ones (magnetization, magnetic field, susceptibility) must be used (see, for example, Huang, 1987; Toda et al., 1992).

[5]This name refers to the celebrated lattice model of a magnet introduced by Ising (1925). This model has a scalar order parameter, and hence its three-dimensional version has the same critical exponents as real fluids.

1966), and experiments confirm, that these inequalities are rigorous equalities. Thus only two critical exponents are truly independent.

The difference between mean-field predictions and experimentally observed critical behavior is striking. In order to understand it, we write the important equation

$$\frac{kTK_T}{V} = \frac{\langle(\delta\rho)^2\rangle}{\langle\rho\rangle^2},$$ (2.67)

where angular brackets denote thermal averaging, and $\langle(\delta\rho)^2\rangle$ is the mean squared density fluctuation in an open volume V. Equation (2.67) relates a thermodynamic derivative (K_T, an observable) to a microscopic fluctuation.[6] In particular, this equation shows that the isothermal compressibility is a direct measure of the fluctuations in density that occur in an open control volume V. It follows from (2.16), (2.25), and (2.67) that density fluctuations are arbitrarily large near the critical point. In mean-field theories molecules are pictured as interacting with a continuum having the fluid's bulk properties. This neglect of fluctuations causes mean-field theories to break down in the fluctuation-dominated critical region (Sengers and Levelt Sengers, 1978).

Two milestones that contributed to the realization that "something was wrong" with mean-field theories when applied near the critical point were Onsager's exact solution of the two-dimensional Ising model (Onsager, 1944), and Guggenheim's observation that the gas-liquid coexistence curve of Ar, Ne, Kr, Xe, N_2, O_2, CO, and CH_4 did not follow the expected classical behavior, $\beta = 1/2$ (see Table 2.2) (Guggenheim, 1945). Onsager's solution yielded nonanalytic critical exponents, a startling result that cannot be explained by a mean-field theory. Guggenheim for his part was able to fit coexistence curves with great accuracy using $\beta = 1/3$ (see Table 2.2). Almost three decades later, Wilson, building upon the deep insights provided by the work of Widom (1965), Kadanoff (1966), and Fisher (1967b, 1969), was finally able to provide a rigorous theory of critical phenomena, the renormalization-group theory (Wilson, 1971; Wilson and Fisher, 1972; Wilson and Kogut, 1974). This theory allows the accurate calculation of critical exponents, and it explains why systems as different as fluids, magnets, and the three-dimensional Ising model have the same critical behavior.

2.2.5 Stability Limit Predictions with Equations of State

Equations of state are frequently used to calculate stability limits of liquids and their mixtures. Such predictions can then be related to experiments. These experiments involve careful penetration into the coexistence region up to a point where a new phase is suddenly formed, no matter how careful the experiment.

---

[6]Equation (2.67) and the general relationship between thermodynamic derivatives and fluctuations are discussed in Sections 2.3.2 and 2.4.

As will be explained in Chapter 3, in the absence of dissolved gases, suspended impurities, or solid surfaces, the appearance of a new phase occurs via the formation of embryos of the stable phase within the bulk metastable liquid (vapor embryos in a superheated liquid, crystal embryos in a supercooled liquid). This process is known as homogeneous nucleation. The rate of formation of critical-sized embryos[7]exhibits a very sharp dependence on the extent of penetration into the coexistence region. Consequently, superheated liquids evolve very suddenly from a condition of apparent stability to one in which the new phase is formed rapidly and spontaneously. This condition is called the superheat limit, and it represents the practical, experimentally attainable stability limit. It is kinetically rather than thermodynamically controlled, and therefore it is not an infinitely sharp stability boundary like the spinodal. The spinodal curve should always lie beyond the observable, kinetically controlled homogeneous nucleation event.

Figures 2.5–2.7 show superheat limits for several substances measured with different techniques, and the corresponding spinodal predictions using the van der Waals, Redlich-Kwong, and Peng-Robinson equations of state (Redlich and Kwong, 1949; Peng and Robinson, 1976; Reid et al., 1987). The Redlich-Kwong and Peng-Robinson equations read, respectively,

$$P = \frac{RT}{v - b} - \frac{aT^{-1/2}}{v(v + b)},$$ (2.68)

$$P = \frac{RT}{v - b} - \frac{a}{v(v + b) + b(v - b)},$$ (2.69)

where v is the molar volume and R is the gas constant (equivalently, v is the molecular volume and R is replaced by k). The a parameter in the Peng-Robinson equation is given by

$$a = a_c[1 + K(1 - T_r^{1/2})]^2,$$ (2.70)

$$K = 0.37464 + 1.54226\omega - 0.2699\omega^2,$$ (2.71)

where ω, the acentric factor (Pitzer, 1955; Pitzer et al., 1955), is a substance-specific measure of deviation from spherical shape, and T_r is the reduced temperature. The parameters a, b, and a_c are obtained from the criticality conditions. Both the Redlich-Kwong and the Peng-Robinson equations are modified van der Waals equations, and are widely used in practical calculations. Spinodal curves result from applying the first condition in Equation (2.37). For the Redlich-Kwong equation they read

$$T_r = 1.1804 \left[\rho_r \left(2 + 0.2599\rho_r\right) \left(\frac{1 - 0.2599\rho_r}{1 + 0.2599\rho_r}\right)^2 \right]^{2/3},$$ (2.72)

---

[7]Subcritical embryos dissolve spontaneously; supercritical embryos grow spontaneously (see Chapter 3).

$$P_r = 3.5412 \left[\left(\frac{\rho_r^5}{2 + 0.2599\rho_r} \right)^{1/3} \left(\frac{1 + 0.2599\rho_r}{1 - 0.2599\rho_r} \right)^{2/3} \right]$$

$$\times \left(\frac{1 - 0.0676\rho_r - 0.5198\rho_r}{1 + 0.0676\rho_r + 0.5198\rho_r} \right), \tag{2.73}$$

and, for the Peng-Robinson equation,

$$\frac{T_r}{f(w, T_r)} = \frac{\rho_r (0.2683 + 0.0711\rho_r)}{[0.0778\rho_r + (0.0861 + 0.0228\rho_r)/(0.2934 - 0.0778\rho_r)]^2}, \tag{2.74}$$

$$\frac{P_r}{f(w, T_r)} = \frac{\rho_r^2 (0.0394 - 0.0028\rho_r^2 - 0.0209\rho_r)}{0.0861 - 0.0061\rho_r^2 + 0.0457\rho_r}, \tag{2.75}$$

where

$$f(w, T_r) = [1 + K(1 - T_r^{1/2})]^2 = \frac{a}{a_c}. \tag{2.76}$$

Note that the Redlich-Kwong spinodal is substance independent, as is also the case for the van der Waals equation [Equations (2.39) and (2.40)]. The references and techniques corresponding to the experimental superheat limits reported in Figures 2.5–2.7 are summarized in Table 2.3.

It can be seen from Figures 2.5–2.7 that the van der Waals equation underpredicts superheat limits. At zero pressure, the van der Waals fluid can be superheated only up to a reduced temperature of 27/32 (= 0.8438...). For the substances in Figures 2.5–2.7, atmospheric pressure corresponds approximately to zero reduced pressure (it represents a reduced pressure of 0.062 for perfluoroheptane, and lower than this number in all other cases shown). Experiments, however, show that liquids can be superheated at atmospheric pressure up to roughly 90% of their critical temperature (Blander and Katz, 1975). The Redlich-Kwong equation underpredicts superheat limits less severely than the van der Waals equation.

Equations of state are useful for estimating the maximum extent to which liquids can be superheated (see, for example, Eberhart, 1976; Eberhart and Schnyders, 1973; Lienhard et al., 1986; Dong and Lienhard, 1986; Reid, 1976; Skripov, 1974; Skripov et al., 1988). As shown in Figures 2.5–2.7, however, the predicted spinodals depend sensitively on the particular equation being used. In most cases, the Peng-Robinson spinodal is a realistic upper boundary within which lie the experimentally attainable superheat limits of a wide variety of liquids.

2.2.6 Continuity and Divergences in Superheated Liquids

There exists a considerable body of literature on the thermophysical properties of superheated liquids. It is mostly associated with Skripov and his collaborators. The bulk of this information has been collected in the first handbook

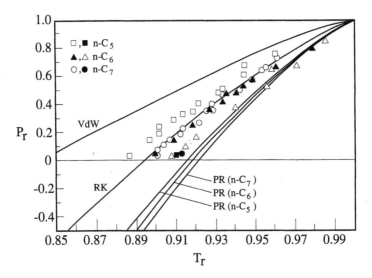

Figure 2.5. Comparison between spinodal curves (vdW = van der Waals; RK = Redlich-Kwong; PR = Peng-Robinson) and experimental superheat limits for n-pentane (n-C$_5$), n-hexane (n-C$_6$), and n-heptane (n-C$_7$). $\square$, $\blacktriangle$, $\bigcirc$, Skripov and Ermakov (1964); $\blacksquare$, Pavlov and Skripov (1965); $\triangle$, Skripov et al. (1965), and Pavlov and Skripov (1970); $\bullet$, Skripov and Pavlov (1970).

of thermophysical properties of superheated liquids (Skripov et al., 1988), to which the reader is referred for tables of surface tension, density, specific heat, viscosity, and thermal conductivity of superheated argon, water, n-hexane, diethyl ether, and benzene. Examples of (P, v, T) measurements for superheated liquids include experiments on n-hexane (Ermakov and Skripov, 1968; Skripov, 1974; Skripov et al., 1988), diethyl ether (Ermakov et al., 1973; Skripov et al., 1988), water and heavy water (Chukanov and Skripov, 1971; Estefeev et al., 1977; Skripov et al., 1988), benzene (Skripov et al., 1988), argon (Baidakov and Skripov, 1978; Skripov et al., 1988), xenon (Baidakov et al., 1988), krypton (Baidakov and Gurina, 1989a), methane (Baidakov and Gurina, 1989b), and oxygen (Baidakov and Gurina, 1985). In this section we discuss the behavior of the thermodynamic properties of superheated liquids, both near and far from the binodal.

A key conclusion from experimental investigations of superheated liquids is the continuity of properties and their derivatives at the binodal. This is illustrated in Figure 2.8, which shows (P, v, T) measurements along isotherms and isochores for superheated water and methane. Similar behavior occurs with all other measured properties (Skripov et al., 1988). All available experimental evidence to date indicates that the properties of superheated liquids are smooth

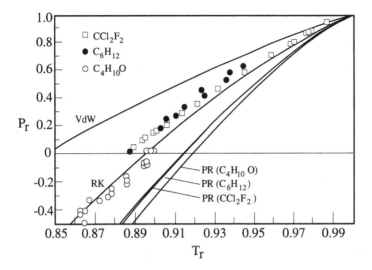

Figure 2.6. Comparison between spinodal curves (vdW = van der Waals; RK = Redlich-Kwong; PR = Peng-Robinson) and experimental superheat limits for dichlorodifluoromethane (CCl_2F_2), cyclohexane (C_6H_{12}), and dimethylether ($C_4H_{10}O$). □, Moore (1959); ●, Skripov (1974); ○, Apfel (1971).

continuations of stable behavior, and that there is no detectable singularity at the binodal, either in the measured properties or in their derivatives.[8]

The extent of penetration into the metastable region is limited by the short lifetime of highly metastable states. This imposes great difficulties on the accurate experimental study of the functionality that characterizes the increase of K_T, c_p, and α_p deep inside the superheated region. We now discuss some experiments aimed at investigating this behavior. The sound velocity is related to thermodynamic quantities by the equations

$$c^2 \rho MW = K_s^{-1},$$ (2.77)

$$K_T = K_s + \frac{vT\alpha_p^2}{c_p},$$ (2.78)

where c is the sound velocity, ρ the molar density, MW the molecular weight, and K_s the adiabatic compressibility,

$$K_s = \left(\frac{-\partial \ln v}{\partial P}\right)_s.$$ (2.79)

---

[8]From a rigorous statistical-mechanical viewpoint, this is not a settled issue; it is discussed in Section 2.4.

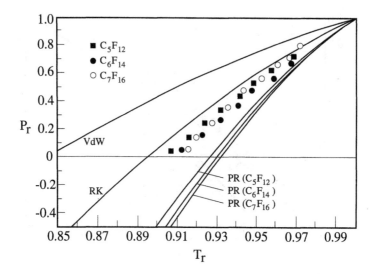

Figure 2.7. Comparison between spinodal curves (vdW = van der Waals; RK = Redlich-Kwong; PR = Peng-Robinson) and experimental superheat limits for perfluoropentane (C_5F_{12}), perfluorohexane (C_6F_{14}), and perfluoroheptane (C_7H_{16}). Data from Ermakov and Skripov (1967).

Figure 2.9 shows the isothermal compressibility of superheated water at atmospheric pressure, obtained by Hareng and Leblond (1980) by sound velocity measurements, using experimental values for the density (Kell, 1975) and thermal expansion coefficient (Kell, 1975; Estefeev et al., 1979), and a combination of experimental and extrapolated heat capacity values. Hareng and Leblond fitted their results to an expression of the form

$$K_T \propto [T_{\text{sp}}(P) - T]^{-\gamma} \tag{2.80}$$

with

$$\gamma = 0.97 \pm 0.05, \quad T_{\text{sp}}(P) = 588 \pm 10 \text{ K}. \tag{2.81}$$

The regressed spinodal temperature is in good agreement with equation of state estimates (Chukanov and Skripov, 1974). The fitted exponent, γ, is in disagreement with the pseudocritical (mean-field) value of $1/2$ [see Equation (2.51)]. Note that the maximum temperature studied by Hareng and Leblond (220°C) is much lower than the measured superheat limit of 302°C (Skripov, 1974). Even if one accepts the validity of an analytic expansion for the Helmholtz energy [Equation (2.41), from which the pseudocritical exponents were derived], this enormous distance between experimental conditions and the assumed underlying singularity severely limits the usefulness of power-law extrapolations.

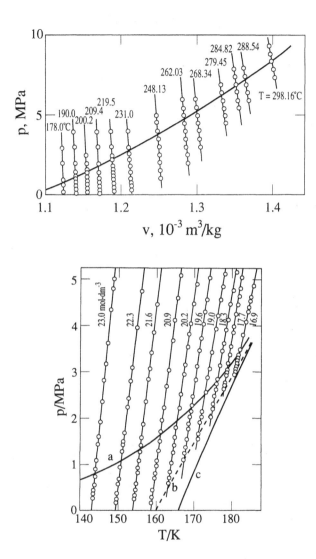

Figure 2.8. (a, top): P, v, T data for superheated water. The dark curve is the saturation line. [Reprinted, with permission, from Skripov et al., *Thermophysical Properties of Liquids in the Metastable (Superheated) State*. Gordon and Breach Science Publishers: New York (1988)] (b, bottom): P, v, T data for superheated methane. a, saturation line; b, locus along which the calculated homogeneous nucleation rate (see Chapter 3) is 6×10^{-3} cm^{-3} sec^{-1}; c, locus along which the calculated homogeneous nucleation rate is 10 cm^{-3} sec^{-1}. [Reprinted, with permission, from Baidakov and Gurina, *J. Chem. Thermodyn.*, 21: 1009 (1989)]

Table 2.3. References and Techniques for Superheat Limits Shown in Figures 2.5–2.7

Substance	Technique[a]	Reference
n-Pentane	Droplet superheat	Skripov and Ermakov (1964)
	Pulse heating	Pavlov and Skripov (1965)
n-Hexane	Droplet superheat	Skripov and Ermakov (1964)
	Pulse heating	Skripov et al. (1965)
	Pulse heating	Pavlov and Skripov (1970)
n-Heptane	Droplet superheat	Skripov and Ermakov (1964)
	Pulse heating	Skripov and Pavlov (1970)
Perfluoropentane	Droplet superheat	Ermakov and Skripov (1967)
Perfluorohexane	Droplet superheat	Ermakov and Skripov (1967)
Perfluoroheptane	Droplet superheat	Ermakov and Skripov (1967)
Cyclohexane	Droplet superheat	Skripov (1974)
Dimethylether	Acoustic levitation	Apfel (1971)
Dichlorodi-		
fluoromethane	Droplet superheat	Moore (1959)

[a] The droplet superheat technique used by Skripov and coworkers and by Moore is very similar to Apfel's experiment described in Section 1.1.2, but allows measurements to be made at higher than atmospheric pressure. Apfel's acoustic levitation technique (see also Section 1.3.2) is an ingenious modification of the droplet superheat method that allows the determination of superheat limits at negative pressure. The pulse heating method utilizes the rapid heating of a thin wire submerged in the liquid under study to attain superheats that exceed those measured with alternative techniques. See also Chapter 3.

If an accurate equation of state is available, Equations (2.77) and (2.78) can be used to calculate the isobaric heat capacity from sound velocity measurements. Skripov et al. (1988) provide comprehensive data on the isobaric heat capacities of argon, water, and heavy water, obtained via sound velocity measurements, and also by extrapolation of data taken in the stable region. Figure 2.10 shows the isobaric heat capacity of superheated argon obtained from speed-of-sound measurements (Baidakov and Skripov, 1978; Skripov et al., 1988), plotted according to Equation (2.49). Spinodal pressures were calculated via the Peng-Robinson equation of state. The closest approach to the calculated spinodal is 5.3 bar at 135 K, 1.7 bar at 140 K, and 0.8 bar at 145 K. The data show a clear increase in the heat capacity as the fluid is decompressed isothermally. The functionality that describes this increase, however, is difficult to ascertain accurately given the scarcity of data at the temperature of closest approach to the calculated spinodal.

Speedy (1982b) analyzed the behavior of K_T, c_p, and α_p for superheated water, and plotted the data as shown in Figure 2.11. The singular temperature, 598 K, is a theoretical estimate due to Skripov (1980). The Helmholtz energy

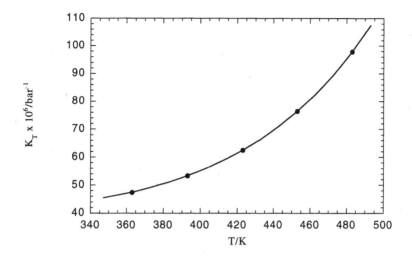

Figure 2.9. Temperature dependence of the isothermal compressibility (K_T) of superheated water at atmospheric pressure. Data from Hareng and Leblond (1980).

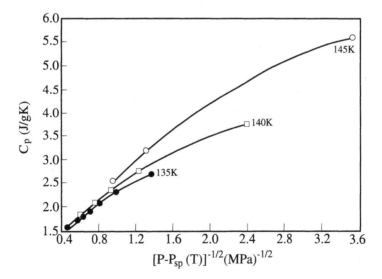

Figure 2.10. Isobaric heat capacity (c_p) of superheated argon, plotted according to Equation (2.49), with spinodal pressures calculated with the Peng-Robinson equation of state. Data from Baidakov and Skripov (1978) and Skripov et al. (1988).

expansion [Equation (2.41)] prescribes not only the pseudocritical exponents but also the proportionality coefficients in Equations (2.45), (2.47), (2.49), and (2.51). Starting from Equation (2.41), one obtains, along an isobar,

$$\frac{C_\alpha}{C_K} = \frac{dP_{\text{sp}}}{dT}, \qquad \frac{C_c}{C_K} = v_{\text{sp}} T_{\text{sp}} \left(\frac{dP_{\text{sp}}}{dT}\right)^2, \tag{2.82}$$

where C_α, C_c, and C_K denote the proportionality coefficients between α_p, c_p, K_T, and $[T_{\text{sp}}(P) - T]^{-1/2}$, respectively; dP_{sp}/dT is the slope of the spinodal's (P, T) projection; and T_{sp} and v_{sp} denote the temperature and specific volume at the spinodal. It follows from this that the limiting slopes of the three lines in Figure 2.11 are not independent. Speedy used this property to draw the compressibility and thermal expansion limiting lines consistently with the heat capacity data. The high-temperature c_p data in Figure 2.11 were estimated from an equation of state (Haar et al., 1980). As with Figure 2.9, experimental values of α_p, K_T, and c_p in Figure 2.11 correspond to a maximum temperature of 220°C, which is much lower than the measured superheat limit of 302°C (Skripov, 1980). The calculated heat capacity shows the linearity predicted by mean-field theory [Equation (2.51)]. The other solid lines are the predicted limiting slopes of K_T and α_p according to Equation (2.82). Figure 2.11 is not inconsistent with mean-field theory.

It is clear from the available experimental data that:

  (i) measured thermophysical properties and their derivatives are continuous at the binodal;
 (ii) K_T, c_p, and α_p increase sharply inside the coexistence region;
(iii) the short lifetimes of metastable systems far removed from the binodal limit the extent of penetration into the coexistence region; hence the true asymptotic behavior of K_T, α_p, and c_p is difficult (if not impossible) to measure experimentally.

2.2.7 The "Pseudospinodal"

Benedek (1969) proposed a method of obtaining spinodal curves near the critical point experimentally. Such curves have come to be known, oddly, as "pseudospinodal." Consider a quantity, such as K_T, that diverges along the spinodal. Then, in the vicinity of the critical point, we may write

$$\Psi = \Psi_0 \left[\frac{T - T_{\text{sp}}(\rho)}{T_c}\right]^{-\gamma} \tag{2.83}$$

(Benedek, 1969; Sorensen and Semon, 1980), where Ψ is the diverging quantity, Ψ_0 and γ are the amplitude and exponent that characterize its behavior along an isochore, and T_c is the critical temperature [for $\Psi = K_T$, $\gamma = 1$, assuming mean-field behavior; see Equation (2.47)]. Starting from an expression such

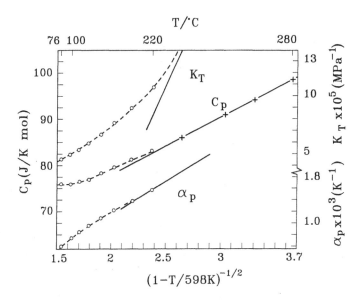

Figure 2.11. Temperature dependence of the isothermal compressibility (K_T), isobaric heat capacity (c_p), and thermal expansion coefficient (α_p) of superheated water at atmospheric pressure (Speedy, 1982b). O, Hareng and Leblond (1980); +, Haar et al. (1980); full lines are limiting α_p and K_T slopes consistent with c_p data, plotted according to Equation (2.82). (Adapted from Speedy, 1982b)

as Equation (2.83), experimental values of Ψ along different isochores should yield the spinodal curve $T_{sp}(\rho)$. The latter can be represented by an expression of the form

$$\frac{|\rho - \rho_c|}{\rho_c} = B_{sp}\left[\frac{T_c - T_{sp}(\rho)}{T_c}\right]^{\beta^+}, \qquad (2.84)$$

where B_{sp} and β^+ are the amplitude and exponent that describe the behavior of the spinodal near the critical point. Internal consistency demands that Ψ_0 and γ be density independent; homogeneity and scaling arguments (Chu et al. 1969; Green et al., 1967; Vicentini-Missoni et al., 1969; Widom, 1965) require that β^+ should be equal to β, the analogous critical exponent for the coexistence curve (see Table 2.2).

Equations (2.83) (with $\Psi = K_T$) and (2.84) imply that

$$\rho\left(\frac{\partial P}{\partial \rho}\right)_T = h\varepsilon^\gamma\left[1 + \frac{a\gamma}{\varepsilon}\Delta\tilde{\rho}^{1/\beta^+} + \cdots\right], \qquad (2.85)$$

where h and a are constants, $\varepsilon = (T - T_c)/T_c$, and $\Delta\tilde{\rho} = |\rho - \rho_c|/\rho_c$ (Chu et al. 1969). If the pressure were an analytic function of density close to the critical point, β^+ should be $1/2$. This is of course what one obtains from a Taylor expansion around the critical point. If $\beta^+ \geq 1/3$ (see Table 2.2), the fourth derivative of the pressure with respect to density is singular along the critical isochore for $\varepsilon > 0$ (Chu et al., 1969). Though this does not violate any physical law, there is no reason to expect such behavior in real fluids.

In their light scattering study of the isobutyric-acid–water mixture, Chu et al., (1969) obtained $\beta^+ = 0.37 \pm 0.04$, by using the mixture analogues of (2.83) and (2.84). This is an interesting result. On the one hand, it validates the idea behind (2.83) and (2.84), especially because $\beta^+ \approx \beta$. On the other hand, the fourth-order singularity casts some doubt on the validity of (2.83) and (2.84). These authors therefore argued in favor of using the term "pseudospinodal" to describe curves obtained by extrapolation of data taken in the stable, as opposed to the metastable, region. Experimentally, however, there is no evidence of discontinuities at the binodal (see Figure 2.8). This contradiction between the existence of a spinodal with $\beta^+ \neq 1/2$ and singularities in the equation of state for $T > T_c$ remains unresolved. We thus encounter a first serious argument against the concept of an infinitely sharp spinodal. Statistical-mechanical and kinetic arguments in favor of a nonsharp stability boundary will be discussed in Section 2.4 and Chapter 3, respectively.

The "pseudospinodal" concept has been used to correlate the near-critical behavior of quantities such as the isochoric heat capacity c_v that exhibit singularities at the critical point but not along the spinodal. The conceptual problems that can arise from this broad interpretation of Equation (2.84) have not always been taken into consideration. They follow from writing the necessary and sufficient conditions for the stability of a pure substance in the form

$$U_{SS} > 0, \tag{2.86}$$

$$U_{SS}U_{VV} - U_{SV}^2 > 0, \tag{2.87}$$

where subscripts denote differentiation [e.g., $U_{SS} = (\partial^2 U/\partial S^2)_{N,V}$] (Lifshitz and Pitaevskii, 1980; see also Appendix 1). Rearranging the natural variables of U from (S, V, N) to (V, S, N) leads to the equivalent stability inequality

$$U_{VV} > 0 \tag{2.88}$$

with (2.87) unchanged. Since U_{VV} and U_{SS} are positive, the inequality (2.87) is always violated first as the spinodal is approached from a metastable region. Therefore, since

$$U_{SS} = \frac{T}{Nc_v}, \quad U_{SS}U_{VV} - U_{SV}^2 = \frac{T}{N^2 v c_v K_T}, \tag{2.89}$$

it follows that the compressibility always diverges before the isochoric heat capacity. The spinodal, in other words, is a locus of diverging compressibility and isobaric heat capacity but finite c_v, except at the critical point. The locus defined by a diverging c_v lies entirely inside the unstable region, except at the critical point, and is different from the spinodal. In this restricted sense, it is reasonable, indeed necessary, to distinguish the different loci, and the term "pseudospinodal" acquires renewed if unintended significance.

A number of papers have been published that are based on the assumption of the existence of a single "pseudospinodal" regardless of the nature of Ψ in Equation (2.83) (Osman and Sorensen, 1980; Abdulagatov and Alibekov, 1982, 1983, 1984, 1985; Lysenkov, 1985; Akhundov et al., 1985; Abdulagatov, 1984). This assumption is incorrect in the case of c_v, and an open question in the case of transport properties. Rykov (1985, 1986) has pointed out the theoretical inconsistencies behind the broad interpretation of the "pseudospinodal" concept. When used correctly, however, the "pseudospinodal" idea is a useful way of estimating the location of spinodal curves, especially for mixtures (Sorensen, 1991; see also Section 2.3.1). However, the issue of the consistency of a "pseudospinodal" defined by (2.83) and (2.84) with the scaling and analyticity properties of the equation of state in the near-critical region has not been resolved, and remains an active area of research (Sorensen and Semon, 1980; Rykov, 1986)

2.2.8 Liquids That Expand When Cooled: The Stability Limit Conjecture

Most liquids show no evidence of approaching a condition of impending loss of stability, in the sense of violating inequalities (2.18)–(2.23), upon supercooling. Their isothermal compressibility and isobaric heat capacity do not increase anomalously below the freezing temperature. Supercooled liquids, however, are metastable with respect to a solid phase, and inequalities (2.18)–(2.23) are not the relevant stability criteria.[9] This is because under the action of applied forces, solids can change their shape as well as their volume; for fluids, shape is an irrelevant variable. Thus for a fixed mass of fluid we write

$$dU = TdS - PdV, \qquad (2.90)$$

whereas for a solid we have instead

$$dU = TdS - V_0\sigma_{ik}du_{ik}, \qquad (2.91)$$

where σ_{ik} and u_{ik} are the stress and strain tensors (Landau and Lifshitz, 1986), and V_0 is the volume of the undeformed body. Whether or not it is possible to generalize the thermodynamic (i.e., macroscopic, continuum) stability criteria

---

[9]The stability of liquids with respect to solids is discussed in Section 2.5 from a microscopic viewpoint.

(2.18)–(2.23) to the case where the new phase is solid is an interesting open problem.

There is at least one important exception to the common behavior of supercooled liquids: in water (and in D_2O), both c_p and K_T increase sharply upon supercooling. This raises two questions: What does thermodynamics say in general about the stability of any liquid whose response functions (compressibility, heat capacity, thermal expansion coefficient) increase upon supercooling? What is the microscopic basis of water's peculiar behavior? In this section we address the former question; a separate section, 4.6, is devoted exclusively to supercooled water. Many of the ideas presented here will be used in Section 4.6 to analyze the phase behavior of supercooled water.

Figures 2.12, 2.13, and 2.14 show the isothermal compressibility, isobaric heat capacity, and thermal expansion coefficient of supercooled water (also D_2O in Figure 2.14) at atmospheric pressure. It can be seen that K_T and c_p increase and α_p becomes large in magnitude upon supercooling. This behavior is in sharp contrast to that of most liquids, whose response functions decrease upon supercooling until freezing or vitrification occurs. The available information can be summarized as follows:[10] at atmospheric pressure K_T, c_p, and the magnitude of α_p continue to increase down to the lowest temperatures at which such measurements have been made ($-38°C$ for c_p; Rasmussen et al., 1973); on the other hand, c_v, calculated using (2.43), does not increase (Angell et al., 1973; Speedy and Angell, 1976). At higher pressures, the trends are similar, but the increases occur at progressively lower temperatures. This is illustrated in Figure 2.15 for D_2O.

A phenomenological interpretation of the anomalous behavior of supercooled water was proposed originally by Speedy (1982a,b), who called it the stability limit conjecture. Later, Debenedetti and co-workers generalized Speedy's theory to any liquid capable of expanding when cooled at constant pressure (D'Antonio and Debenedetti, 1987; Debenedetti and D'Antonio, 1988). The salient feature of the stability limit conjecture, shown schematically in Figure 2.16, is the continuous spinodal curve bounding the superheated and supercooled regions. At point e, the liquid attains its maximum tensile strength, and the (P, T) projection of the spinodal changes the sign of its slope (retracing spinodal). Thermodynamic consistency (D'Antonio and Debenedetti, 1987; Debenedetti and D'Antonio, 1988) demands that this occur whenever a locus of diverging compressibility (ce) intersects a line along which the thermal expansion coefficient vanishes (fae). Between curves fae and fe, the liquid's thermal expansion coefficient is negative. This is shown by isochores g and h ($\rho_g > \rho_h$). The former becomes tangent to the spinodal with negative slope (isochoric cooling with pressure increase); the latter with positive slope. Thus the change in slope of the spinodal follows from the fact that it is an envelope of

---

[10] See Section 4.6 for a detailed discussion of the thermophysical properties of supercooled water.

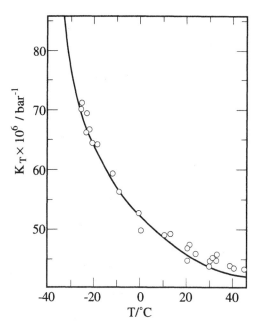

Figure 2.12. Temperature dependence of the isothermal compressibility (K_T) of super-cooled water at atmospheric pressure. The solid line is a fit, $K_T = 29.65 \times 10^{-6}$ (bar^{-1}) $(T - 228)^{-0.349}$. (Adapted from Speedy and Angell, 1976)

isochores (Section 2.2.2), the slope of isochores vanishing along *fae*. If, starting anywhere along *fe*, the temperature is increased isobarically, the density will increase, reach a maximum along *fae*, and decrease thereafter: line *fae* is a locus of density maxima.

In Speedy's treatment, the continuous spinodal was obtained by several procedures. In the 0–100 °C range, Speedy used a density-explicit Taylor expansion about the limit of stability,

$$P(T) = P_s(T) + \sum_j (1/j!) \left(\partial^j P/\partial \rho^j\right)_{T,P=P_s} (\rho - \rho_s)^j$$

$$= P_s(T) + \sum_j C_j(T) (\rho - \rho_s)^j, \qquad (2.92)$$

and fitted the temperature-dependent coefficients to an empirical but accurate equation of state for water (Chen et al., 1977). In Equation (2.92), the subscript *s* denotes the spinodal. At lower temperatures, Speedy used power-law extrapolations of transport property measurements, of the form given by Equation (2.51). Debenedetti and D'Antonio (1988) showed that the same overall

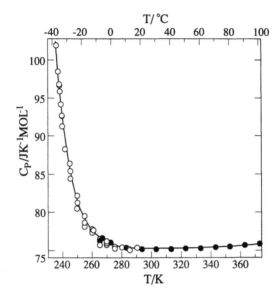

Figure 2.13. Temperature dependence of the isobaric heat capacity (c_p) of supercooled water at atmospheric pressure. ●, literature data; ○, data of Angell and co-workers (Adapted from Angell et al., 1982)

picture applies in principle to any fluid capable of exhibiting density anomalies $(\alpha_p < 0)$. Their treatment was based exclusively on thermodynamic consistency arguments. Specifically, the scenario depicted in Figure 2.16 was deduced by assuming that the Helmholtz energy is analytic at the spinodal and that the locus of density maxima has finite slope in the (P, T) plane;[11] by neglecting the possibility of density minima; and by taking into account that the spinodal is an envelope of isochores, and that isochores cannot intersect in the (P, T) plane. From this phenomenological viewpoint, then, it is indeed possible for a supercooled liquid to lose stability by becoming infinitely compressible. The supercooled liquid spinodal, a sharply defined low-temperature limit below which the liquid cannot exist, can only appear when a liquid can contract when heated isobarically. The locus of density maxima causes the (P, T) projection of the spinodal to retrace, giving rise to a tensile strength maximum and to a continuous stability boundary. The two intersections of the locus of density maxima with the spinodal (f, e) define the ranges of temperature and pressure within which the liquid exhibits density anomalies. In the absence of density anomalies, the liquid's tensile strength increases monotonically upon cooling,

---

[11]This is a restrictive assumption. It will be shown in Section 4.6 that an altogether different scenario results if this assumption is not made.

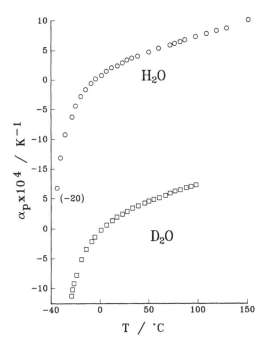

Figure 2.14. Temperature dependence of the thermal expansion coefficient (α_p) of supercooled water and D_2O at atmospheric pressure. (Adapted from Hare and Sorensen, 1986)

that is to say, the superheated liquid spinodal continues into the negative pressure region without changing the sign of its slope.

In the case of water, which expands upon freezing below ca. 2 kbar, the initial portion of the melting curve originating at the triple point has a negative slope in (P, T) projection. Hence supercooling can be caused by isothermal expansion. This perturbation is opposed by the liquid's cohesion, as is superheating. Provided a metastable liquid can exhibit density anomalies below the melting temperature, it will be expanded, not compressed, when supercooled. Curve *fe* (Figure 2.16) is therefore a locus of diverging compressibility, but it is not brought about by compression at high density.

The most interesting result of the stability limit conjecture is to relate density anomalies to loss of stability upon supercooling. Because this conclusion follows from general thermodynamic arguments, it is relevant not just to water and heavy water but to any liquid that expands when cooled at constant pressure, such as supercooled SiO_2 (Angell and Kanno, 1976). In model fluids, density anomalies often result when the repulsive core of the spherically symmetric pair potential has two inflection points (or a sudden change in the core diameter, if the repulsion is hard). This behavior is known as core softening

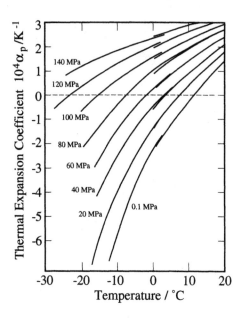

Figure 2.15. Temperature and pressure dependence of the thermal expansion coefficient (α_p) of supercooled D_2O. (Adapted from Kanno and Angell, 1980.) The high-temperature portions of the isobars are from Emmett and Millero (1975).

(Debenedetti et al., 1991; see also Hemmer and Stell, 1970; Stell and Hemmer, 1972; Kincaid et al., 1976a,b; Kincaid and Stell, 1977, 1978; Young and Alder, 1977, 1979; Stillinger and Weber, 1978). The effective pair potential of many liquid metals is core softened (see, e.g., Yokoyama and Ono, 1985; Hoshino et al., 1987; March, 1987, 1990; Hafner, 1987; Iida and Guthrie, 1993). Anomalous increases in the response functions have been measured for several supercooled liquid metals (Perepezko and Paik, 1984), in particular for Te and Te alloys (Tsuchiya, 1991a,b,c, 1992, 1993, 1994; Kakinuma and Ohno, 1987; Hosokawa et al., 1993; Kakinuma et al., 1993). Several elements expand upon freezing, that is to say, their phase diagrams include a negatively sloped melting curve within some pressure range. They include Rb, Cs, Ba, Ga, Si, C, Ge, P, Sb, Bi, S, Se, Te, Ce, Eu, and Pu (Young, 1991). This phenomenon is closely related to the appearance of density anomalies in the liquid phase. Thus the stability limit conjecture suggests that interesting connections may exist between supercooled water, heavy water, and SiO_2, on the one hand, and supercooled liquid metals, elements that expand upon freezing, and core-softened model fluids, on the other. What these systems have in common is that they can adopt low-density, energetically favored configurations. Hence they expand when

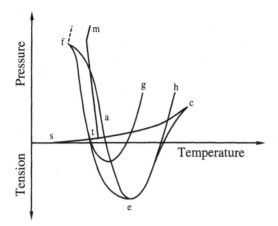

Figure 2.16. Schematic representation of Speedy's stability limit conjecture. st is the sublimation curve, tc is the boiling curve, and tm is the melting curve. g and h are isochores ($\rho_g > \rho_h$), c and t are the critical and triple points, fae is the locus of density maxima, and cef is the continuous spinodal bounding the superheated, supercooled, and simultaneously superheated-supercooled states.

cooled. This can be achieved either by forming tetrahedrally coordinated open networks that collapse into denser and more disordered arrangements through input of thermal or mechanical energy (H_2O, D_2O, SiO_2), or by core softening, which can cause pressure and temperature to become negatively correlated (Debenedetti el al., 1991). The experimental and computational investigation of supercooled liquids that expand when cooled is an interesting and virtually unexplored area.

The stability limit conjecture has one important limitation: it assumes that the phase with respect to which the supercooled liquid loses stability is isotropic. This is because the stability criteria $K_T > 0$, $c_p > 0$ result from considering fluctuations that are macroscopic in extent (they involve a large number of molecules), but small in intensity. Such fluctuations involve the appearance of homogeneous and isotropic regions with different intensive properties than the bulk fluid, but which are locally at equilibrium. Hence the fluid's stability is being tested with respect to the appearance of another isotropic (fluid) phase (Appendix 1). In a supercooled liquid, on the other hand, the relevant fluctuations are with respect to an ordered phase, and this is ignored in the stability limit conjecture. Thus a more complete theory would presumably lead to a liquid-solid limit of stability preceding the retracing locus fe.[12] As mentioned above, thermodynamic stability criteria of liquids with respect to ordered

---

[12]Recently, Sastry el al. (1993) have obtained such a solid-liquid spinodal in a waterlike lattice model. This work is discussed in Section 4.6.

phases have not been derived. In this sense, the stability limit conjecture is not incorrect but incomplete. It provides one possible explanation for the growth in the response functions; it shows the relation between this growth and density anomalies; but it predicts that the response functions should diverge, even though the supercooled liquid is not metastable with respect to the vapor. The missing liquid-solid spinodal should not be confused with the homogeneous nucleation locus, which is the kinetically controlled and practically attainable limit of supercooling: in a more complete thermodynamic theory of the stability limits of supercooled liquids, this missing spinodal should be an absolute limit, albeit not with respect to an isotropic phase.

The above discussion has focused on the stability limit conjecture as it applies in general to liquids with density anomalies. In the specific case of water, alternative interpretations of the stability problem exist. They are discussed in Section 4.6.

2.2.9 Metastable Phase Equilibrium

The coexistence of metastable phases is known as metastable phase equilibrium. A pure substance can, in principle, exhibit several types of two-phase metastability: vapor-solid coexistence above the triple point,[13] solid-liquid coexistence below the triple point, and liquid-vapor coexistence below the triple point, of which the latter two have been observed (Roedder, 1967; Kraus and Greer, 1984; Henderson and Speedy, 1987; D'Antonio, 1989). In addition, different solid forms of a substance can show metastable coexistence [for example, in water, ice III–ice IX, ice III–ice Ih (Hobbs, 1974)]. Consider a temperature T_α, intermediate between the triple and critical points (Figure 2.17). At T_α, there is equilibrium between the solid and liquid phases at a pressure P_1, and between the liquid and vapor phases at $P_3 < P_1$. At point 2, the solid and the vapor have the same temperature, pressure, and chemical potential (equilibrium); at fixed temperature and pressure, the two-phase system will relax spontaneously to a single-phase liquid (metastability). This can be seen from the fact that a vertical line through 2 intersects the liquid branch in the (μ, P) plane, resulting in a condition of lower chemical potential at fixed T and P.

The middle diagrams in Figure 2.17 show stable three-phase equilibrium at the triple point. Consider now a temperature T_β, lower than the triple point temperature. At T_β, there is phase equilibrium between the solid and vapor phases at a pressure P_6. Additional intersections of μ vs. P curves correspond to metastable phase equilibrium between the solid and subtriple liquid phases under tension (7; $P_7 < 0$), and between the vapor and subtriple liquid phases (5). A subtriple liquid is a metastable liquid below its triple point temperature

---

[13]For liquids that expand when they freeze, such as water, metastable solid-liquid coexistence occurs below the triple point pressure but above the triple point temperature (see Figure 1.14).

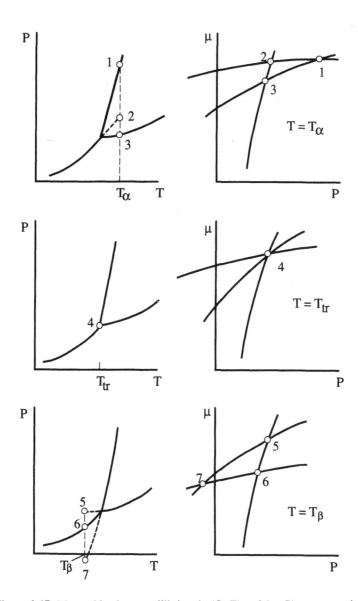

Figure 2.17. Metastable phase equilibrium in (P, T) and (μ, P) representations.

and pressure (D'Antonio, 1989). It is not necessary for P_7 to be negative: the solid and liquid phases can coexist in metastable equilibrium at positive pressures for T_β sufficiently close to the triple point.

In the (μ, P) plane, a vertical line descending from a condition of two-phase metastability intersects the stable equilibrium state that would result from a relaxation at fixed temperature and pressure. The same construction in (a, v) coordinates (specific Helmholtz energy, specific volume) identifies the stable condition following a relaxation at fixed temperature and volume. For example, a subtriple liquid in metastable equilibrium with the vapor phase (5) will solidify completely if constrained in temperature and pressure. Likewise, a metastable liquid-vapor system (5) will relax to a solid-vapor mixture if constrained in temperature and volume, as will a metastable solid-liquid system (7).

If a clean, deaerated liquid with positive thermal expansion coefficient is cooled inside a rigid container which it fills completely, its pressure will decrease. Assuming the container to be perfectly rigid, the pressure drop can be very pronounced [ca. 8.5 bar K^{-1} for benzene at its triple point density (D' Antonio, 1989)]. Isochoric cooling is in fact one of the commonly used methods for generating tension in liquids (Trevena, 1987), and can lead to substantial penetration into the subtriple region (D'Antonio, 1989). Relaxations away from metastability can be very fast relative to characteristic heat transfer times. When this is the case, they can be treated as adiabatic; furthermore, if the container is rigid, the system performs no mechanical work. During such a phase transition the system is therefore, for all practical purposes, isolated. The process can be depicted conveniently with an energy-volume diagram (D'Antonio, 1989).

Figure 2.18 is a schematic projection of the phase diagram of a pure substance onto the (U, V) plane. Line LCV is the vapor-liquid binodal; C is the critical point. The pairs of lines $(s_1 S; vV)$, and $(s_2 S; lL)$ are the solid-vapor and solid-liquid binodals. Points S, L, and V represent coexisting solid, liquid, and vapor phases at the triple point. At equilibrium, points inside triangle SLV represent mixtures of solid, liquid, and vapor phases at the triple temperature and pressure; the specific energy (volume) of any such point is a linear combination of the solid, liquid, and vapor triple point energy (volume), each weighted by the appropriate mass fraction. Consider now the isochoric cooling of an initially stable liquid at its triple point density. This process is represented by the vertical line α. Point 1 is a metastable subtriple liquid. If relaxation occurs from this condition, the system will evolve towards three-phase coexistence at the triple point. For a truly isolated system, the overall specific energy and volume of the resulting three-phase mixture are identical to those of the starting condition (point 1). In the laboratory, this phenomenon involves the simultaneous boiling and freezing of a subtriple liquid (Hayward, 1971; D'Antonio, 1989). Point 2 represents the limiting condition beyond which no stable liquid phase remains following relaxation at constant energy and volume. If the subtriple liquid can be successfully cooled to point 3 and relaxes from this condition, it does so towards a solid-vapor mixture, the mass fractions of solid and vapor phases being proportional to the distances from point 3 along the tie line SV to the vapor and solid binodals, respectively. Such tie lines connect coexisting phases with the same temperature, pressure, and chemical potential.

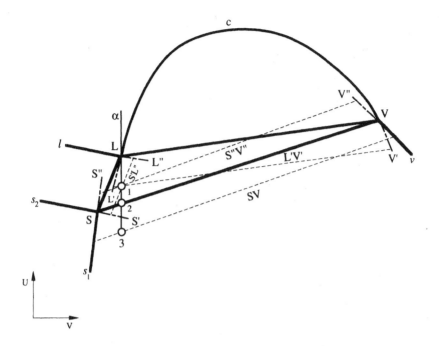

Figure 2.18. Energy-volume projection of the phase diagram of a pure substance show-ing metastable phase equilibrium. See text for explanation.

The relaxation of a subtriple liquid towards stable equilibrium does not al-ways occur in one step. Lines $S'L''$, $S''V''$, and $L'V'$ in Figure 2.18 are tie lines connecting solid-liquid, solid-vapor, and liquid-vapor phases. The end points of these lines lie on the metastable continuations of the respective binodals; the tie lines therefore connect pairs of states in metastable equilibrium with each other. If a subtriple liquid having the specific energy and volume corre-sponding to point 1 relaxes towards two-phase metastability without interacting with its surroundings (i.e., the system is isolated), it can, in principle, do so in one of three ways: by freezing partially to a solid-liquid mixture (end points of $S'L''$), by boiling partially to a liquid-vapor mixture (end points of $L'V'$), and by freezing and boiling to a vapor-solid mixture ($S''V''$). If the system is isolated during the transition, it will relax by increasing its entropy. The final state (that is to say, the condition of maximum entropy) cannot be metastable: the subtriple liquid represented by point 1 will always relax to the triple point if isolated. Nevertheless, the relaxation can occur in steps, towards conditions of two-phase metastability. Observations of this kind include partial boiling of subtriple benzene to a metastable vapor-liquid mixture (D'Antonio, 1989), and partial freezing of subtriple water to a metastable liquid-solid mixture (Schubert

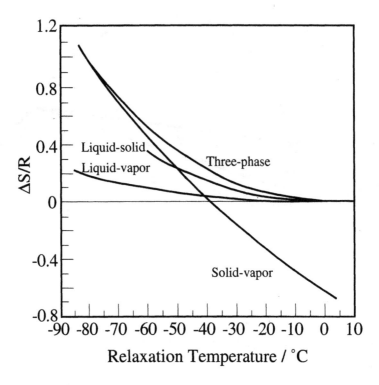

Figure 2.19. Calculated molar entropy changes towards two- and three-phase states that accompany the relaxation (at constant energy and volume) of subtriple benzene at its liquid triple-point density, as a function of the temperature just prior to relaxation. (Adapted from D'Antonio, 1989)

and Lingenfelter, 1970). Whether a particular transition towards two-phase metastability can occur depends on the accompanying entropy change. Figure 2.19 shows the calculated entropy changes that accompany the relaxation, at constant volume and energy, of subtriple benzene at its triple point density, as a function of the temperature just prior to relaxation [the triple point of benzene occurs at 278.7 K and 0.048 bar (Rowlinson and Swinton, 1982)]. The largest entropy change is always towards a stable equilibrium condition. Partial boiling to a metastable liquid-vapor mixture is always accompanied by an entropy increase, and complete freezing and boiling to a solid-vapor mixture is only accompanied by entropy increase after substantial cooling. The termination of the liquid-solid line occurs when the liquid phase in metastable equilibrium with the solid becomes unstable due to the intersection of the metastable liquid binodal (LL'') with the superheated liquid spinodal. Calculations leading to Figure 2.19 (D'Antonio, 1989) are based on an empirical equation of state (Wenzel and Schmidt, 1980) that models the solid phase as a compressed fluid.

Figure 2.20 shows the phase diagram of a model of C_{60} (buckminster-fullerene) obtained by computer simulation (Hagen et al., 1993). The inter-molecular potential was constructed by assuming that carbon atoms on two different C_{60} molecules interact through a Lennard-Jones potential, and by av-eraging over all the relative orientations of the C_{60} molecules (Girifalco, 1992). The remarkable feature of this calculated phase diagram is the absence of a stable liquid phase: the entire vapor-liquid binodal is metastable. The critical point of this metastable phase transition lies below the sublimation curve. In a molecule like C_{60}, the effective range of intermolecular attractions is deter-mined by carbon-carbon interactions, whereas the repulsive core is fixed by the size of the large C_{60} cage (Hagen et al., 1993). Thus this molecule has a much larger ratio of repulsive core to attractive well range than small-molecule systems. That this feature may cause the phase behavior shown in Figure 2.20 is suggested (Hagen et al., 1993) by analogy with colloidal suspensions (Russel et al., 1989), where similar phase behavior occurs if nonadsorbing polymer is added under conditions such that the range of the polymer-induced attractions is smaller than one third of the radius of the colloidal particles. Though the phase behavior shown in Figure 2.20 is based on a model system, and not on actual C_{60}, it raises the interesting question of the relationship between short-ranged attractions and vapor-liquid metastability in real systems.

2.3 PHENOMENOLOGICAL APPROACH:
STABILITY OF FLUID MIXTURES

The phenomenological treatment yields the general stability criterion Equation (2.15). It will now be applied to systems with more than one component. The goal is to determine spinodal curves for fluid mixtures, and to understand the differences between metastability in pure substances and in mixtures.

2.3.1 Binary Mixtures

For a binary mixture ($n = 2$), the general stability criterion reads

$$\left(\frac{\partial \xi_3}{\partial X_3}\right)_{\xi_1,\xi_2,X_4} > 0 \tag{2.93}$$

(Beegle et al., 1974a,b). When, starting from a stable or metastable condition, Equation (2.93) first becomes an equality, the mixture has reached a limit of

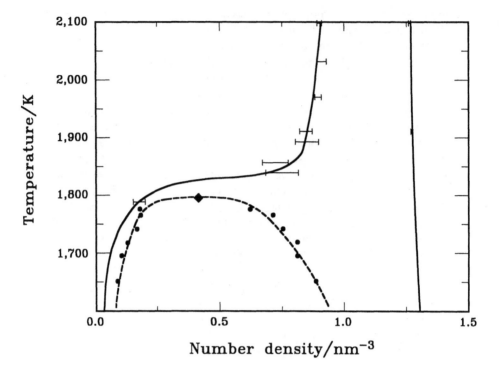

Figure 2.20. Computed phase diagram of a model of C_{60}. The solid lines are the branches of the stable solid-fluid binodal. The dashed line is the metastable vapor-liquid binodal. The dots are simulation results, the diamond is the estimated critical point, and the dashed line is a fit through the points. [Reprinted, with permission, from Hagen et al., *Nature*, 365: 425 (1993). Copyright 1993, Macmillan Magazines Limited]

stability. The spinodal curve is therefore given by the condition

$$\left(\frac{\partial \xi_3}{\partial X_3}\right)_{\xi_1, \xi_2, X_4} = 0. \tag{2.94}$$

For a binary mixture, there are four independent extensive variables X_i: S, V, N_1, and N_2, where N_i denotes the number of moles (or molecules) of species i. Therefore the stability criterion can be written in 4! equivalent ways. However, those arrangements of the independent variables that differ only in the ordering of N_1 and N_2 will yield symmetric expressions. Furthermore, as can be seen from Equation (2.94), the ordering of the first two variables is irrelevant. In practice, expressions with chemical potentials as independent variables are seldom used. This eliminates all orderings in which the first two variables are mole numbers. Therefore the stability criterion for a binary mixture reads

$$\left(\frac{\partial \mu_1}{\partial N_1}\right)_{T, P, N_2} > 0 \tag{2.95}$$

or, equivalently,

$$\left(\frac{\partial \mu_1}{\partial x_1}\right)_{T,P} > 0, \tag{2.96}$$

where x_1 is the mole fraction of component 1. The spinodal is the locus along which, starting from a stable or metastable condition, the inequality (2.96) is first violated,

$$\left(\frac{\partial \mu_1}{\partial x_1}\right)_{T,P} = 0. \tag{2.97}$$

As with pure substances, there exists a correspondence between stability coefficients and fluctuations. One example of this correspondence is Equation (2.67). It can be expressed in general form by the fluctuation identity

$$\left(\frac{\partial \xi_{n+1}}{\partial X_{n+1}}\right)_{\xi_1,\xi_2,\dots,\xi_n,X_{n+2}} = \frac{kT}{\langle (\delta X_{n+1})^2 \rangle_{X_{n+2}}} \tag{2.98}$$

(Debenedetti, 1985; Panagiotopoulos and Reid, 1986), which relates a stability coefficient to the mean square fluctuation of an extensive variable X_{n+1} within a region whose boundaries are so defined as to contain a fixed amount of X_{n+2}. For a binary mixture, the above equation becomes

$$\left(\frac{\partial \xi_3}{\partial X_3}\right)_{\xi_1,\xi_2,X_4} = \frac{kT}{\langle (\delta X_3)^2 \rangle_{X_4}}, \tag{2.99}$$

which, for the ordering (S, N_2, N_1, V), reads

$$\left(\frac{\partial \mu_1}{\partial N_1}\right)_{T,\mu_2,V} = \frac{kT}{\langle (\delta N_1)^2 \rangle_V}. \tag{2.100}$$

The left-hand side is a stability coefficient: it vanishes at the spinodal. The corresponding diverging fluctuation is thus a composition fluctuation. Upon switching labels between the two components, we obtain an identical equation with subscripts 1 and 2 interchanged. Therefore the molecular mechanism that causes a binary mixture to become unstable is the unbounded growth of composition fluctuations.

Consider a binary liquid mixture whose components are not fully miscible across the entire composition range. The change in Gibbs energy associated with mixing N_1 molecules of liquid 1 and N_2 molecules of liquid 2 at constant temperature and pressure is given by

$$\frac{G_{\text{mix}} - G}{N} = \Delta g = x_1(\mu_1 - g_1) + (1 - x_1)(\mu_2 - g_2), \tag{2.101}$$

where $N = N_1 + N_2$, g_1 and g_2 are the chemical potentials (per molecule) of the pure components at the mixture's temperature and pressure, G_{mix} is the mixture's Gibbs energy, and G is the sum of pure-component Gibbs energies at the mixture's temperature and pressure. Figure 2.21 is a schematic representation of the isothermal and isobaric Δg vs. x_1 relationship for a binary liquid mixture whose components are not fully miscible at the temperature and pressure of interest. It is analogous to the Helmholtz energy vs. volume diagram, Figure 2.1, used to describe fluid metastability in pure substances, in that it assumes analytic continuation of a thermodynamic function (Δg in this case) into an unstable region. The meaning of Figure 2.21 follows from the two thermodynamic relations

$$\mu_1 - g_1 = (1 - x_1)\left(\frac{\partial \Delta g}{\partial x_1}\right)_{T,P} + \Delta g, \qquad (2.102)$$

$$x_1\left(\frac{\partial \mu_1}{\partial x_1}\right)_{T,P} + (1 - x_1)\left(\frac{\partial \mu_2}{\partial x_1}\right)_{T,P} = 0. \qquad (2.103)$$

A tangent to curve $acdeb$ defines two y intercepts. By virtue of Equation (2.102), the left intercept is the quantity $(\mu_2 - g_2)$; the right intercept, $(\mu_1 - g_1)$. Points a and b share a common tangent. The chemical potential of each component is thus equal at a and b, and these points represent two coexisting phases in equilibrium. Points intermediate between a and c and between b and d are metastable. Mixtures within these two composition ranges will split spontaneously into phases a and b at the given temperature and pressure, the process being accompanied by a decrease in Gibbs energy. The final amounts of a and b are in the ratio β/α for a mixture such as e. At points c and d, the curvature changes sign. It follows from Equation (2.103) that

$$\left(\frac{\partial^2 \Delta g}{\partial x_1^2}\right)_{T,P} = \left(\frac{1}{1 - x_1}\right) \cdot \left(\frac{\partial \mu_1}{\partial x_1}\right)_{T,P}. \qquad (2.104)$$

Therefore points c and d are limits of stability. By identical reasoning, points between c and d are unstable ($\partial \mu_1/\partial x_1 < 0$). This type of instability with respect to concentration fluctuations is called diffusional or material instability.

Some possible binodal and spinodal loci defined by pairs of points such as ab and cd (Figure 2.21) at different temperatures and constant pressure are shown in Figure 2.22. In case (a), immiscibility disappears above the upper critical solution temperature (UCST). In case (c), immiscibility also disappears below the lower critical solution temperature (LCST). Mixing is favored entropically; hence this contribution to the mixture's free energy predominates at high temperature. This explains the UCST. In order for a mixture to exhibit a LCST, strong association forces, such as hydrogen bonding, must exist between the two components. Such forces, which favor mixing, predominate at low temperatures. In Figure 2.22, the strength of association forces between unlike components increases from left to right. A detailed discussion of liquid

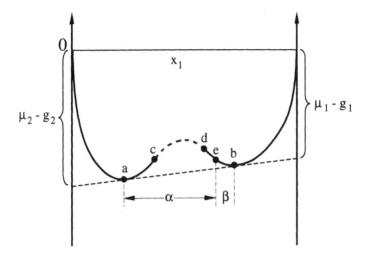

Figure 2.21. Composition dependence of the specific Gibbs energy change of mixing (Δg) for a binary liquid mixture whose components are not completely miscible at the given temperature and pressure. x_1 is a mole fraction; phases a and b are in equilibrium; points c and d are limits of stability; states between c and d are unstable.

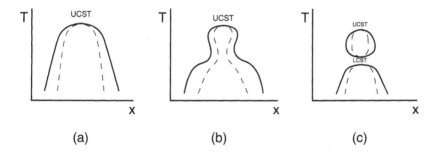

Figure 2.22. Possible binodal (full) and spinodal (dashed) loci at constant pressure for binary liquid mixtures exhibiting immiscibility. UCST and LCST denote the upper and lower critical solution temperatures. Association forces (e.g., hydrogen bonds) between unlike molecules increase in strength from left to right.

mixture phase behavior is beyond the scope of this chapter but can be found in books by Rowlinson and Swinton (1982), and by Walas (1985), and in an interesting descriptive article by Walker and Vause (1987).

A convenient way of studying the stability of mixtures experimentally is by light scattering (e.g., Chu et al., 1969; Kojima et al., 1975; Sorensen, 1988).

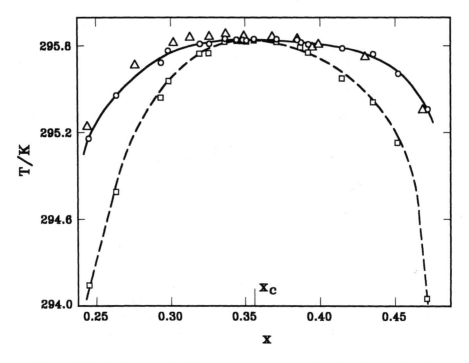

Figure 2.23. Binodal and spinodal curves (the latter determined by light scattering) for the binary liquid mixture n-hexane–n-tetradecafluorohexane. The triangles are data of Gaw and Scott (1971). (Adapted from Pozharskaya et al., 1984)

An example is given in Figure 2.23, which shows spinodal and binodal curves for the mixture n-hexane–n-tetradecafluorohexane (Pozharskaya et al., 1984). The zero-angle intensity of scattered light due to concentration fluctuations I is given by

$$\lim_{q \to 0} I(q) = I(0) = A + B \left[\left(\frac{\partial \mu_1}{\partial N_1} \right)_{T,P,N2} \right]^{-1}, \qquad (2.105)$$

$$q = \frac{4\pi}{\lambda} \sin \left(\frac{\theta}{2} \right). \qquad (2.106)$$

A and B depend on mixture characteristics such as refractive index and compressibility; λ is the wavelength of light in the scattering medium, and θ is the scattering angle (Chu et al., 1968, 1969; Chu and Schoenes, 1968). Near the spinodal, therefore, $I(0)^{-1}$ vanishes as

$$I(0)^{-1} \propto \left(\frac{\partial \mu_1}{\partial x_1} \right)_{T,P}. \qquad (2.107)$$

Since the stability coefficient vanishes at the spinodal, we can write

$$\left(\frac{\partial \mu_1}{\partial x_1}\right)_{T,P} = C \left[\frac{T - T_{sp}(x_1)}{T_c}\right]^{\gamma}, \tag{2.108}$$

where C and γ are the amplitude and exponent that characterize the vanishing of the stability coefficient as the spinodal is approached by changing the temperature isobarically at constant composition, and T_c is the mixture's critical temperature at the given pressure. Measurements of the temperature dependence of the zero-angle scattered light intensity at constant composition and pressure yield $T_{sp}(x)$ by extrapolation to the condition $I^{-1} = 0$. Equation (2.108) is the binary mixture analogue of the pseudospinodal Equation (2.83).

Both the UCST and the LCST are mixture critical points. They are the only stable states along a mixture spinodal. Therefore, they satisfy simultaneously the spinodal condition [Equation (2.97)] and the stability conditions

$$\left(\frac{\partial^2 \mu_1}{\partial N_1^2}\right)_{T,P,N_2} = 0, \qquad \left(\frac{\partial^3 \mu_1}{\partial N_1^3}\right)_{T,P,N_2} > 0 \tag{2.109}$$

(Appendix 1; Section 2.2.1). As a mixture approaches the spinodal, it becomes progressively less stable towards composition fluctuations. However, at the limit of stability, the mixture has a finite compressibility and isobaric heat capacity. Thus material stability is violated before mechanical stability (Modell and Reid, 1983).

We now consider the calculation of mixture spinodals from pressure-explicit equations of state. To this end, we write

$$\left(\frac{\partial \mu_1}{\partial N_1}\right)_{T,P,N_2} = \left(\frac{\partial \mu_1}{\partial N_1}\right)_{T,V,N_2} - V K_T \left[\left(\frac{\partial P}{\partial N_1}\right)_{T,V,N_2}\right]^2. \tag{2.110}$$

Equation (2.110) is useful for practical calculations because the independent variables in the right-hand side (T, V, N_1, N_2) are the natural arguments of pressure-explicit equations of state. Thus, if the mixture's equation of state $P = P(T, V, N_1, N_2)$ is known, the spinodal is obtained from the condition

$$\left(\frac{\partial \mu_1}{\partial N_1}\right)_{T,V,N_2} - V K_T \left[\left(\frac{\partial P}{\partial N_1}\right)_{T,V,N_2}\right]^2 = 0 \tag{2.111}$$

in conjunction with the identity

$$\left(\frac{\partial \mu_1}{\partial N_1}\right)_{T,V,N_2} = \int_V^{\infty} \left(\frac{\partial^2 P}{\partial N_1^2}\right)_{T,V,N_2} dV + \frac{kT}{N_1} \tag{2.112}$$

(Modell and Reid, 1983).

Metastability in binary mixtures is not confined to systems that exhibit immiscibility. Binary liquids can be superheated, and binary vapors supercooled, just like their single-component counterparts. The thermodynamic limit of superheat is the locus along which Equation (2.97) is satisfied. These stability limits are thermodynamically identical to the spinodal curves of systems exhibiting liquid-liquid immiscibility, even though their phenomenological description suggests a similarity with spinodals in single-component vapor-liquid equilibrium.

Figures 2.24 and 2.25 show kinetic and thermodynamic superheat limits of n-pentane–n-hexane and of benzene-cyclohexane mixtures at atmospheric pressure (Holden and Katz, 1978).[14] As with pure components, the attainable penetration into the metastable region is kinetically rather than thermodynamically controlled. Consequently, accurate equations of state give stability boundaries that can only be approached in actual experiments. The spinodals in Figures 2.24 and 2.25 were calculated using the Peng–Robinson equation of state (Peng and Robinson, 1976). Other examples of studies of kinetic superheat limits in binary mixtures[15] include the systems ethane–n-butane and n-propane–isopropane (Porteous and Blander, 1975); n-propane–n-butane (Blander and Katz, 1975); n-pentane–n-hexane (Skripov, 1974); n-pentane–cyclohexane (Eberhart et al., 1975); n-pentane–n-heptane (Skripov and Kukushkin, 1961); n-pentane–n-dodecane (Eberhart et al., 1975); n-pentane–n-hexadecane (Blander et al., 1971; Eberhart et al., 1975); n-hexane–cyclohexane (Holden and Katz, 1978); n-hexane–benzene (Holden and Katz, 1978); n-hexane–hexafluorobenzene (Skripov et al., 1988); benzene–cyclohexane (Skripov et al., 1988); benzene–carbon tetrachloride (Jalaluddin and Sinha, 1962); benzene-ethanol (Skripov et al., 1988); benzene-acetone (Skripov et al., 1988); and acetone-butanol (Skripov et al., 1988).

The n-pentane–n-hexane superheat limits (Figure 2.24) are mole fraction averages of the pure component values. In this case, both components are miscible in all proportions, and their mixtures exhibit small deviations from ideality. The benzene-cyclohexane superheat limits, on the other hand, exhibit a nonmonotonic dependence on the mixture's composition (Figure 2.25). Though miscible in all proportions, these two components form mixtures that deviate markedly from ideality. In particular, the (P, T) projection of the locus of mixture critical points shows a minimum (Rowlinson and Swinton, 1982). Note that in both cases the calculated spinodal lies beyond the experimental and theoretical kinetic limits of superheat.

Figure 2.26 shows the (P, T) projection of the limits of stability for the mixture n-pentane–n-hexane, calculated with the Peng-Robinson equation of state. C_1 (n-pentane) and C_2 (n-hexane) are the pure-component critical points.

---

[14]See Sections 3.1.1 and 3.1.4 for the calculation of kinetic limits of superheat (full line in Figures 2.24 and 2.25).

[15]These investigations are discussed in detail in Section 3.1.4.

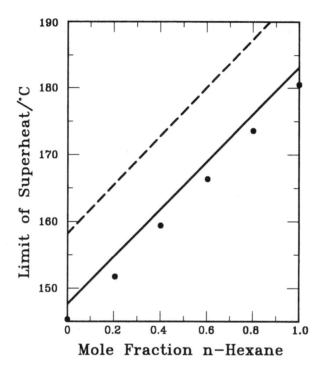

Figure 2.24. Thermodynamic and kinetic limits of superheat of the mixture n-pentane + n-hexane at atmospheric pressure. The points are measurements, using the droplet superheat technique (Holden and Katz, 1978). The full line is a kinetic calculation, using a nucleation rate of 10^4 cm^{-3} sec^{-1} (Holden and Katz, 1978). The dashed line is the spinodal calculated with the Peng-Robinson equation of state.

Curves labeled V and L are vapor and liquid spinodals, respectively. The four intermediate curves are for n-hexane mole fractions of 0.2, 0.4, 0.6, and 0.8, increasing in the direction of the arrow. Line $C_1 C_2$ is the locus of mixture critical points. A constant-pressure line intersects the liquid spinodals at temperatures that increase linearly with hexane mole fraction: this is the behavior shown in Figure 2.24. The four mixture spinodal curves of Figure 2.26 are stability limits along which the supercooled vapor or superheated liquid becomes materially but not mechanically unstable. The experimentally inaccessible locus of mechanical stability limits along which the mixture's compressibility diverges lies to the right of curves L and to the left of curves V. Thus, upon isobaric superheating, a given mixture will reach a condition of material instability, L, while its compressibility is finite. Figure 2.27 shows the limits of stability for the mixture benzene-cyclohexane, calculated with the Peng-Robinson equation

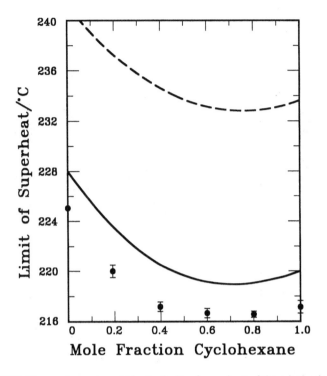

Figure 2.25. Thermodynamic and kinetic limits of superheat of the mixture benzene + cyclohexane at atmospheric pressure. The points are measurements, using the droplet superheat technique (Holden and Katz, 1978). The full line is a kinetic calculation, using a nucleation rate of 10^4 cm^{-3} sec^{-1} (Holden and Katz, 1978). The dashed line is the spinodal calculated with the Peng-Robinson equation of state.

of state. Note that there is now a pressure range within which the intersection of a constant-pressure line with the liquid spinodals occurs at temperatures that do not increase monotonically with benzene's mole fraction. This is the behavior shown in Figure 2.25.

Figure 2.28 (Peters et al., 1995) shows an example of metastable phase equilibria in a binary mixture (propane-triphenylmethane). Loci of stable three-phase coexistence are shown in Figure 2.28(a), where l_1 denotes a propane-rich liquid phase, l_2 a triphenylmethane-rich liquid phase, g the gas phase, and s_B solid triphenylmethane. Q is the point where the four phases coexist; K is an upper critical end point, where the difference between l_1 and l_2 disappears. In Figure 2.28(b), the metastable continuation of the l_1l_2g locus for $T < T_Q$ is shown. The (P, x) projection of the metastable phase coexistence at 330 K is shown in Figure 2.28(c) (x denotes the mole fraction of triphenylmethane): the entire liquid dome, including the UCEP (upper critical end point), is metastable.

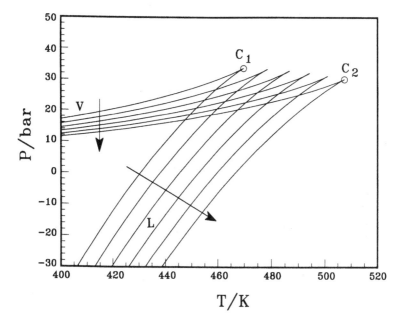

Figure 2.26. Limits of stability for n-pentane + n-hexane mixtures calculated with the Peng-Robinson equation of state. C_1 (n-pentane) and C_2 (n-hexane) are the pure-component critical points. V and L are vapor and liquid spinodals for mixtures of n-hexane mole fractions 0, 0.2, 0.4, 0.6, 0.8, and 1, respectively, increasing in the direction of the arrow.

The horizontal line where the liquid-liquid dome terminates denotes the pressure at which the three fluid phases (l_1, l_2, g) coexist; the compositions of the coexisting liquid phases are given by the intersection of the binodal with the horizontal three-phase coexistence locus; g is pure propane. By plotting the pressure at which the three fluid phases coexist as a function of temperature, one obtains the metastable three-phase locus $l_1 l_2 g$ at $T < T_Q$ in Figure 2.28(b).

2.3.2 Multicomponent Mixtures

The general stability criterion for an n-component mixture reads

$$\left(\frac{\partial \xi_{n+1}}{\partial X_{n+1}} \right)_{\xi_1, \xi_2, \dots, \xi_n, X_{n+2}} > 0 \qquad (2.15)$$

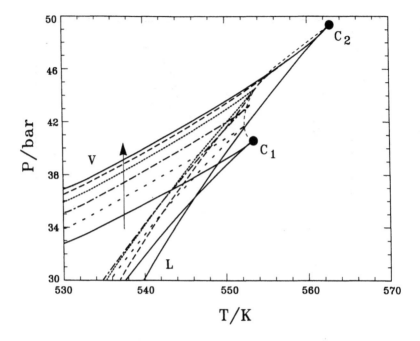

Figure 2.27. Limits of stability for benzene + cyclohexane mixtures calculated with the Peng-Robinson equation of state. C_1 (cyclohexane) and C_2 (benzene) are the pure-component critical points. V and L denote the vapor and liquid spinodals. The full lines are cyclohexane and benzene spinodals, terminating at C_1 and C_2, respectively. The dashed spinodals are for mixtures with benzene mole fractions 0.2, 0.4, 0.6, and 0.8, increasing in the direction of the arrow for the vapor spinodals. Note the crossing of liquid spinodals. The dashed curve joining C_1 and C_2 is the locus of mixture critical points.

(Beegle et al., 1974a,b). The locus along which the inequality is first violated, starting from a stable or metastable condition, defines the spinodal curve

$$\left(\frac{\partial \xi_{n+1}}{\partial X_{n+1}}\right)_{\xi_1,\xi_2,\ldots,\xi_n,X_{n+2}} = 0. \tag{2.16}$$

The independent variables in Equation (2.16) may not always be convenient for practical calculations involving mixtures with more than two components. As an example, consider a ternary mixture ($n = 3$), and order the extensive variables (S, V, N_1, N_2, N_3). Then the equation of the spinodal is

$$\left(\frac{\partial \mu_2}{\partial N_2}\right)_{T,P,\mu_1,N_3} = 0, \tag{2.113}$$

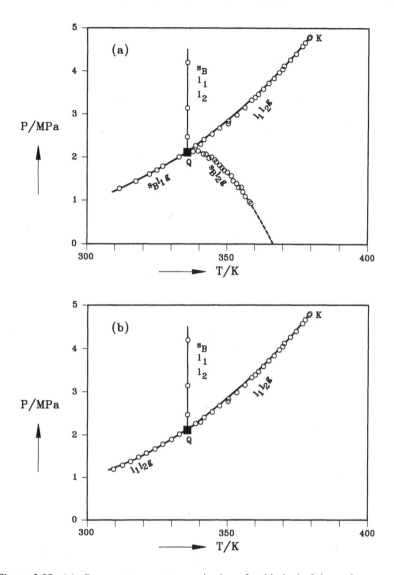

Figure 2.28. (a): Pressure-temperature projection of stable loci of three-phase coexistence in the system propane + triphenylmethane. (b): Three-phase coexistence loci including the metastable line $l_1 l_2 g$ at $T < T_Q$. [Reprinted, with permission, from Peters et al., *Fluid Phase Equil.*, 109: 99 (1995)] *Continued next page.*

which requires an expression of the form $\mu_2 = \mu_2(T, P, N_2, N_3, \mu_1)$. In such cases, it is convenient to express the stability criteria in determinant form (Modell and Reid, 1983; Kumar and Reid, 1986; Panagiotopoulos and Reid, 1986).

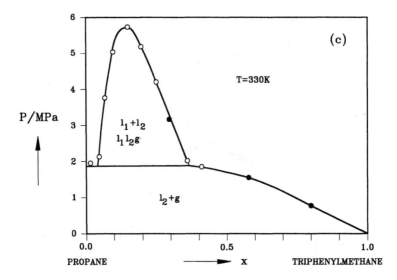

Figure 2.28. *Continued.* Pressure-composition projection of metastable phase coexistence at 330 K (c). K is an upper critical end point. At Q, four phases coexist. l_1 denotes the propane-rich liquid phase; l_2, the triphenylmethane-rich liquid phase; g is the gas phase; s_B is solid triphenylmethane. [Reprinted, with permission, from Peters et al., *Fluid Phase Equil.*, 109: 99 (1995)]

A system is stable or metastable when the following inequality is satisfied:

$$L_i > 0 \qquad\qquad (2.114)$$

and the equation of the spinodal reads

$$L_i = 0, \qquad\qquad (2.115)$$

where L_i is a determinant defined by

$$L_i = \begin{vmatrix} y^{(i)}_{i+1,i+1} & \cdots & y^{(i)}_{i+1,n+1} \\ \cdot & \cdots & \cdot \\ \cdot & \cdots & \cdot \\ \cdot & \cdots & \cdot \\ y^{(i)}_{n+1,i+1} & \cdots & y^{(i)}_{n+1,n+1} \end{vmatrix}, \quad 0 \le i \le n \qquad (2.116)$$

and $y^{(i)}$ is the ith Legendre transform of the energy

$$y^{(i)} = U - \sum_{j=1}^{i} \xi_j X_j = y^{(i)}(\xi_1, \xi_2, \ldots, \xi_i, X_{i+1}, \ldots, X_{n+2}) \qquad (2.117)$$

$$dy^{(i)} = -\sum_{j=1}^{i} X_j d\xi_j + \sum_{j=i+1}^{n+2} \xi_j dX_j \qquad (2.118)$$

(Beegle et al., 1974a; Modell and Reid, 1983; Callen, 1985). In equation (2.116), $y_{j,k}^{(i)}$ denotes a second-order partial derivative, for example,

$$y_{i+1,i+1}^{(i)} = \left(\frac{\partial^2 y^{(i)}}{\partial X_{i+1}^2}\right)_{\xi_1,\dots,\xi_i,X_{i+2},\dots,X_{n+2}} = \left(\frac{\partial \xi_{i+1}}{\partial X_{i+1}}\right)_{\xi_1,\dots,\xi_i,X_{i+2},\dots,X_{n+2}}. \qquad (2.119)$$

For any given ordering of the independent variables (e.g., S, V, N_1, N_2, and N_3 for a ternary mixture), (2.116) defines $(n+1)$ equivalent criteria.

As an example of the application of the determinant form of the stability criterion, consider the calculation of a ternary mixture's spinodal, assuming knowledge of the temperature, pressure, and composition dependence of the chemical potentials [i.e., $\mu_1 = f(T, P, x_2, x_3)$, which can always be written as $\mu_1 = g(T, P, N_1, N_2, N_3)$]. Then, choosing $i = 2$ and ordering the extensive variables as (S, V, N_1, N_2, N_3),

$$y^{(2)} = U - TS + PV = G(T, P, N_1, N_2, N_3), \qquad (2.120)$$

$$dy^{(2)} = -SdT + VdP + \mu_1 dN_1 + \mu_2 dN_2 + \mu_3 dN_3, \qquad (2.121)$$

$$
\begin{aligned}
L_2 &= \begin{vmatrix} y_{33}^{(2)} & y_{34}^{(2)} \\ y_{43}^{(2)} & y_{44}^{(2)} \end{vmatrix} \\
&= \left(\frac{\partial \mu_1}{\partial N_1}\right)_{T,P,N_2,N_3} \left(\frac{\partial \mu_2}{\partial N_2}\right)_{T,P,N_1,N_3} - \left[\left(\frac{\partial \mu_1}{\partial N_2}\right)_{T,P,N_1,N_3}\right]^2 \\
&= 0. \qquad (2.122)
\end{aligned}
$$

Mole-fraction-explicit expressions, as well as determinants suitable for use with pressure-explicit equations of state, can be found in the text by Modell and Reid (1983). Equations (2.114)–(2.116) are completely general, and can also be applied to single-component and binary systems.

Criticality criteria can also be expressed as partial derivatives:

$$\left(\frac{\partial \xi_{n+1}}{\partial X_{n+1}}\right)_{\xi_1,\dots,\xi_n,X_{n+2}} = 0, \quad \left(\frac{\partial^2 \xi_{n+1}}{\partial X_{n+1}^2}\right)_{\xi_1,\dots,\xi_n,X_{n+2}} = 0,$$

$$\left(\frac{\partial^3 \xi_{n+1}}{\partial X_{n+1}^3}\right)_{\xi_1,\dots,\xi_n,X_{n+2}} > 0, \qquad (2.123)$$

or, equivalently

$$L_i = 0 \qquad (2.115)$$

(Beegle et al., 1974b; Reid and Beegle 1977; Modell and Reid, 1983), and simultaneously

$$M_i = 0, \tag{2.124}$$

where M_i is a determinant defined by

$$M_i = \begin{vmatrix} y^{(i)}_{i+1,i+1} & \cdots & y^{(i)}_{i+1,n+1} \\ \vdots & \cdots & \vdots \\ y^{(i)}_{n,i+1} & \cdots & y^{(i)}_{n,n+1} \\ \left(\frac{\partial L_i}{\partial X_{i+1}}\right)_{\xi_1,\dots,\xi_i,X_{i+2},\dots,X_{n+2}} & \cdots & \left(\frac{\partial L_i}{\partial X_{n+1}}\right)_{\xi_1,\dots,\xi_i,X_{i+1},\dots,X_n,X_{n+2}} \end{vmatrix}. \tag{2.125}$$

To illustrate the use of (2.124), consider a ternary mixture. Choosing $i=2$ and ordering the extensive variables as (S, V, N_1, N_2, N_3), the spinodal line satisfies (2.122), and the criticality condition is

$$M_2 = \begin{vmatrix} \left(\frac{\partial \mu_1}{\partial N_1}\right)_{T,P,N_2,N_3} & \left(\frac{\partial \mu_1}{\partial N_2}\right)_{T,P,N_1,N_3} \\ \left(\frac{\partial L_2}{\partial N_1}\right)_{T,P,N_2,N_3} & \left(\frac{\partial L_2}{\partial N_2}\right)_{T,P,N_1,N_3} \end{vmatrix} = 0, \tag{2.126}$$

where

$$\begin{aligned} \left(\frac{\partial L_2}{\partial N_1}\right)_{T,P,N_2,N_3} &= \left(\frac{\partial^2 \mu_1}{\partial N_1^2}\right)_{T,P,N_2,N_3} \left(\frac{\partial \mu_2}{\partial N_2}\right)_{T,P,N_1,N_3} \\ &+ \left(\frac{\partial \mu_1}{\partial N_1}\right)_{T,P,N_2,N_3} \left(\frac{\partial^2 \mu_1}{\partial N_1 \partial N_2}\right)_{T,P,N_3} \\ &- 2\left(\frac{\partial \mu_1}{\partial N_2}\right)_{T,P,N_1,N_3} \left(\frac{\partial^2 \mu_1}{\partial N_1 \partial N_2}\right)_{T,P,N_3}, \end{aligned} \tag{2.127}$$

$$\begin{aligned} \left(\frac{\partial L_2}{\partial N_2}\right)_{T,P,N_1,N_3} &= \left(\frac{\partial^2 \mu_1}{\partial N_1 \partial N_2}\right)_{T,P,N_3} \left(\frac{\partial \mu_2}{\partial N_2}\right)_{T,P,N_1,N_3} \\ &+ \left(\frac{\partial \mu_1}{\partial N_1}\right)_{T,P,N_2,N_3} \left(\frac{\partial^2 \mu_1}{\partial N_2^2}\right)_{T,P,N_1,N_3} \\ &- 2\left(\frac{\partial \mu_1}{\partial N_2}\right)_{T,P,N_1,N_3} \left(\frac{\partial^2 \mu_1}{\partial N_2^2}\right)_{T,P,N_1,N_3}. \end{aligned} \tag{2.128}$$

Often it is desired to calculate spinodal curves and critical points starting from a pressure-explicit equation of state, $P = P(T, V, N_1, \dots, N_n)$ (e.g., Peng and Robinson, 1977; Baker and Luks, 1980; Heidemann and Khalil, 1980; Michelsen and Heidemann, 1981; Michelsen, 1982a,b; Nagarajan et al., 1991).

Then, using the ternary case as an example, the criticality condition reads, choosing $i = 1$ and after ordering the extensive variables as $(S, V, N_1, N_2 N_3)$,

$$
L_1 = \begin{vmatrix} y_{22}^{(1)} & y_{23}^{(1)} & y_{24}^{(1)} \\ y_{32}^{(1)} & y_{33}^{(1)} & y_{34}^{(1)} \\ y_{42}^{(1)} & y_{43}^{(1)} & y_{44}^{(1)} \end{vmatrix}
$$

$$
= \begin{vmatrix} -\left(\frac{\partial P}{\partial V}\right)_{T,N_1,N_2,N_3} & -\left(\frac{\partial P}{\partial N_1}\right)_{T,V,N_2,N_3} & -\left(\frac{\partial P}{\partial N_2}\right)_{T,V,N_1 N_3.} \\ -\left(\frac{\partial P}{\partial N_1}\right)_{T,V,N_2,N_3} & \left(\frac{\partial \mu_1}{\partial N_1}\right)_{T,V,N_2,N_3} & \left(\frac{\partial \mu_1}{\partial N_2}\right)_{T,V,N_1,N_3} \\ -\left(\frac{\partial P}{\partial N_2}\right)_{T,V,N_1,N_3} & \left(\frac{\partial \mu_1}{\partial N_2}\right)_{T,V,N_1,N_3} & \left(\frac{\partial \mu_2}{\partial N_2}\right)_{T,V,N_1,N_3} \end{vmatrix} = 0,
$$

$$(2.129)$$

$$
M_1 = \begin{vmatrix} y_{22}^{(1)} & y_{23}^{(1)} & y_{24}^{(1)} \\ y_{32}^{(1)} & y_{33}^{(1)} & y_{34}^{(1)} \\ \left(\frac{\partial L_1}{\partial V}\right)_{T,N_1,N_2,N_3} & \left(\frac{\partial L_1}{\partial N_1}\right)_{T,V,N_2,N_3} & \left(\frac{\partial L_1}{\partial N_2}\right)_{T,V,N_1,N_3} \end{vmatrix} = 0. \quad (2.130)
$$

The chemical potential derivatives in (2.129) and (2.130) can be expressed in a form useful for equation-of-state calculations, using the identity (see 2.112)

$$
\left(\frac{\partial \mu_i}{\partial N_j}\right)_{T,V,N[j]} = \frac{kT}{N_i}\delta_{ij} + \int_V^\infty \left(\frac{\partial^2 P}{\partial N_i \partial N_j}\right)_{T,V,N_k} dV, \quad (2.131)
$$

where δ_{ij} is Kronecker's delta, and $N[j]$ denotes constancy of all mole numbers except N_j. Equation (2.129), by itself, defines the spinodal. The theoretical and numerical aspects of the calculation of mixture critical points are discussed in the excellent review by Heidemann (1994).

The general fluctuation relationship

$$
\left(\frac{\partial \xi_{n+1}}{\partial X_{n+1}}\right)_{\xi_1,\xi_2,...,\xi_n,X_{n+2}} = \frac{kT}{\langle(\delta X_{n+1})^2\rangle_{X_{n+2}}} \quad (2.132)
$$

can always be arranged so that X_{n+2} is V and X_{n+1} is N_i. In light of (2.16), this implies that the mechanism of instability in any mixture, regardless of the number of components, is the growth of concentration fluctuations (material instability).

2.4 CRITIQUE OF THE PHENOMENOLOGICAL APPROACH: METASTABILITY AND STATISTICAL MECHANICS

The behavior of bulk matter is a consequence of molecular motion. Hence a rigorous theory must arrive at a description of bulk behavior starting from the forces between molecules. Statistical mechanics provides the link between the molecular and bulk domains. Central to the theory is the partition function (see Appendix 3), a sum over all assignments of positions and momenta to the molecules that constitute the system under study (Hill, 1956). Now we may be interested in studying an isolated system, or one whose temperature, volume, and mass are fixed, or a system with given temperature and pressure. To each of these choices there corresponds a different partition function. Consider first a fixed region inside a liquid, into and out of which molecules are free to move. Let the volume under consideration be V. Also, let the temperature imposed by the surrounding fluid be T. We choose V such that the average number of molecules contained in it, $\langle N \rangle$, is large [of the order of Avogadro's number, $\langle N \rangle \approx O(10^{23})$]. The properties of the fluid region under consideration can be computed from the grand partition function Ξ,

$$\Xi(T, V, \mu) = \sum_N e^{\beta \mu N} \sum_i e^{-\beta E_i} \qquad (2.133)$$

(Appendix 3; Hill, 1956), where $\beta = 1/kT$, and k is Boltzmann's constant. The first summation is over all possible values of N, and the second over all possible assignments of positions and momenta to the N molecules in V. In the above equation, μ is the chemical potential imposed by the surrounding fluid, and E_i is the energy associated with the ith assignment of positions and momenta. The connection with thermodynamics follows from the equation

$$kT \ln \Xi(\beta \mu, \beta, V) = PV \qquad (2.134)$$

where P is the pressure.

The average number of molecules in V is given by

$$\left(\frac{\partial \ln \Xi}{\partial \beta \mu} \right)_{\beta, V} = \frac{\sum_N N e^{N \beta \mu} \sum_i e^{-\beta E_i}}{\sum_N e^{N \beta \mu} \sum_i e^{-\beta E_i}} = \langle N \rangle. \qquad (2.135)$$

In general, it is impossible to enumerate the configurations accessible to a macroscopic system, and Ξ cannot be calculated exactly. Nevertheless, it is not necessary to obtain the function $\Xi(\beta \mu, \beta, V)$ exactly for the purpose of the present discussion. Instead, we differentiate Equation (2.135) once more, to obtain

$$\left(\frac{\partial^2 \ln \Xi}{\partial (\beta \mu)^2} \right)_{\beta, V} = \langle N^2 \rangle - \langle N \rangle^2 \equiv \langle (\delta N)^2 \rangle, \qquad (2.136)$$

where angular brackets denote average values inside V, and $\delta N = N - \langle N \rangle$. Thus the curvature of the $\ln \Xi$ vs. $\beta \mu$ relationship is equal to the mean squared fluctuation in the number of molecules present in V about the average value $\langle N \rangle$. The derivative in the left-hand side of Equation (2.136) has thermodynamic significance; this can be seen from the identity

$$\left(\frac{\partial^2 \beta P V}{\partial (\beta \mu)^2} \right)_{\beta, V} = \frac{kTV}{v^2} K_T, \qquad (2.137)$$

where $v = 1/\rho = V/\langle N \rangle$. Comparison of Equations (2.134), (2.136), and (2.137) yields

$$\frac{kT K_T}{V} = \frac{\langle (\delta\rho)^2 \rangle}{\langle \rho \rangle^2}, \tag{2.138}$$

where we have used the fact that in an open system with fixed volume, $\delta N = V\delta\rho$. Finally, we write

$$\left(\frac{\partial^2 A}{\partial V^2}\right)_{T,N} = -\left(\frac{\partial P}{\partial V}\right)_{T,N} = \frac{1}{V K_T} = \frac{kT}{V^2} \cdot \frac{\langle \rho \rangle^2}{\langle (\delta\rho)^2 \rangle}. \tag{2.139}$$

Segment ef of Figure 2.1 is therefore unphysical. The isothermal compressibility is inherently positive [Equation (2.138)] and so, therefore, is the curvature of the Helmholtz energy–volume relationship [Equation (2.139)]. This conclusion is independent of the type of molecular interaction that exists in the system under study, since this information was not invoked in the previous arguments. Note that this conclusion is more profound than the mere recognition of the mechanical instability of states along ef: we have shown that any microscopic model of matter yields a Helmholtz energy that does not depend on volume as shown by curve $befb'$ in Figure 2.1. This type of functionality, including the unphysical portion ef, is obtained only when approximate techniques for solving the partition function (or empirical equations of state) are used, and the system is implicitly constrained to having a uniform density within the coexistence region.

Consider now a closed system of given volume and temperature. The appropriate partition function Q (the canonical partition function) and its connection to thermodynamics are given by

$$Q(N, V, T) = \exp(-\beta A) = \sum_i \exp(-\beta E_i) \tag{2.140}$$

(Appendix 3; Hill, 1956), where A is the Helmholtz energy, and the summation is over all distinct configurations (assignments of positions and velocities to all molecules) accessible to N molecules contained in a fixed volume V, and maintained at temperature T through contact with a large heat bath (Hill, 1956).

If we impose a temperature and overall density that fall inside a two-phase region (e.g., vapor-liquid), the equilibrium state will be a combination of the coexisting phases at the given temperature (we are considering for simplicity a pure substance). It is clear that to study metastable states we need to impose restrictions that prevent the system from undergoing the phase transition that will lead to equilibrium.

Metastable states cannot be obtained from an unconstrained treatment of equilibrium, such as would result from the literal application of the usual statistical mechanical formalism [for example, performing the summation in Equation (2.140)]. This is because a metastable state is not a condition of maximum entropy for an isolated system; or, equivalently, of minimum Helmholtz energy for a closed system of fixed volume and temperature, or of minimum Gibbs energy for a closed system of fixed temperature and pressure. Hence in the thermodynamic limit metastable states make an insignificant contribution to the summation in Equation (2.140). What is needed in order to reconcile metastability with rigorous statistical mechanics is to restrict the summation in Equation (2.140) to those microscopic configurations in which the spatial distribution of molecules is "reasonably uniform." The partition function would then be evaluated exactly, but over a restricted set of configurations. It is important to distinguish the usual artificial simplification whereby the system's density is strictly uniform inside the coexistence region (e.g., the van der Waals theory) from this proposed rigorous calculation, in which only configurations exceeding a threshold (and as yet unspecified) non-uniformity are excluded. Approximate lattice theories are illustrative of the way in which mathematical approximations give rise to continuous loops spanning the unstable region (van der Waals loops). In the cluster-based quasichemical approximation, for example (Guggenheim and McGlashan, 1951), the system is treated as a random mixture of basic clusters. Because all possible configurations of the molecules in such clusters must be enumerable, the size of the clusters is too small to allow macroscopically inhomogeneous configurations to contribute to the partition function.

We have shown that the exact Helmholtz energy is a convex function of volume. In the previous paragraph, we have gone further still in the direction of questioning the rigorous basis of metastable states such as those that result in the van der Waals theory. For we are now not only denying a microscopic basis to a continuous curve spanning the coexistence region via an unstable portion (e.g., ef in Figure 2.1), but also saying that even states lying on convex portions of the isotherm are inconsistent with the unrestricted statistical-mechanical formalism if they lie above the common tangent construction that yields the equilibrium state. Thus the rigorous statistical mechanics of metastability is the statistical mechanics of constrained systems (Reiss, 1975; Stillinger, 1995).

In the laboratory, one studies metastable liquids by imposing kinetic barriers that prolong the system's lifetime (see Chapter 3). In contrast, the calculation of the thermophysical properties of a metastable system requires that we impose analytical barriers (constraints) that block access to regions of phase space[16] where the system is nonuniform. This means restricting the summation in the

---

[16]The motion of a system of N molecules with m degrees of freedom per molecule ($m = 3$ for an atom; $m = 6$ for a perfectly rigid polyatomic molecule) can be represented in a $2Nm$-dimensional space consisting of mN generalized coordinates $(q_1, q_2, q_3, \ldots, q_{mN})$ and mN generalized momenta $(p_1, p_2, p_3, \ldots, p_{mN})$. This space is called phase space (Toda et al., 1992). The instantaneous state of a system is represented by a point in phase space. The time evolution of the instantaneous states of the system defines an orbit in phase space.

partition function to configurations where the system is reasonably uniform.[17] While rigorous, this approach suffers from an obvious limitation: in most cases, the partition function is impossible to calculate exactly. Consequently, if configurations cannot be exhaustively enumerated, neither can they be selectively eliminated. Computers are a promising tool for studying metastable liquids because constraints can be easily implemented in simulations (e.g., Corti and Debenedetti, 1994, 1995; Stillinger, 1995). To date, however, this promising approach has not been given the detailed attention it merits.

The most important attempt at a rigorous thermodynamic treatment of metastability to date is the penetrating work of Lebowitz and Penrose (Lebowitz and Penrose, 1966; Penrose and Lebowitz, 1971, 1987). These authors first investigated the conditions under which the Maxwell construction for vapor-liquid equilibrium becomes exact (Lebowitz and Penrose, 1966). They considered a van der Waals–type fluid whose molecules interact in pairwise fashion through a potential (v) that can be decomposed into short-range repulsive (q) and longer-ranged attractive (ϕ) components,

$$v(r) = q(r) + \gamma^{\nu}\phi(\gamma r), \qquad (2.141)$$

where γ^{-1} is a measure of the range of the potential, and ν the dimensionality of the space under consideration. Lebowitz and Penrose showed that in the limit where the attractive potential becomes infinitely long ranged $(\gamma^{-1} \to \infty)$, the system's Helmholtz energy density $(a = A/V)$ becomes

$$a(\rho, 0+) = \lim_{\gamma \to 0} a(\rho, \gamma) = \text{CE}\left[a_0(\rho) + \frac{1}{2}\alpha\rho^2\right], \qquad (2.142)$$

where a_0 is the Helmholtz energy density of the system when $\phi = 0$ (purely repulsive interactions), and

$$\alpha = \gamma^{\nu}\int \phi(\gamma r)\, d^{\nu}r. \qquad (2.143)$$

In Equation (2.142), CE denotes a convex envelope, that is to say, the convex function whose value nowhere exceeds that of its argument. Note that the reference Helmholtz energy density a_0 is a convex function of ρ; however, since α is negative (ϕ attractive), the quantity in brackets is not necessarily a convex function of ρ. For low enough temperatures, it has an intermediate concave

---

[17]A metastable system will evolve irreversibly towards a different equilibrium state only if localized nonuniformities formed initially by spontaneous molecular fluctuations exceed a critical size (for example, if large enough droplets are formed in a supercooled vapor). Such a critical-sized nonuniformity is called the critical nucleus. The calculation of the critical size beyond which the formation of a new phase occurs spontaneously is discussed in Sections 3.1.1 and 3.1.2. Thus a system is "reasonably" uniform when it contains no nonuniformity whose size exceeds that of the critical nucleus.

portion. The equation of state is then given by the application of Maxwell's construction to the function $P_0 + (\alpha\rho^2)/2$, where P_0 is the pressure of the reference system whose molecules interact via the purely repulsive potential q at the given density.

The proof of Equation (2.142) involves dividing the physical size V into regions of size ω, such that

$$V^{1/\nu} \gg \gamma^{-1} \gg \omega^{1/\nu} \gg r_0, \qquad (2.144)$$

where r_0 is the diameter of the molecular repulsive core. The limits in Equation (2.144) are to be taken sequentially, that is to say, $V \to \infty$; $\gamma^{-\nu} \to \infty$; $\omega \to \infty$. Thus, in the limit where the attractive contribution to the pair potential becomes infinitely long ranged, it is possible to define a length scale $\omega^{1/\nu}$ that is much longer than the range of the repulsive core, yet much shorter than the range of the attractive forces. Lebowitz and Penrose were able to show that a system for which this separation of length scales is possible exhibits a first order phase transition. The quantitative description of this transition follows from applying the Maxwell construction to the van der Waals–type equation of state that results from adding a mean-field attractive correction to the pressure exerted by the purely repulsive reference system. Of course, this complete separation of length scales is not possible in real systems.

Important results concerning metastability in the vapor-liquid transition for systems with long-ranged attractive forces were also derived by Penrose and Lebowitz (1971). They characterized a metastable state as one satisfying three conditions: (i) only one phase is present (condition of rough uniformity); (ii) a system that starts in such a state is likely to take a long time to leave it; (iii) once the system leaves the metastable state, it is unlikely to return. These authors showed that a restricted set of configurations can be defined satisfying the above conditions if, at a given density ρ,

$$a_0(\rho) + \frac{1}{2}\alpha\rho^2 > a(\rho, 0+),$$

$$\left(\frac{d^2 a_0}{d\rho^2}\right) + 2\alpha > 0. \qquad (2.145)$$

Note that, since $\alpha < 0$, Equation (2.145) is more restrictive than the requirement of positive curvature for the function $a_0 + (\alpha\rho^2)/2$. The specification of metastable states involves once more a division of the total volume V into equal cells of size ω. Then, a subset R of possible realizations of the system is defined such that in every case the density in each cell falls between two values, $\rho-$ and $\rho+$. The latter are chosen so that the second inequality in (2.145) holds for all $\rho- < \rho < \rho+$. This subset of configurations constitutes the restricted ensemble appropriate to the calculation of thermodynamic properties of metastable states. Penrose and Lebowitz showed that a rigorous upper bound on the rate λ at which the system escapes the set of restricted configurations R can be written as

$$\lambda \leq \text{const} \times V\omega^{-1/\nu} \exp[-C\omega + f(\omega)]$$
$$= \text{const} \times \omega^{-1/\nu} \exp[\ln V - C\omega + f(\omega)], \tag{2.146}$$

where λ has units of reciprocal time, C is positive as long as ρ, $\rho-$, and $\rho+$ are on the same locally convex branch of $a_0 + (\alpha\rho^2)/2$, and $f(\omega)$ increases with ω more slowly than ω in itself. Therefore λ can be made arbitrarily small by appropriate choices of V, ω, and γ provided that

$$V^{1/\nu} \gg \gamma^{-1} \gg \omega^{1/\nu} \gg r_0 \ln V. \tag{2.147}$$

Under these conditions, the free energy density of the metastable state, computed by performing the summation prescribed in Equation (2.140) over the restricted ensemble R, is given by $a_0 + (\alpha\rho^2)/2$.

The separation of the intermolecular potential into short-ranged repulsive and arbitrarily long-ranged attractive components allows the definition of a scale $\omega^{1/\nu}$. By virtue of (2.147), this length scale satisfies $\omega^{1/\nu} \gg r_0 \ln V$, as well as $\omega^{1/\nu} \ll \gamma^{-1}$. The former inequality guarantees a small escape rate [see Equation (2.146)] by ensuring that, ω being macroscopic, there are always enough molecules in every cell so as to make a violation of the uniformity condition $\rho- < \rho < \rho+$ very unlikely. The second inequality ensures that phase transitions due to the long-ranged attraction (which occur over a length scale γ^{-1}) are suppressed in the restricted ensemble. Hence for the system to be in a configuration belonging to the set R is in itself an improbable event under conditions where the phase transition is the equilibrium state for the unrestricted system. Thus, $\gamma^{-1} \gg \omega^{1/\nu}$ guarantees uniformity [condition (i)]; $\omega^{1/\nu} \gg \ln V$ guarantees a small escape rate [condition (ii)]; and $V^{1/\nu} \gg \gamma^{-1}$ guarantees the thermodynamic limit [in which a metastable state has negligible probability of occurrence in an unconstrained ensemble, hence, once left, this condition is highly unlikely to recur, and condition (iii) is met]. All of these inequalities must apply simultaneously, rather than sequentially; thus the difference between (2.144) and (2.147).

The separation of length scales on which the above treatment is based is not in general possible for real systems. Nevertheless, the work of Lebowitz and Penrose is of great importance. It shows the limiting conditions necessary for the van der Waals theory to be exact; it is an example of the successful use of restricted ensembles[18] and it gives pragmatic guidelines for defining metastable states in a way that is amenable to statistical-mechanical treatment. Detailed proofs of the existence of Lebowitz-Penrose metastability in the Ising

---

[18]Elkoshi et al. (1985) were able to prove metastability rigorously, including negative pressures, in a one-dimensional system of hard rods constrained to have a given concentration of holes.

ferromagnet and in the Widom-Rowlinson model [a binary system with hard-core repulsion between unlike particles and no interaction between like particles (Widom and Rowlinson, 1970)] have been provided by Capocaccia et al. (1974), and by Cassandro and Oliveri (1977), respectively.

In 1967, Fisher showed that an idealized model of a vapor at sufficiently subcritical temperatures exhibits a mathematical singularity at the condensation point. This finding prompted the question of whether the equilibrium properties of metastable states can be calculated by analytic continuation of thermodynamic functions into the coexistence region. Fisher considered a vapor in the vicinity of the condensation point. At sufficiently low density (and hence temperature), it can be pictured as a collection of isolated, noninteracting molecules. Also present are clusters of a few molecules, held together loosely by attractive forces. Because of the low density, these clusters are also treated as noninteracting. One then has an ideal gas of noninteracting clusters, for which the equation of state can be written as

$$\beta P = \sum_{j=1}^{\infty} \rho_j = \sum_{j=1}^{\infty} q_j z^j V^{-1} \qquad (2.148)$$

(Fisher, 1967a,b), where ρ_j is the number density of clusters composed of j molecules, V is the system's total volume, z is the bulk vapor-phase activity [$z = \exp(\beta\mu)$, with μ the bulk vapor-phase chemical potential], and q_j is the configurational canonical partition function for a j-molecule cluster (i.e., the partition function that arises from summing $\exp[-\beta E_{pot,i}(j)]$ over the possible spatial arrangements of the j molecules in the cluster, where $E_{pot,i}(j)$ is the potential energy of the ith spatial arrangement of such molecules). Large clusters resemble droplets of the liquid phase, and this model is therefore called the droplet model. This picture involving embryos of the new phase (for example liquid droplets in a supercooled vapor; bubbles in a superheated liquid; crystallites in a supercooled liquid) underlies most of the kinetic models to be presented in Chapter 3. Approximating the configurational partition function by its maximum term, we have

$$q_j \approx V\Omega(\Sigma) \exp(\beta j \epsilon_0) \exp(-\beta \epsilon_s \Sigma) \qquad (2.149)$$

(Fisher, 1967b), where Ω is the number of configurations accessible to j molecules with fixed center of mass and surface area Σ; $-\epsilon_0$ is a characteristic short-ranged binding energy per molecule; ϵ_s is a specific surface energy arising due to binding energy loss at the surface; and Σ is the most probable surface area of a j-molecule cluster. For Σ and Ω Fisher wrote

$$\Sigma = a_0 j^m, \quad 0 < m < 1,$$

$$\Omega \approx \frac{g_0 \lambda^\Sigma}{\Sigma^{\tau/m}}, \qquad (2.150)$$

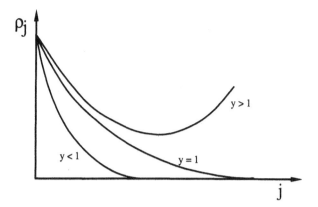

Figure 2.29. Concentration of clusters composed of j molecules in a nonsaturated vapor $(y < 1)$, at condensation $(y = 1)$, and in a supersaturated vapor $(y > 1)$. The portion of the supersaturated curve beyond the minimum is unphysical. $y = \exp(\delta\mu/kT)$, where $\delta\mu$ is the difference between the bulk vapor-phase chemical potential and its value at saturation. (Adapted from Fisher, 1967a)

with m approaching unity for compact clusters. The factor $(\Sigma^{\tau/m})^{-1}$ in the expression for Ω corrects the numerator due to the fact that the surface is topologically closed, and the exponent τ satisfies $\tau/m > 2$ (Fisher, 1967b). The equation of state therefore reads

$$\beta P \approx \frac{g_0}{a_0^{\tau/m}} \sum_{j=1}^{\infty} y^j j^{-\tau} x^{j^m} \qquad (2.151)$$

with

$$y = z \exp(\beta\epsilon_0), \quad x = \exp[\beta a_0 (wT - \epsilon_s)], \quad w = k \ln \lambda. \qquad (2.152)$$

For $T < \epsilon_s/w$, x is less than 1. At low density (and hence low z), y is also small. Under these conditions, comparison of Equations (2.148), (2.151), and (2.152) shows that ρ_j decreases rapidly to 0 as j increases. With $x < 1$, the decay of ρ_j to zero becomes slower as y approaches unity. However, when $y > 1$, ρ_j no longer decays monotonically to zero. Instead, it decreases at first, reaches a minimum at $j = j^*$, and increases thereafter (Figure 2.29). The unbounded increase in the concentration of large clusters implies the occurrence of condensation. Therefore we identify the condition $y = 1$ with condensation, and we can write $y = \exp(\beta\delta\mu)$, where $\delta\mu$ is the difference between the bulk vapor-phase chemical potential and its value at condensation. The vanishing of the surface energy term $wT - \epsilon_s$ corresponds to the critical point ($T_c = \epsilon_s/w$).

j^* is given (approximately) by

$$j^* \approx \left[\frac{ma_0(\epsilon_s - Tw)}{\delta\mu} \right]^{1/(1-m)}. \tag{2.153}$$

Clusters of size j^* (critical clusters) play an essential role in the theory of nucleation discussed in Chapter 3. There, it will be shown that the formation of critical clusters represents a free energy barrier that must be overcome by the system before the new phase, the liquid in the present case, can be formed. Clusters containing $j < j^*$ molecules evaporate spontaneously, and those with $j > j^*$ grow spontaneously. The goal of nucleation theory is the calculation of the rate of formation of critical clusters, or nucleation rate.

The density and its derivatives with respect to pressure are obtained from (2.151) by successive differentiation with respect to the activity z (Fisher, 1967a)

$$\pi^{(0)} = \beta P = \frac{g_0}{a_0^{\tau/m}} \sum_{j=1}^{\infty} y^j j^{-\tau} x^{j^m}, \tag{2.154}$$

$$\pi^{(1)} = \rho = z \left(\frac{\partial \pi^{(0)}}{\partial z} \right)_\beta = \frac{g_0}{a_0^{\tau/m}} \sum_{j=i}^{\infty} y^j j^{1-\tau} x^{j^m}, \tag{2.155}$$

$$\pi^{(2)} = \rho^2 k T K_T = z \left(\frac{\partial \pi^{(1)}}{\partial z} \right)_\beta = \frac{g_0}{a_0^{\tau/m}} \sum_{j=1}^{\infty} y^j j^{2-\tau} x^{j^m}, \tag{2.156}$$

$$\vdots$$

$$\pi^{(n)} = z \left(\frac{\partial \pi^{(n-1)}}{\partial z} \right)_\beta = \frac{g_0}{a_0^{\tau/m}} \sum_{j=1}^{\infty} y^j j^{n-\tau} x^{j^m}. \tag{2.157}$$

Thus, the pressure and all its derivatives with respect to z are well defined at the condensation point, $y = 1$. For $y > 1$, however, when $\delta\mu > 0$ and the supercooled vapor is metastable, the series (2.151) does not converge. Thus the model's thermodynamic functions exhibit a so-called essential singularity at the condensation point, beyond which they cannot be continued analytically (Fisher, 1967a,b). This singularity is extremely weak: it would not be possible to measure it experimentally since all thermodynamic quantities are well behaved at the condensation point. Although the droplet model is not rigorous, and it applies only to low-density vapors, its implications are quite provocative. In particular, it suggests that it is not possible to describe metastable states rigorously by analytic continuation of thermodynamic quantities into the coexistence region.

Fisher's arguments have been given a more rigorous basis by Isakov (1984). For the d-dimensional lattice-gas model ($d \geq 2$; single occupancy, nearest-neighbor attraction), Isakov showed that the nth derivatives of the Gibbs energy at the phase transition satisfy

$$\frac{\partial^n G}{\partial P^n} \approx A b^n (n!)^{d/(d-1)}, \tag{2.158}$$

where A and b are temperature dependent, d is the dimensionality, and the expression is valid for large n (Fisher, 1990). Thus one cannot construct a convergent Taylor series about the first-order phase transition point because the higher-order coefficients of the series are arbitrarily large.

The existence of an essential singularity at the binodal has only been proved for the lattice gas. Fisher's droplet arguments, though plausible, are not completely rigorous. Nevertheless, it is reasonable to expect that such a singularity would exist at any first-order phase transition, and that it would be so weak as to be impossible to detect experimentally. Approaching the binodal from the single-phase region, no constraints are needed to describe single-phase systems. Suddenly, at the binodal and beyond, constraints are needed to describe a metastable single-phase system. Thus it is reasonable to associate the singularity with the fact that further continuation of the single-phase equation of state into the coexistence region requires the sudden introduction of constraints. The weakness of the singularity, on the other hand, is due to the fact that the constraints are infinitely weak at the binodal, where the critical nucleus is infinitely large (see Section 3.1.1). Under such circumstances, a constrained system is indistinguishable from an unconstrained one. Though these arguments are qualitative, they point to the theoretical significance of the Fisher-Isakov singularity as an indication that the usual statistical-mechanical formalism must be modified through the introduction of constraints in order to describe metastability.

Langer (1967) investigated the possibility of evaluating the properties of suppercooled vapors rigorously by using complex variable calculus to continue the free energy analytically around the singularity at the phase transition. The treatment yields the mathematical equivalent of calculating the properties of the metastable vapor in the droplet model by truncating the series in Equation (2.148) at j^*, and writing its equation of state as

$$\beta P = \sum_{j=1}^{j^*} \rho_j. \tag{2.159}$$

The continuation procedure yields a complex βP, the real part of which is given formally by Equation (2.159), and the imaginary part by

$$i\beta P = \frac{i\rho_1 \sqrt{\pi k T \sigma^3}}{6(\delta\mu)^2} \exp\left[-\frac{4\sigma^3}{27(\delta\mu)^2 kT}\right] \tag{2.160}$$

(Langer, 1980), where ρ_1 is the bulk density of the metastable vapor, σ is the surface tension, $\delta\mu$ is the difference in chemical potential between the supercooled and the saturated vapor at the same temperature, and $i = \sqrt{-1}$. This imaginary part of βP is related to the rate of decay of the metastable state, a topic that we discuss in Chapter 3.

In subsequent work, Langer generalized the above ideas (Langer, 1974, 1980). His treatment is based on a coarse-grained Helmholtz energy,

$$A\{\rho\} = -kT \ln \sum \exp(-\beta E), \tag{2.161}$$

where the summation is over all microscopic configurations consistent with the assignment of a specific density distribution, hence the notation $A\{\rho\}$. Formally, one constructs a coarse-grained Helmholtz energy by dividing the system into cells, and specifying density distributions that vary slowly over distances comparable to the cell size. The cell size should be large compared to the molecular volume, yet small enough to prevent phase separation from occurring inside any given cell. The grand partition function is then written as a functional integral over all possible density distributions,

$$\Xi(T, V, \mu) = \int \delta\rho \exp(-\beta\Psi\{\rho\}) = \exp(\beta P V),$$

$$\Psi = A\{\rho\} - \mu \int \rho(r)\, d^3r, \tag{2.162}$$

where $\delta\rho$ denotes functional integration. Consider a vapor at fixed temperature, and let μ and μ_c correspond to the vapor's chemical potential and the chemical potential at saturation. When $\delta\mu (= \mu - \mu_c) < 0$ (stable system), Ψ is minimum at a particular density, the equilibrium density corresponding to the imposed temperature and chemical potential. The grand partition function can then be formally evaluated by expanding Ψ about the uniform density. Of course, Ψ also has a local minimum corresponding to a uniform metastable state at the same temperature and chemical potential; for example, a superheated liquid when the stable state is an unsaturated vapor. However, this local minimum in Ψ corresponding to a uniform metastable state at the same temperature and chemical potential makes a negligible contribution to the grand partition function. When $\delta\mu > 0$ (metastable system), there are also two uniform densities where Ψ is minimum (ρ_0, ρ_{liq}). These correspond to the metastable supercooled vapor (ρ_0) and the stable unsaturated liquid (ρ_{liq}) at the same temperature. The uniform metastable state makes a negligible contribution to the unconstrained grand partition function; however, it is precisely this contribution that can be interpreted as determining the analytic continuation of $\Xi(\mu)$ (Langer, 1980). The metastable contribution can be calculated because, in addition to the local minima at ρ_0 and ρ_{liq}, Ψ has a saddle point in the vicinity of ρ_0; at this saddle point the vapor's density is everywhere constant except for the appearance of a critical nucleus of the new phase. The mathematical procedure required to compute the free energy of the uniform metastable state involves a functional expansion of $\Psi(\rho)$ about the saddle point, and integration of the grand partition function from ρ_0 to the saddle point, followed by integration in the complex plane (Langer, 1980). This results in a complex free energy, the real part of which gives the equilibrium properties of the uniform metastable state, and the imaginary part of which is proportional to its lifetime. Integration in the complex plane, and hence a complex free energy, arise because one wishes to continue analytically a function around a singular point; the singularity arises because the integral that defines Ξ, evaluated by perturbation about a uniform density distribution, is divergent for $\mu > \mu_c$ for the particular choice of uniform state that we are interested in here (that is to say, the metastable vapor). The topic of analytic continuations at first-order phase transitions and the connection between imaginary free energies and decay kinetics has been extensively investigated by Schulman and co-workers (Newman and Schulman, 1977; McCraw and Schulman, 1978; Newman and Schulman, 1980; Privman and Schulman, 1982a,b; Roepstorff and Schulman, 1984). The applicability to realistic systems of the concept of a complex free energy has yet to be established (Penrose and Lebowitz, 1987).

The statistical mechanics of metastable (constrained) systems is still in its infancy. Much has been learned about the limitations of the phenomenological approach to metastability, but very little progress has been made on the more difficult task of translating this criticism into useful results for realistic systems. A good starting point for a theory of constrained systems would be to use cluster-based methods for evaluating the partition function (Guggenheim and McGlashan, 1951; Kikuchi, 1951; Kurata et al., 1953) to investigate the effect of varying the cluster size, and of constraining the density microscopically (in the clusters), on the predicted behavior inside the coexistence region. One such study has been done by Kikuchi (1967). Given the ease with which constraints can be imposed in computer simulations (Corti and Debenedetti, 1994, 1995), it is surprising that this powerful tool has not been more widely used to study metastable liquids. At present, equations of state and approximate theories of mixtures are the only available tools for estimating the thermodynamic properties of metastable liquids. Improved theoretical and computational tools are clearly needed.

2.5 STABILITY OF LIQUIDS WITH RESPECT TO CRYSTALLINE SOLIDS

The inequalities (2.18)–(2.23) describe the stability of a fluid with respect to the formation of another fluid phase. The potentially destabilizing density fluctuations in this case are isotropic. The relevant fluctuations in the case of supercooled liquids, on the other hand, are periodic density fluctuations whose wavelength corresponds to the lattice constant of the stable crystalline solid. To test for stability, we study the response of the density to an external potential $U(\mathbf{r})$,

$$\delta\rho(\mathbf{r}_1) = \int d\mathbf{r}_2 \left[\frac{\delta\rho(\mathbf{r}_1)}{\delta U(\mathbf{r}_2)} \right]_{U=0} U(\mathbf{r}_2), \qquad (2.163)$$

where the symbol $\delta\rho/\delta U$ inside the integral denotes functional differentiation (Lovett, 1977). Expressing this quantity in terms of the pair correlation function $g(r)$ (Percus, 1964) and taking the Fourier transform of (2.163) gives

$$kT\delta\rho(\mathbf{k}) = -\rho S(\mathbf{k}) U(\mathbf{k}), \qquad (2.164)$$

where $\delta\rho(\mathbf{k})$ and $U(\mathbf{k})$ are Fourier-transformed quantities, and $S(\mathbf{k})$ is the structure factor,

$$S(\mathbf{k}) = 1 + \rho \int d\mathbf{r} \exp(i\mathbf{k} \cdot \mathbf{r}) [g(r) - 1] = [1 - \rho c(\mathbf{k})]^{-1}, \qquad (2.165)$$

where $c(\mathbf{k})$ is the Fourier transform of the direct correlation function (Appendix 3). Thus the liquid becomes unstable with respect to periodic fluctuations of wave vector $\mathbf{k}$ when

$$\rho c(\mathbf{k}) = 1. \qquad (2.166)$$

Several theoretical studies have addressed this type of instability (e.g., Schneider et al., 1970; Haus and Meijer, 1975; Lovett, 1977; Vicsek, 1980; Robledo, 1980; Lovett and Buff, 1980). Experimental evidence on its existence, however, remains inconclusive (Bosio and Windsor, 1975; Suck et al., 1981).

As shown by Lovett (1977), the instability criterion (2.166) is identical to the one that results from searching for spatially nonuniform, solidlike solutions for the density, for a given liquid-phase direct correlation function. The basic equation in this equivalent approach is the lowest-order member of the Born-Green integral equation hierarchy for the molecular correlation functions

$$\frac{d\ln\rho(\mathbf{r}_1)}{d\mathbf{r}_1} = \int d\mathbf{r}_2 c(\mathbf{r}_1, \mathbf{r}_2) \frac{d\rho(\mathbf{r}_2)}{d\mathbf{r}_2} \qquad (2.167)$$

(Hill, 1956). The analysis consists of searching for nonuniform solutions for $\rho(\mathbf{r})$, assuming knowledge of $c(\mathbf{r})$ at the given state point (ρ, T). Since the uniform density is always a solution, the situation of interest is one in which an additional solidlike solution appears at high enough density. This bifurcation point coincides with the condition (2.166) (Lovett, 1977). There is a substantial literature on this topic, originating in Kirkwood's work (e.g., Kirkwood and Monroe, 1940, 1941; Kirkwood, 1951; Kunkin and Frisch, 1969; Weeks et al., 1970; Raveché and Stuart, 1975, 1976; Raveché and Kayser, 1978; Müller-Krumbhaar and Haus, 1978; Kozak, 1979; Feijoo and Rahman, 1982; Bagchi et al., 1983a,b; Cerjan et al., 1985). These and other aspects of the theory of the liquid-solid transition are discussed in two reviews by Haymet (1987a,b).

Kirkwood's interpretation of the bifurcation point as signaling the coexistence of the solid and liquid phases has been gradually abandoned in favor of the correct viewpoint in which bifurcation signals a limit of stability.

References

Abdulagatov, I.M. 1984."Analytical Description and Calculation of the Transport Properties near the Critical Point by the Pseudo-Spinodal Curve Method." *Russ. J. Phys. Chem.* 58: 1492.

Abdulagatov, I.M., and B.G. Alibekov. 1982."Equation of State of n-Hexane Allowing for Scale Theory Effects near the Critical Point." *Russ. J. Phys. Chem.* 56: 1612.

Abdulagatov, I.M., and B.G. Alibekov. 1983."The 'Pseudospinodal' Curve in the Description of the Scale Characteristics of the Behavior of a Substance near the Critical Point." *Russ. J. Phys. Chem.* 57: 285.

Abdulagatov, I.M., and B.G. Alibekov. 1984."Spinodal Equation." *Russ. J. Phys. Chem.* 58: 595.

Abdulagatov, I.M., and B.G. Alibekov. 1985."Relation Between the 'Pseudospinodal' Hypothesis and the 'Linear Model' of Critical Phenomena." *High Temp.* 23: 378.

Akhundov, T.S., I.M. Abdulagatov, and R.T. Akhundov. 1985."Analysis of the Behavior of the Viscosity of Benzene and Toluene close to the Critical Point on the Basis of the 'Pseudospinodal-Curve' Method." *High Temp.* 23: 73.

Andrews, T. 1869."On the Continuity of the Gaseous and Liquid States of Matter." *Phil. Trans. Roy. Soc. London.* 159: 575.

Angell, C.A., and H. Kanno. 1976."Density Maxima in High-Pressure Supercooled Water and Liquid Silicon Dioxide." *Science.* 193: 1121.

Angell, C.A., J. Shuppert, and J.C. Tucker. 1973."Anomalous Properties of Supercooled Water. Heat Capacity, Expansivity, and Proton Magnetic Resonance Chemical Shift from 0 to $-38°$." *J. Phys. Chem.* 77: 3092.

Angell, C.A., M. Oguni, and W.J. Sichina. 1982. "Heat Capacity of Water at Extremes of Supercooling and Superheating." *J. Phys. Chem.* 86: 998.

Apfel, R.E. 1971."A Novel Technique for Measuring the Strength of Liquids." *J. Acoust. Soc. Amer.* 49: 145.

Bagchi, B., C. Cerjan, and S.A. Rice. 1983a. "Contribution to the Theory of Freezing." *J. Chem. Phys.* 79: 5595.

Bagchi, B., C. Cerjan, and S.A. Rice. 1983b. "A Study of the Freezing Transition in the Lennard-Jones System." *J. Chem. Phys.* 79: 6222.

Baidakov, V.G. and T.A. Gurina. 1985. "(P, ρ, T) of Superheated Liquid Oxygen." *J. Chem. Thermodyn.* 17: 131.

Baidakov, V.G., and T.A. Gurina. 1989a. "Thermodynamic Properties of Metastable Liquefied Inert Gases. Part 2. P, ρ, T Properties of Superheated Krypton." *Physica B.* 160: 221.

Baidakov, V.G. and T.A. Gurina. 1989b. "An Experimental Study of the Equation of State of Superheated Liquid Methane." *J. Chem. Thermodyn.* 21: 1009.

Baidakov, V.G., and V.P. Skripov. 1978. "Thermodynamic Properties of Liquid Argon in the Metastable (Superheated) State." Preprint No. TF-001/7801. Ural Scientific Center Press, Academy of Sciences of the USSR, Sverdlovsk.

Baidakov, V.G., A.M. Rubshtein, V.R. Pomortsev, and I.J. Sulla. 1988. "The Equation of State of Metastable Liquid Xenon Near the Critical Point." *Phys. Lett. A.* 131: 119.

Baker, L.E., and K.D. Luks. 1980. "Critical Point and Saturation Pressure Calculations for Multipoint Systems." *Soc. Petr. Eng. J.* 20: 15.

Beegle, B.L., M. Modell, and R.C. Reid. 1974a. "Legendre Transforms and their Applications in Thermodynamics." *Amer. Inst. Chem. Eng. J.* 20: 1194.

Beegle, B.L., M. Modell, and R.C. Reid. 1974b. "Thermodynamic Stability Criterion for Pure Substances and Mixtures." *Amer. Inst. Chem. Eng. J.* 20: 1200.

Benedek, G.B. 1969. "Optical Mixing Spectroscopy, with Applications to Problems in Physics, Chemistry, Biology and Engineering." In *Polarisation, Matière et Rayonnement; Volume Jubilaire en l'Honneur d' Alfred Kastler.* Presses Universitaires de France: Paris.

Binder, K. 1987. "Theory of First-Order Phase Transitions." *Rep. Prog. Phys.* 50: 783.

Binney, J.J., N.J. Dowrick, A.J. Fisher, and M.J. Newman. 1992. *The Theory of Critical Phenomena. An Introduction to the Renormalization Group.* Chap. 1. Clarendon Press: Oxford.

Blander, M., and J.L. Katz. 1975. "Bubble Nucleation in Liquids." *Amer. Inst. Chem. Eng. J.* 21: 833.

Blander, M., D. Hengstenberg, and J.L. Katz. 1971. "Bubble Nucleation in n-Pentane, n-Hexane, n-Pentane+Hexadecane Mixtures, and Water." *J. Phys. Chem.* 75: 3613.

Boiko, V.G., A.V. Chalyj, and H.J. Moegel. 1984. "Pseudocritical Exponents of Metastable Fluids in the Mean-Field Theory." Preprint No. ITP-84-119E. Institute of Theoretical Physics, Academy of Sciences of Ukraine SSR, Kiev.

Borick, S.S., and P.G. Debenedetti. 1993. "Equilibrium, Stability, and Density Anomalies in a Lattice Model with Core-Softening and Directional Bonding." *J. Phys. Chem.* 97: 6292.

Bosio, L., and C.G. Windsor. 1975. "Observation of a Metastability Limit in Liquid Gallium." *Phys. Rev. Lett.* 35: 1652.

Bragg, W.L., and E.J. Williams. 1934. "The Effect of Thermal Agitation on Atomic Arrangements in Alloys." *Proc. Roy. Soc. A* 145: 699.

Callen, H.B. 1985. *Thermodynamics and an Introduction to Thermostatistics.* 2nd ed. Chap. 5. Wiley: New York.

Capocaccia, D., M. Cassandro, and E. Olivieri. 1974. "A Study of Metastability in the Ising Model." *Commun. Math. Phys.* 39: 185.

Cassandro, M., and E. Oliveri. 1977. "A Rigorous Study of Metastability in a Continuous Model." *J. Stat. Phys.* 17: 229.

Cerjan, C., B. Bagchi, and S.A. Rice. 1985. "A Comment on the Consistency of Truncated Nonlinear Integral Equation Based Theories of Freezing." *J. Chem. Phys.* 83: 2376.

Chen, C.-T., R.A. Fine, and F.J. Millero. 1977. "The Equation of State of Pure Water Determined from Sound Speeds." *J. Chem. Phys.* 66: 2142.

Chu, B., and F.J. Schoenes. 1968. "Diffusion Coefficient of the Isobutyric Acid–Water System in the Critical Region." *Phys. Rev. Lett.* 21: 6.

Chu, B., F.J. Schoenes, and W.P. Kao. 1968. "Spatial and Time-Dependent Concentration Fluctuations of the Isobutyric Acid–Water System in the Neighborhood of its Critical

Mixing Point." *J. Amer. Chem. Soc.* 90: 3042.

Chu, B., F.J. Schoenes, and M.E. Fisher. 1969."Light Scattering and Pseudospinodal Curves: The Isobutyric Acid–Water System in the Critical Region." *Phys. Rev.* 185: 219.

Chukanov, V.N., and V.P. Skripov. 1971. "Specific Volumes of Severely Superheated Water." *High Temp.* 9: 672.

Chukhanov, V.N., and V.P. Skripov. 1974. "Specific Volumes of Metastable Water." In *Proceedings of the 8th International Conference on the Properties of Water and Steam.* Vol. 1, p. 512.

Compagner, A. 1974. "On Pseudocritical Exponents at Endpoints of Metastable Branches." *Physica.* 72: 115.

Corti, D.S., and P.G. Debenedetti. 1994. "A Computational Study of Metastability in Vapor-Liquid Equilibrium." *Chem. Eng. Sci.* 49: 2717.

Corti, D.S., and P.G. Debenedetti. 1995. "Metastability and Constraints: A Study of the Superheated Lennard-Jones Liquid in the Void-Constrained Ensemble." *Ind. Eng. Chem. Res.* 34: 3573.

D'Antonio, M.C. 1989. "A Thermodynamic Investigation of Tensile Instabilities and Sub-Triple Liquids." Ph.D. Thesis, Princeton University, Princeton, N.J.

D'Antonio, M.C., and P.G. Debenedetti. 1987. "Loss of Tensile Strength in Liquids Without Property Discontinuities: A Thermodynamic Analysis." *J. Chem. Phys.* 86: 2229.

Debenedetti, P.G. 1985. "On the Relationship Between Principal Fluctuations and Stability Coefficients." *J. Chem. Phys.* 84: 1778.

Debenedetti, P.G., and M.C. D'Antonio. 1988. "Stability and Tensile Strength of Liquids Exhibiting Density Maxima." *Amer. Inst. Chem. Eng. J.* 34: 447.

Debenedetti, P.G., V.S. Raghavan, and S.S. Borick. 1991. "Spinodal Curve of Some Supercooled Liquids." *J. Phys. Chem.* 95: 4540.

Dong, W., and J.H. Lienhard, 1986. "Corresponding States Correlation of Saturated and Metastable Properties." *Can. J. Chem. Eng.* 64: 158.

Eberhart, J.G. 1976. "The Thermodynamic and Kinetic Limits of Superheat of a Liquid." *J. Colloid Interf. Sci.* 56: 262.

Eberhart, J.G., and H.C. Schnyders. 1973. "Application of the Mechanical Stability Condition to the Prediction of the Limit of Superheat for Normal Alkanes, Ether, and Water." *J. Phys. Chem.* 23: 2730.

Eberhart, J.G., W. Kremsner, and M. Blander. 1975. "Metastability Limits of Superheated Liquids: Bubble Nucleation Temperatures of Hydrocarbons and their Mixtures." *J. Colloid Interf. Sci.* 50: 369.

Elkoshi, Z., H. Reiss, and A.D. Hammerich. 1985. "One-Dimensional Rigorous Hole Theory of Fluids: Internally Constrained Ensembles." *J. Stat. Phys.* 41: 685.

Emmett, R.T., and F.J. Millero. 1975. "Specific Volume of Deuterium Oxide from 2° to 40° and 0 to 1000 Bars Applied Pressure."*J. Chem. Eng. Data.* 20: 351.

Ermakov, G.V., and V.P. Skripov. 1967. "Saturation Line, Critical Parameters, and the Maximum Degree of Superheating of Perfluoro-Paraffins." *Russ. J. Phys. Chem.* 41: 39.

Ermakov, G.V., and V.P. Skripov. 1968. "Experimental Determination of the Specific Volumes of a Superheated Liquid." *High Temp.* 6: 86.

Ermakov, G.V., V.G. Baidakov, and V.P. Skripov. 1973. "Density of Superheated Ethyl

Ether and the Limited Stability of the Liquid State." *Russ. J. Phys. Chem.* 47: 582.

Estefeev, V.N., V.N. Chukanov, and V.P. Skripov. 1977. "Specific Volumes of Super-heated Water." *High Temp.* 15: 550.

Estefeev, V.N., V.P. Skripov, and V.N. Chukanov. 1979. "Experimental Determination of the Speed of Ultrasound in Superheated Ordinary and Heavy Water." *High Temp.* 17: 252.

Feijoo, L., and A. Rahman. 1982. "A Study of Spatially Nonuniform Solutions of the First BBGKY Equation." *J. Chem. Phys.* 77: 5687.

Fisher, M.E. 1967a. "The Theory of Condensation and the Critical Point." *Physics.* 3: 255.

Fisher, M.E. 1967b. "The Theory of Equilibrium Critical Phenomena." *Rep. Prog. Phys.* 30: 615.

Fisher, M.E. 1969. "Rigorous Inequalities for Critical-Point Correlation Exponents." *Phys. Rev.* 180: 594.

Fisher, M.E. 1990. "Phases and Phase Diagrams." In *Proceedings of the Gibbs Symposium, Yale University, May 15–17, 1989*, D.G. Caldi and G.D. Mostow, eds. American Mathematical Society: Providence R.I.; American Institute of Physics: New York.

Gaw, W.J., and R.L. Scott. 1971. "Volume Changes in the Critical Region." *J. Chem. Thermodyn.* 3: 335.

Gibbs, J.W. 1875–76 and 1877–78. "On the Equilibrium of Heterogeneous Substances." *Trans. Conn. Acad.* III: 108 (Oct. 1875–May 1876) and III: 343 (May 1877–July 1878).

Gibbs, J.W. 1961. *The Scientific Papers of J. Willard Gibbs, Ph.D., LL.D. I. Thermodynamics*, pp. 219–331. Dover: New York.

Girifalco, L.A. 1992. "Molecular Properties of C_{60} in the Gas and Solid Phases." *J. Phys. Chem.* 96: 858.

Green, M.S., M. Vicentini-Missoni, and J.M.H. Levelt Sengers. 1967. "Scaling Law Equation of State for Gases in the Critical Region." *Phys. Rev. Lett.* 18: 1113.

Griffiths, R.B. 1965. "Thermodynamic Inequality near the Critical Point for Ferromagnets and Fluids." *Phys. Rev. Lett.* 14: 623.

Guggenheim, E.A. 1945. "The Principle of Corresponding States." *J. Chem. Phys.* 13: 253.

Guggenheim, E.A., and M.C. McGlashan. 1951. "Statistical Mechanics of Regular Mixtures." *Proc. Roy. Soc. A.* 206: 335.

Haar, L., J.S. Gallagher, and G.S. Kell. 1980. "Thermodynamic Properties for Fluid Water." In *Water and Steam. Their Properties and Current Industrial Applications. Proceedings of the 9th International Conference on the Properties of Steam*, J. Straub and K. Scheffler, eds. p. 69. Pergamon Press: Oxford.

Hafner, J. 1987. *From Hamiltonians to Phase Diagrams. The Electronic and Statistical Mechanical Theory of sp-Bonded Metals and Alloys.* Chaps. 1 and 2. Springer-Verlag: Berlin.

Hagen, M.H.J., E.J. Meijer, G.C.A.M. Mooij, D. Frenkel, and H.N.W. Lekkerkerker. 1993. "Does C_{60} Have a Liquid Phase?" *Nature.* 365: 425.

Hare, D.E., and C.M. Sorensen. 1986. "Densities of Supercooled H_2O and D_2O in 25 μm Glass Capillaries." *J. Chem. Phys.* 84: 5085.

Hareng, M., and J. Leblond. 1980. "Brillouin Scattering in Superheated Water." *J. Chem. Phys.* 73: 622.

Haus, J.W., and P.H.E. Meijer. 1975. "Static Aspects of a Model for Metastable Fluid States." *Physica*. 80A: 313.

Haymet, A.D.J. 1987a. "Freezing." *Science*. 236: 1076.

Haymet, A.D.J. 1987b. "Theory of the Equilibrium Liquid-Solid Transition." *Annu. Rev. Phys. Chem.* 38: 89.

Hayward, A.T. 1971."Negative Pressure in Liquids: Can It Be Harnessed to Serve Man?" *Amer. Sci.* 59: 434.

Heidemann, R.A. 1994. "The Classical Theory of Critical Points." In *Supercritical Fluids. Fundamentals for Application*, E. Kiran and J.M.H. Levelt Sengers, eds., p. 39. NATO Advanced Study Institute Series E, vol. 273. Kluwer Academic Publishers: Dordrecht.

Heidemann, R.A., and A.M. Khalil. 1980. "The Calculation of Critical Points." *Amer. Inst. Chem. Eng. J.* 26: 769.

Hemmer, P.C., and G. Stell. 1970. "Fluids with Several Phase Transitions." *Phys. Rev. Lett.* 24: 1284.

Henderson, S.J., and R.J. Speedy. 1987. "Melting Temperature of Ice at Positive and Negative Pressures." *J. Phys. Chem.* 91: 3069.

Hill, T.L. 1956. *Statistical Mechanics. Principles and Selected Applications*. McGraw-Hill: New York.

Hobbs, P.V. 1974. *Ice Physics*. Chap. 1. Clarendon Press: Oxford.

Holden, B.S., and J.L. Katz. 1978. "The Homogeneous Nucleation of Bubbles in Superheated Binary Liquid Mixtures." *Amer. Inst. Chem. Eng. J.* 24: 260.

Hoshino, K., C.H. Leung, I.L. McLaughlin, S.M.M. Rahman, and W.H. Young. 1987. "Pair Potential Trends from the Evidence of Observed Liquid-Metal Structure Factors." *J. Phys. F. Met. Phys.* 17: 787.

Hosokawa, S., S. Yamada, and K. Tamura. 1993. "Density Measurement for Liquid Se-Te Mixtures at High Temperatures and Pressures." *J. Non-Cryst. Sol.* 156–158: 708.

Huang, K. 1987. *Statistical Mechanics*. 2nd ed. Chap. 14. John Wiley and Sons: New York.

Iida, T., and R.I.L. Guthrie. 1993. *The Physical Properties of Liquid Metals*. Chap. 2. Clarendon Press: Oxford.

Isakov, S.N. 1984. "Nonanalytic Features of the First Order Phase Transition in the Ising Model." *Commun. Math. Phys.* 95: 427.

Ising, E. 1925. "Beitrag zur Theorie des Ferromagnetismus." *Z. Phys.* 31: 253.

Jalaluddin, A.K., and D.B. Sinha. 1962. "Maximum Superheat of Binary Liquid Mixtures." *Ind. J. Phys.* 36: 312.

Josephson, B.D. 1967a. "Inequality for the Specific Heat. I. Derivation." *Proc. Phys. Soc.* 92: 269.

Josephson, B.D. 1967b. "Inequality for the Specific Heat. II. Application to Critical Phenomena." *Proc. Phys. Soc.* 92: 276.

Kadanoff, L.P. 1966. "Scaling Laws for Ising Models Near T_c." *Physics*. 2: 263.

Kakinuma, F., and S. Ohno. 1987. "Heat Capacity of Liquid Se-Te Mixtures." *J. Phys. Soc. Jpn.* 56: 619.

Kakinuma, F., S. Ohno, and K. Suzuki. 1993. "Specific Heat of Liquid S-Te Mixtures." *J. Non-Cryst. Sol.* 156–158: 691.

Kanno, H., and C.A. Angell. 1979. "Water: Anomalous Compressibilities to 1.9 kbar and Correlation with Supercooling Limits." *J. Chem. Phys.* 70: 4008.

Kanno, H., and C.A. Angell. 1980. "Volumetric and Derived Thermal Characteristics of Liquid D_2O at Low Temperatures and High Pressures." *J. Chem. Phys.* 73: 1940.

Kell, G.S. 1975. "Density, Thermal Expansivity, and Compressibility of Liquid Water from 0° to 150°C: Correlations and Tables for Atmospheric Pressure and Saturation Reviewed and Expressed on 1968 Temperature Scale." *J. Chem. Eng. Data.* 20: 97.

Kikuchi, R. 1951. "A Theory of Cooperative Phenomena." *Phys. Rev.* 81: 988.

Kikuchi, R. 1967. "Cooperative Phenomena in the Triangular Lattice." *J. Chem. Phys.* 47: 1664.

Kincaid, J.M., and G. Stell. 1977. "Isostructural Phase Transitions Due to Core Collapse. III. A Model for Solid Mixtures." *J. Chem. Phys.* 67: 420.

Kincaid, J.M., and G. Stell. 1978. "Structure Factor of a One-Dimensional Shouldered Hard-Sphere Fluid." *Phys. Lett.* 65A: 131.

Kincaid, J.M., G. Stell, and C.K. Hall. 1976a. "Isostructural Phase Transitions Due to Core Collapse. I. A One-Dimensional Model." *J. Chem. Phys.* 65: 2161.

Kincaid, J.M., G. Stell, and E. Goldmark. 1976b. "Isostructural Phase Transitions Due to Core Collapse. II. A Three-Dimensional Model with a Solid-Solid Critical Point." *J. Chem. Phys.* 65: 2172.

Kirkwood, J.G., 1951. "Crystallization as a Cooperative Phenomenon." In *Phase Transformations in Solids*, R. Smoluchowski, J. E. Mayer, and W. A. Weyl. eds., p. 67. Wiley: New York.

Kirkwood, J.G., and E. Monroe. 1940. "On the Theory of Fusion." *J. Chem. Phys.* 8: 845.

Kirkwood, J.G., and E. Monroe. 1941. "Statistical Mechanics of Fusion." *J. Chem. Phys.* 9: 514.

Kojima, J., N. Kuwahara, and M. Kaneko. 1975. "Light Scattering and Pseudospinodal Curve of the System Polystyrene-Cyclohexane in the Critical Region." *J. Chem. Phys.* 63: 333.

Kozak, J.J. 1979. "Nonlinear Problems in the Theory of Phase Transitions." *Adv. Chem. Phys.* 40: 229.

Kraus, G.F., and S.C. Greer. 1984. "Vapor Pressures of Supercooled H_2O and D_2O" *J. Phys. Chem.* 88: 4781.

Kumar, S.K., and R.C. Reid. 1986. "Derivation of the Relationships between Partial Derivatives of Legendre Transforms." *Amer. Inst. Chem. Eng. J.* 32: 1224.

Kunkin, W., and H.L. Frisch. 1969. "Comment on the Kirkwood Instability." *J. Chem. Phys.* 50: 1817.

Kurata, M., R. Kikuchi, and T. Watari. 1953. "A Theory of Cooperative Phenomena. III. Detailed Discussions of the Cluster Variation Method." *J. Chem. Phys.* 21: 434.

Landau, L.D. 1965. "Zur Theorie der Phasenumwandlungen, I." In *Collected Papers of L.D. Landau*. D. ter Haar, ed. Pergamon Press: London.

Landau, L.D. and E.M. Lifshitz. 1986. *Theory of Elasticity*. Vol. 7 of Course of Theoretical Physics. 3rd ed., Chap. 1. Pergamon Press: Oxford.

Langer, J.S. 1967. "Theory of the Condensation Point." *Ann. Phys.* 41: 108.

Langer, J.S. 1974. "Metastable States." *Physica.* 73: 61.

Langer, J.S. 1980. "Kinetics of Metastable State." In *Systems Far from Equilibrium*, L. Garrido, ed. Springer-Verlag: Berlin.

Lebowitz, J.L., and O. Penrose. 1966. "Rigorous Treatment of the van der Waals–

Maxwell Theory of the Liquid-Vapor Transition." *J. Math. Phys.* 7: 98.

Le Guillou, J.-C., and J. Zinn-Justin. 1985. "Accurate Critical Exponents from the ε-Expansion." *J. Phys. Lett.* 46: L-137.

Levelt Sengers, J.M.H. 1991. "Thermodynamics of Solutions Near the Solvent's Critical Point." In *Supercritical Fluid Technology. Reviews in Modern Theory and Applications*, T.J. Bruno and J.F. Ely eds., chap. 1. CRC Press: Boca Raton, FL.

Lienhard, J.H., N. Shamsundar, and P.O. Biney. 1986. "Spinodal Lines and Equations of State: A Review." *Nucl. Eng. Design.* 95: 297.

Lifshitz, E.M., and L.P. Pitaevskii. 1980. *Statistical Physics. Part I.* Vol. 5 of Course of Theoretical Physics, by L.D. Landau and E.M. Lifshitz. Chaps. 2 and 14. Pergamon Press: Oxford.

Lovett, R. 1977. "On the Stability of a Fluid Toward Solid Formation." *J. Chem. Phys.* 66: 1225.

Lovett, R., and F.P. Buff. 1980. "Phase Instability and the Direct Correlation Function Integral Equation." *J. Chem. Phys.* 72: 2425.

Lysenkov, V.F. 1985. "The 'Pseudo-spinodal' Hypothesis and the Scale Equation of State for the Critical Region." *Russ. J. Phys. Chem.* 59: 502.

McCraw, R.J., and L.S. Schulman. 1978. "Metastability in the Two-Dimensional Ising Model." *J. Stat. Phys.* 18: 293.

March, N.H. 1987. "Structure and Forces in Liquid Metals and Alloys." *Can. J. Phys.* 65: 219.

March, N.H. 1990. *Liquid Metals.* Chaps. 3–5. Cambridge University Press: Cambridge.

Michelsen, M.L. 1982a. "The Isothermal Flash Problem. Part I. Stability." *Fluid Phase Equil.* 9: 1.

Michelsen, M.L. 1982b. "The Isothermal Flash Problem. II. Phase-Split Calculation." *Fluid Phase Equil.* 9: 21.

Michelsen, M.L., and Heidemann, R.A. 1981. "Calculation of Critical Points from Cubic Two-Constant Equations of State." *Amer. Inst. Chem. Eng. J.* 27: 521.

Modell, M., and R.C. Reid. 1983. *Thermodynamics and its Applications.* 2nd ed. Prentice-Hall: Englewood Cliffs, N.J.

Moore, G.R. 1959. "Vaporization of Superheated Drops in Liquids." *Amer. Inst. Chem. Eng. J.* 5: 458.

Müller-Krumbhaar, H., and J. W. Haus. 1978. "On the Bifurcation Analysis of the Liquid-Solid Instability in Hard-Sphere Systems." *J. Chem. Phys.* 69: 5219.

Nagarajan, N.R., A.S. Cullick, and A. Griewank. 1991. "New Strategy for Phase Equilibrium and Critical Point Calculations by Thermodynamic Energy Analysis. Part II. Critical Point Calculations." *Fluid Phase Equil.* 62: 211.

Newman, C.M., and L.S. Schulman. 1977. "Metastability and the Analytic Continuation of Eigenvalues." *J. Math. Phys.* 18: 23.

Newman, C.M., and L.S. Schulman. 1980. "Complex Free Energies and Metastable Lifetimes." *J. Stat. Phys.* 23: 131.

Onsager, L. 1944. "Crystal Statistics. I. A Two-Dimensional Model with an Order-Disorder Transition." *Phys. Rev.* 65: 177.

Osman, J., and C.M. Sorensen. 1980. "Experimental Evidence for the Universality of the Pseudospinodal." *J. Chem. Phys.* 73: 4142.

Panagiotopoulos, A.Z., and R.C. Reid. 1986. "On the Relationship Between Pairwise Fluctuations and Thermodynamic Derivatives." *J. Chem. Phys.* 85: 4650.

Pavlov, P.A., and V.P. Skripov. 1965. "Boiling of a Liquid with Pulsed Heating. 1. Hot-Wire Method of Experiment." *High Temp.* 3: 97.

Pavlov, P.A., and V.P. Skripov. 1970. "Kinetics of Spontaneous Nucleation in Strongly Heated Liquids." *High Temp.* 8: 540.

Peng, D.-Y., and D.B. Robinson. 1976. "A New Two-Constant Equation of State." *Ind. Eng. Chem. Fundam.* 15: 59.

Peng, D.-Y., and D.B. Robinson. 1977. "A Rigorous Method for Predicting the Critical Properties of Multicomponent Systems from an Equation of State." *Amer. Inst. Chem. Eng. J.* 23: 137.

Penrose, O., and J.L. Lebowitz. 1971. "Rigorous Treatment of Metastable States in the van der Waals–Maxwell Theory." *J. Stat. Phys.* 3: 211.

Penrose, O., and J.L. Lebowitz. 1987. "Towards a Rigorous Molecular Theory of Metastability." In *Fluctuation Phenomena*, E.W. Montroll and J.L. Lebowitz, eds., chap. 5. North-Holland: Amsterdam.

Percus, J.K. 1964. "The Pair Distribution Function in Classical Statistical Mechanics." In *The Equilibrium Theory of Classical Fluids*, H.L. Frisch and J.L. Lebowitz, eds., chap. II-3. Benjamin: New York.

Perepezko, J.H., and J.S. Paik. 1984. "Thermodynamic Properties of Undercooled Liquid Metals." *J. Non-Cryst. Sol.* 61–62: 113.

Peters, C.J., J.L. de Roo, and J. de Swaan Arons. 1995. "Phase Equilibria in Binary Mixtures of Propane and Triphenylmethane." *Fluid Phase Equil.* 109: 99.

Pitzer, K.S. 1955. "The Volumetric and Thermodynamic Properties of Fluids. I. Theoretical Basis and Virial Coefficients." *J. Amer. Chem. Soc.* 77: 3427.

Pitzer, K.S., D.Z. Lippmann, R.F. Curl, Jr., C.M. Huggins, and D.E. Petersen. 1955. "The Volumetric and Thermodynamic Properties of Fluids. II. Compressibility Factor, Vapor Pressure, and Entropy of Vaporization." *J. Amer. Chem. Soc.* 77: 3433.

Porteous, W., and M. Blander. 1975. "Limits of Superheat and Explosive Boiling of Light Hydrocarbons, Halocarbons, and Hydrocarbon Mixtures." *Amer. Inst. Chem. Eng. J.* 21: 560.

Pozharskaya, G.I., N.L. Kasapova, V.P. Skripov, and Y. D. Kolpakov. 1984. "The Spinodal Approximation by the Method of Scattering of Light in (n-hexane + n-tetradecafluorohexane)." *J. Chem. Thermodyn.* 16: 267.

Privman, V., and L.S. Schulman. 1982a. "Analytic Properties of Thermodynamic Functions at First-Order Phase Transitions." *J. Phys. A. Math. Gen.* 15: L231.

Privman, V., and L.S. Schulman. 1982b. "Analytic Continuation at First-Order Phase Transitions." *J. Stat. Phys.* 29: 205.

Rasmussen, D.H., A.P. MacKenzie, C.A. Angell, and J.C. Tucker. 1973. "Anomalous Heat Capacities of Supercooled Water and Heavy Water." *Science.* 181: 342.

Raveché, H.J. and R.F. Kaiser. 1978. "Towards a Molecular Theory of Freezing: the Equation of State and Free Energy from the First BBGKY Equation." *J. Chem. Phys.* 68: 3632.

Raveché, H.J., and C.A. Stuart. 1975. "Towards a Molecular Theory of Freezing." *J. Chem. Phys.* 63: 1099.

Raveché, H.J., and C.A. Stuart. 1976. "Towards a Molecular Theory of Freezing. II. Study of Bifurcation as a Function of Density." *J. Chem. Phys.* 65: 2305.

Redlich, O., and J.N.S. Kwong. 1949. "On the Thermodynamics of Solutions. V: An Equation of State. Fugacities of Gaseous Solutions". *Chem. Rev.* 44: 234.

Reid, R.C. 1976. "Superheated Liquids." *Amer. Sci.* 64: 146.

Reid, R.C., and B.L. Beegle. 1977. "Critical Point Criteria in Legendre Transform Notation." *Amer. Inst. Chem. Eng. J.* 23: 726.

Reid, R.C., J.M. Prausnitz, and B.E. Poling. 1987. *The Properties of Gases and Liquids.* 4th ed. Chap. 3. McGraw-Hill: New York.

Reiss, H. 1975. "The Existence of the Spinodal, an Incompletely Solved Problem in the Thermodynamics of Solids." *Ber. Bunsenges. Phys. Chem.* 79: 943.

Robledo, A. 1980. "The Liquid-Solid Transition for the Hard Sphere System from Uniformity of the Chemical Potential." *J. Chem. Phys.* 72: 1701.

Roedder, E. 1967. "Metastable Superheated Ice in Liquid-Water Inclusions Under High Negative Pressure." *Science.* 155: 1413.

Roepstorff, G., and L.S. Schulman. 1984. "Metastability and Analyticity in a Dropletlike Model." *J. Stat. Phys.* 34: 35.

Rowlinson, J.S., ed. 1988. *J.D. van der Waals: On the Continuity of Gaseous and Liquid States.* North-Holland: Amsterdam.

Rowlinson, J.S., and F.L. Swinton. 1982. *Liquids and Liquid Mixtures.* 3rd ed. Butterworths: London.

Rushbrooke, G.S. 1963. "On the Thermodynamics of the Critical Region for the Ising Problem." *J. Chem. Phys.* 39: 842.

Russel, W.B., D.A. Saville, and W.R. Schowalter. 1989. *Colloidal Dispersions.* Chap. 10 Cambridge University Press: Cambridge.

Rykov, V.A. 1985. "Definition of the 'Pseudo-spinodal' Curve through the Thermodynamic Equalities $(\partial T/\partial s)_v = 0$ and $(\partial v/\partial p)_T = 0$." *Russ. J. Phys. Chem.* 59: 1743.

Rykov, V.A. 1986. "The Pseudo-spinodal Curve Hypothesis." *Russ. J. Phys. Chem.* 60: 476.

Sastry, S., F. Sciortino, and H.E. Stanley. 1993. "Limits of Stability of the Liquid Phase in a Lattice Model with Water-Like Properties." *J. Chem. Phys.* 98: 9863.

Schneider, T., R. Brout, H. Thomas, and J. Feder. 1970. "Dynamics of the Liquid-Solid Transition." *Phys. Rev. Lett.* 25: 1423.

Schubert, G. and R.E. Lingenfelter. 1970. "Superheated Ice Formed by the Freezing of Superheated Water." *Science.* 168: 469.

Sengers, J., and J.M.H. Levelt Sengers. 1978. "Critical Phenomena in Classical Fluids." In *Progress in Liquid Physics*, C.A. Croxton, ed., chap. 4. Wiley: New York.

Sengers, J., and J.M.H. Levelt Sengers. 1986. "Thermodynamic Behavior of Fluids Near the Critical Point." *Annu. Rev. Phys. Chem.* 37: 189.

Skripov, V.P. 1966. "Important Property of the Spinodal." *High Temp.* 4: 757.

Skripov, V.P. 1974. *Metastable Liquids.* Israel Program for Scientific Translations: Jerusalem; Wiley: New York.

Skripov, V.P. 1980. "Metastable States of Water-Superheating, Supercooling and the Boundary of Thermodynamic Stability." In *Water and Steam. Their Properties and Current Industrial Applications. Proceedings of the 9th International Conference on the Properties of Steam*, J. Straub and K. Scheffler, eds., p. 261. Pergamon Press: Oxford.

Skripov, V.P., and G.V. Ermakov. 1964. "Pressure Dependence of the Limiting Superheating of a Liquid." *Russ. J. Phys. Chem.* 38: 208.

Skripov, V.P., and V.I. Kukushkin. 1961. "Apparatus for Observing the Superheating

Limits of Liquids." *Russ. J. Phys. Chem.* 35: 1393.

Skripov, V.P., and P.A. Pavlov. 1970. "Explosive Boiling of Liquids and Fluctuation Nucleus of Formation." *High Temp.* 8: 782.

Skripov, V.P., P.A. Pavlov, and E.N. Sinitsyn. 1965. "Heating of Liquids to Boiling by a Pulsating Heat Supply." *High Temp.* 3: 670.

Skripov, V.P., E.N. Sinitsyn, P.A. Pavlov, G.V. Ermakov, G.N. Muratov, N.V. Bulanov, and V.G. Baidakov. 1988. *Thermophysical Properties of Liquids in the Metastable (Superheated) State.* Gordon and Breach: New York.

Sorensen, C.M. 1988. "Dynamic Light Scattering Study of Tetrahydrofuran and Water Solutions." *J. Phys. Chem.* 92: 2367.

Sorensen, C.M. 1991. "Comparison of the Pseudospinodal to the Transition from Metastability to Instability in a Binary Liquid Mixture." *J. Chem. Phys.* 94: 8630.

Sorensen, C.M., and M.D. Semon. 1980. "Scaling Equation of State Derived from the Pseudospinodal." *Phys. Rev. A.* 21: 340.

Speedy, R.J. 1982a. "Stability-Limit Conjecture. An Interpretation of the Properties of Water." *J. Phys. Chem.* 86: 982.

Speedy, R.J. 1982b. "Limiting Forms of the Thermodynamic Divergences at the Conjectured Stability Limits in Superheated and Supercooled Water." *J. Phys. Chem.* 86: 3002.

Speedy, R.J., and C.A. Angell. 1976. "Isothermal Compressibility of Supercooled Water and Evidence for a Thermodynamic Singularity at $-45°C$." *J. Chem. Phys.* 65: 851.

Stanley, H.E. 1971. *Introduction to Phase Transitions and Critical Phenomena.* Oxford University Press: New York.

Stell, G. and P.C. Hemmer. 1972. "Phase Transitions Due to Softness of the Potential Core." *J. Chem. Phys.* 56: 4274.

Stillinger, F.H. 1995. "Statistical Mechanics of Metastable Matter: Superheated and Stretched Liquids." *Phys. Rev. E.* 52: 4685.

Stillinger, F.H., and T.A. Weber. 1978. "Study of Melting and Freezing in the Gaussian Core Model by Molecular Dynamics Simulation." *J. Chem. Phys.* 68: 3837.

Suck, J.-B., J.H. Perepezko, I. E. Anderson, and C.A. Angell. 1981. "Temperature Dependence of the Dynamic Structure Factor for Supercooled $Sn_{1-x}Pb_x$ Alloys: A Test of Instability Theories for the Liquid-Solid Phase Transition." *Phys. Rev. Lett.* 47: 424.

Toda, M., R. Kubo, and N. Saitô. 1992. *Statistical Physics I. Equilibrium Statistical Mechanics.* 2nd ed. Springer-Verlag: Berlin.

Trevena, D.H. 1987. *Cavitation and Tension in Liquids.* Chap. 3. Adam Hilger: Bristol.

Tsuchiya, Y. 1991a. "The Anomalous Negative Thermal Expansion and the Compressibility Maximum of Molten Ge-Te Alloys." *J. Phys. Soc. Jpn.* 60: 227.

Tsuchiya, Y. 1991b. "Isothermal Compressibility and Thermodynamics of Structural Transitions in the Liquid Se-Te System." *J. Phys. Soc. Jpn.* 60: 960.

Tsuchiya, Y. 1991c. "Thermodynamic Evidence for a Structural Transition of Liquid Te in the Supercooled Region." *J. Phys. Cond. Matt.* 3: 3163.

Tsuchiya, Y. 1992. "Phase Equilibria in the Liquid Sulphur-Tellurium System: Structural Changes and Two-Melt Phase Separation." *J. Phys. Cond. Matt.* 4: 4335.

Tsuchiya, Y. 1993. "AC Calorimetry of the Thermodynamic Transition in Liquid $Ge_{15}Te_{85}$." *J. Non-Cryst. Sol.* 156–158: 704.

Tsuchiya, Y. 1994. "The Thermodynamics of Structural Changes in the Liquid Sulphur-

Tellurium System: Compressibility and Ehrenfest's Relations." *J. Phys. Cond. Matt.* 6: 2451.

van der Waals, J.D. 1873. "De Continuitet van den Gas-en Vloeistoftoestand." Ph.D. thesis, University of Leiden, Holland.

Vicentini-Missoni, M., J.M.H. Levelt-Sengers, and M.S. Green. 1969. "Thermodynamic Anomalies of CO_2, Xe, and He^4 in the Critical Region." *Phys. Rev. Lett.* 22: 389.

Vicsek, T. 1980. "Simple Variational Method for the Closures of the Ornstein-Zernike Equation: A Study of Liquid Instability." *Physics.* 1024: 523.

Walas, S.M. 1985. *Phase Equilibria in Chemical Engineering.* Chap. 7. Butterworth: Boston.

Walker, S., and C.A. Vause. 1987. "Reappearing Phases." *Sci. Amer.* 256 (5): 98.

Weeks, J.D., S.A. Rice, and J.J. Kozak. 1970. "Analytic Approach to the Theory of Phase Transitions." *J. Chem. Phys.* 52: 2416.

Wenzel, H., and G. Schmidt. 1980. "A Modified van der Waals Equation of State for the Representation of Phase Equilibria between Solids, Liquids and Gases." *Fluid Phase Equil.* 5: 3.

Widom, B. 1965. "Equation of State in the Neighborhood of the Critical Point." *J. Chem. Phys.* 43: 3898.

Widom, B., and J.S. Rowlinson. 1970. "New Model for the Study of Liquid-Vapor Phase Transitions." *J. Chem. Phys.* 52: 1670.

Wilson, K.G. 1971. "Renormalization Group and Critical Phenomena. I. Renormalization Group and the Kadanoff Scaling Picture." *Phys. Rev. B.* 4: 3174.

Wilson, K. G. and M.E. Fisher. 1972. "Critical Exponents in 3.99 Dimensions." *Phys. Rev. Lett.* 28: 240.

Wilson, K.G., and J. Kogut. 1974. "The Renormalization Group and the ϵ Expansion." *Phys. Rep. (Phys. Lett. C).* 12: 75.

Yokoyama, I., and S. Ono. 1985. "Effective Interatomic Pair Potentials in Liquid Polyvalent Metals from Observed Structure Data." *J. Phys. F. Met. Phys.* 15: 1215.

Young, D.A. 1991. *Phase Diagrams of the Elements.* University of California Press: Berkeley.

Young, D.A., and B.J. Alder. 1977. "Melting-Curve Extrema from a Repulsive 'Step' Potential." *Phys. Rev. Lett.* 38: 1213.

Young, D.A., and B.J. Alder. 1979. "Studies in Molecular Dynamics. XVII. Phase Diagrams for 'Step' Potentials in Two and Three Dimensions." *J. Chem. Phys.* 70: 473.

3

Kinetics

The relaxation of metastable liquids towards stable equilibrium involves the formation of one or more new phases. The characteristic rates and mechanisms of the initial stage of this process are the main subjects of this chapter. Knowledge of the rate at which a system evolves towards stable equilibrium is essential to any investigation of metastable liquids, because such systems can only be studied over intervals that are short compared to the characteristic time for the appearance of a new phase.

In most practical circumstances suspended and dissolved impurities, as well as imperfectly wetted solid boundaries, provide preferential sites for the formation of a new phase. This process is known as heterogeneous nucleation. However, in the absence of impurities or solid surfaces, small embryos of the new phase are formed within the bulk metastable liquid. This is an activated process: a free energy barrier must be overcome in order to form embryos of a critical size, beyond which the new phase grows spontaneously.[1] This fundamental mechanism of phase transformation is known as homogeneous nucleation. Similarly to other activated processes, the rate at which critical-sized embryos are formed is extraordinarily sensitive to the height of the free energy barrier, or, equivalently, to the extent of penetration into the metastable region. As a result, carefully purified metastable liquids evolve suddenly from a condition of apparent stability to one of catastrophic growth of a new phase, and homogeneous nucleation is the mechanism that determines the practical, attainable limits of liquid superheating or supercooling.

Whereas metastable systems relax by nucleation, unstable systems do so by spinodal decomposition. Spinodal decomposition will occur if, for example, a binary mixture of near-critical composition is rapidly cooled to a subcritical temperature. This second fundamental mechanism of phase separation occurs spontaneously: no free energy barriers must be overcome. Spinodal decomposition involves the growth of fluctuations of small amplitude that exceed a critical wavelength. The relevant fluctuating property is density for pure substances, and composition for mixtures. Thus metastable systems relax by the activated growth of localized fluctuations of large amplitude, whereas unstable

---

[1] Heterogeneous nucleation is also an activated process.

systems do so by the spontaneous growth of long-wavelength fluctuations of small amplitude.

Homogeneous nucleation is treated in Section 3.1. Included there are discussions of the classical theory of nucleation (Section 3.1.1), of nucleation in superheated and supercooled liquids (Sections 3.1.4 and 3.1.5), and of the attainment of steady state conditions during nucleation (Section 3.1.6). Though still widely used, the classical theory rests on questionable assumptions, the critical reconsideration of which is the basis of rigorous treatments of the energetics of embryo formation (Section 3.1.2), and of kinetic nucleation theories (Section 3.1.3). Spinodal decomposition is discussed in Section 3.2. The transition between nucleation and spinodal decomposition, with emphasis on unsolved aspects of this problem, is addressed in Section 3.3. Finally, heterogeneous nucleation is discussed in Section 3.4.

3.1 HOMOGENEOUS NUCLEATION

3.1.1 Classical Nucleation Theory

The so-called classical nucleation theory originated with the work of Volmer and Weber (1926) (Abraham, 1974). These authors were the first to argue that the nucleation rate should depend exponentially on the reversible work of embryo formation. More quantitative treatments date back to the work of Farkas (1927), who laid the foundation for subsequent developments (Becker and Döring, 1935; Zeldovich, 1942; Frenkel, 1955). There are several excellent pedagogical treatments of classical nucleation theory (McDonald, 1962, 1963; Andres, 1965; Abraham, 1974; Friedlander, 1977; Springer, 1978; Seinfeld, 1986). In spite of its shortcomings, this theory still constitutes the basis of most modern treatments of nucleation processes. The theory was originally developed for droplet condensation from supercooled vapors. Extensions to bubble nucleation in superheated liquids and crystallization in supercooled liquids are discussed in Sections 3.1.4 and 3.1.5, respectively.

In order for a new phase to appear, an interface must be formed. In the absence of suspended impurities, this occurs via the formation of small, localized embryos of the new phase within a bulk metastable phase. Initially, embryos are formed as a result of spontaneous density or composition fluctuations. Nucleation theory aims at quantifying the net rate at which embryos grow to a critical size, beyond which the new phase forms spontaneously. Consider a supercooled vapor. As a result of density fluctuations small embryos of the liquid phase (droplets) are constantly being formed and destroyed. We assume that there is no molecular association in the metastable vapor, an excellent approximation away from criticality, and that the concentration of droplets is small. Under these conditions, droplets can only grow or shrink as a result of single-molecule events. Then $J(n)$, the difference between the rate at which droplets containing n molecules are formed by single-molecule condensation onto $(n-1)$-molecule droplets and the rate at which they are destroyed by single-molecule evaporation, can be written as

$$J(n) = f(n-1)F(n-1)\beta(n-1) - f(n)F(n)\alpha(n), \qquad (3.1)$$

where $J(n)$ has dimensions of (volume^{-1} × time^{-1}); $f(n)$ and $f(n-1)$ are the concentrations of n- and $(n-1)$-molecule droplets; $F(n)$ and $F(n-1)$ are their respective surface areas; $\beta(n-1)$ is the flux per unit time and area of single molecules onto the $(n-1)$-molecule droplet (i.e., molecules arriving at the droplet-vapor interface from the vapor); and $\alpha(n)$ is the flux of single molecules leaving the n-molecule droplet. Of the two rate coefficients, β can be calculated from kinetic theory, but α is not known in general. To overcome this difficulty, equilibrium considerations are invoked. Assuming that an equilibrium distribution of droplets can be established in the bulk metastable phase,[2] we can write, because of microscopic reversibility,

$$N(n-1)F(n-1)\beta = N(n)F(n)\alpha, \qquad (3.2)$$

where we have assumed that the single-molecule fluxes are independent of embryo size, and $N(n-1)$ and $N(n)$ are the equilibrium concentrations of droplets composed of $n-1$ and n molecules, respectively. It is also assumed that the values of α and β do not change in going from an equilibrium to a nonequilibrium situation. Using (3.2) to solve for α, and substituting into (3.1),

$$J(n) = \beta F(n-1)N(n-1)\left[\frac{f(n-1)}{N(n-1)} - \frac{f(n)}{N(n)}\right]. \qquad (3.3)$$

It follows from Equation (3.1) that

$$\frac{\partial f(n,t)}{\partial t} = J(n) - J(n+1). \qquad (3.4)$$

Therefore a time-invariant population of droplets is established when J becomes independent of n, that is to say, when the total rate of formation of n-molecule droplets by single-molecule addition to $(n-1)$-molecule droplets plus single-molecule subtraction from $(n+1)$-molecule droplets equals the total n-molecule droplet rate of disappearance by single-molecule addition to, and subtraction from, n-molecule droplets. The theory assumes that such a steady state is rapidly established.[3] Rearranging Equation (3.3) but with J now independent of n,

$$\frac{J}{\beta F(n-1)N(n-1)} = \frac{f(n-1)}{N(n-1)} - \frac{f(n)}{N(n)}, \qquad (3.5)$$

[2]This is a reasonable assumption as long as the characteristic time for the establishment of an equilibrium embryo distribution is short compared to the lifetime of the metastable phase.

[3]This assumption is discussed in detail in Section 3.1.6.

and summing from $n = 2$ to $n = \Lambda$, where Λ is a large but otherwise as yet undefined number, we obtain

$$J = \frac{\frac{f(1)}{N(1)} - \frac{f(\Lambda+1)}{N(\Lambda+1)}}{\sum\limits_{n=1}^{\Lambda} \frac{1}{\beta F(n) N(n)}}. \qquad (3.6)$$

Equation (3.6) implies that the steady-state nucleation rate J can be calculated from knowledge of the equilibrium droplet distribution $N(n)$, the rate of arrival of single molecules β, and the ratio of actual to equilibrium concentrations of single molecules, and of very large droplets. Since the bulk metastable phase exists overwhelmingly as single molecules, the equilibrium and actual single-molecule concentrations are indistinguishable. Furthermore, the theory describes events that take place at the onset of a phase transition, and provides no information on the system's progress beyond the initial, embryonic stage of phase separation represented by the fluctuation-driven formation of droplets (Lifshitz and Pitaevskii, 1981). On this time scale, it is reasonable to expect that $f(n)$ will vanish for sufficiently large n: otherwise, significant amounts of the liquid phase will have already formed. Since, as will be shown below, $N(n)$ does not vanish for large n, Equation (3.6) becomes

$$J = \frac{1}{\sum\limits_{n=1}^{\Lambda} \frac{1}{\beta F(n) N(n)}}. \qquad (3.7)$$

Thus, because of lack of knowledge about α, the kinetic problem of calculating the nucleation rate has been transformed into the thermodynamic problem of evaluating the equilibrium droplet distribution. This calculation requires knowledge of the energetics of embryo formation, which we now consider.

The minimum (reversible) work needed to form an embryo of n molecules can be written as[4]

$$W_{min} = \sigma F + \left(P - P'\right) V' + n \left[\mu'\left(T, P'\right) - \mu(T, P)\right], \qquad (3.8)$$

where σ is the surface tension, F the interfacial area between the embryo and the bulk phase, P the bulk phase pressure, P' the pressure inside the embryo, V' the embryo's volume, and μ' and μ the chemical potentials in the embryo and in the bulk phase, respectively (Reiss, 1970; Katz and Blander, 1973). For an incompressible embryo (a good approximation for nucleation in a supercooled vapor away from the critical point), we can write

$$\mu'(T, P') - \mu'(T, P) = v'(P' - P), \qquad (3.9)$$

---

[4]Equation (3.8) is derived in Appendix 2.

where v' is the volume per molecule in the embryo phase. Equation (3.8) then becomes

$$W_{\min} = \sigma F + n\left[\mu'(T, P) - \mu(T, P)\right] = \sigma F + n\Delta\mu, \qquad (3.10)$$

where $\Delta\mu$ (< 0) is now the difference between the chemical potentials in the stable and metastable phases at bulk conditions (T, P). Since F is proportional to $n^{2/3}$, (3.10) has the form

$$W_{\min} = an^{2/3} - bn = cr^2 - dr^3, \qquad (3.11)$$

where a, b, c, and d are positive numbers, and r is the radius of the embryo. This function is sketched in Figure 3.1. It has a maximum at n^*, where

$$n^* = \left(\frac{2a}{3b}\right)^3. \qquad (3.12)$$

For spherical embryos, n^* is given by

$$n^* = \frac{32\pi}{3}\left[\frac{\left(v'\right)^{2/3}\sigma}{(-\Delta\mu)}\right]^3 \qquad (3.13)$$

and the radius of the embryo having n^* molecules by

$$r^* = \frac{2\sigma v'}{(-\Delta\mu)}. \qquad (3.14)$$

The minimum (reversible) work needed to form an n^*-molecule embryo is

$$W_{\min} = \frac{4a^3}{27b^2} = \frac{16\pi}{3}\left[\frac{v'\sigma^{3/2}}{(-\Delta\mu)}\right]^2. \qquad (3.15)$$

An embryo containing n^* molecules is called a critical-sized embryo or, equivalently, a critical nucleus. The minimum (reversible) work of embryo formation is maximized for this particular value of n. The critical nucleus is therefore in unstable equilibrium (Gibbs, 1875–76, 1877–78, 1961). Embryos containing fewer than n^* molecules shrink spontaneously (work can be recovered from this process), while embryos larger than n^* grow spontaneously. Thus, in order for the new phase to be formed, the system must first overcome a free energy barrier and form a critical nucleus. Thereafter, the new phase grows spontaneously. The rate at which critical nuclei are formed is the homogeneous nucleation rate. Note that the free energy cost associated with the formation of critical nuclei decreases as $(\Delta\mu)^2$, and the size of the critical nucleus, n^*,

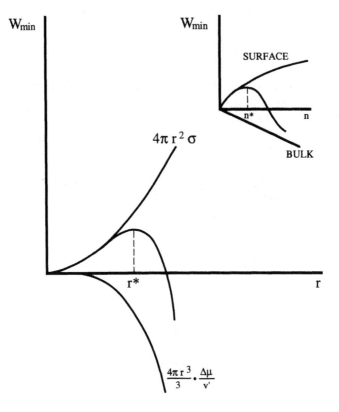

Figure 3.1. Surface and bulk contributions to the reversible work of formation of an incompressible embryo in the classical nucleation picture.

as $(-\Delta\mu)^3$. This means that, as the extent of penetration into the metastable region increases, critical nuclei become smaller, and so does the free energy barrier that must be overcome to form them.

Assuming that an equilibrium embryo distribution can be established (Frenkel, 1955), it should satisfy

$$N(n) \propto \exp\left[-\frac{W(n)}{kT}\right],\qquad(3.16)$$

where $W(n)$ is the reversible (minimum) work needed to form an n-molecule embryo. Since there should, in principle, be no work associated with the formation of a single-molecule embryo, the proportionality constant should equal the number density of unassociated molecules in the bulk metastable phase. We can therefore write

$$N(n) = N_{tot} \exp\left[-\frac{W(n)}{kT}\right],$$ (3.17)

where N_{tot} is now the total number density of the bulk metastable phase.

A more rigorous route to (3.16) and (3.17) takes into account explicitly the sequential formation of embryonic liquid droplets in a supersaturated vapor (Reiss, 1970). In this approach, embryos are treated as distinct molecular species of an ideal gas mixture, and the unimolecular steps leading to their formation are considered to be at equilibrium,

$$a_1 + a_1 = a_2,$$
$$a_2 + a_1 = a_3,$$
$$\vdots$$ (3.18)
$$a_{n-1} + a_1 = a_n,$$

where a_n denotes an embryo containing n molecules. We can therefore write

$$\mu_n = n\mu_1,$$ (3.19)

where μ_n denotes the chemical potential of an n-molecule embryo (droplet), considered as a distinct molecular species, and μ_1 is the chemical potential per molecule in the bulk vapor. Because of the assumption of ideality, it follows that

$$\mu_n = \lambda_n(T, v) + kT \ln \frac{N(n)}{\sum N(n)},$$ (3.20)

where λ_n is the chemical potential of an n-sized embryo when this is the only "species" present, v is the bulk specific volume, and $N(n)$ is the concentration of n-molecule droplets. Solving for $N(n)$,

$$N(n) = \left[\sum N(n)\right] \exp\left[\frac{n\mu_1 - \lambda_n}{kT}\right].$$ (3.21)

This implies

$$W(n) = \lambda_n - n\mu_1.$$ (3.22)

For condensation, the liquid embryos can be considered incompressible. Therefore, from Equations (3.10) and (3.22),

$$\lambda_n = n\mu_1 + \sigma F(n) + n(\mu - \mu_1) = \sigma F(n) + n\mu,$$ (3.23)

where μ is the chemical potential per molecule in the liquid at the vapor's bulk conditions. Thus the chemical potential of a drop containing n molecules, considered as a distinct molecular species, differs from the chemical potential of n molecules in the liquid (at the bulk conditions prevailing in the metastable vapor) by the surface term. If the number of large embryos is small and the vapor is unassociated, $\Sigma N(n)$ differs negligibly from the vapor's number density, and we recover Equation (3.17).

The preexponential factor in Equation (3.17) was arrived at by noting that there can be no work associated with the formation of a single-molecule embryo. However, none of the expressions for the reversible work of embryo formation discussed above vanishes when $n=1$. This inconsistency is due to the use of macroscopic, thermodynamic arguments to calculate the work of formation of molecular-size objects.[5] In particular, the derivation of Equation (3.8) hinges on the assumption that the radius of curvature of the interface is large with respect to molecular dimensions (Gibbs, 1875–76, 1877–78, 1961; see Appendix 2). This is known as the capillarity approximation.

Note that even though the predictions for $N(1)$ and $N(n \gg n^*)$ are unphysical[6] the limits invoked in going from (3.6) to (3.7) are not. Regardless of the predicted value for $N(1)$, it is reasonable to assume that the ratio $f(1)/N(1)$ cannot differ significantly from unity: since the metastable phase is unassociated by hypothesis, this is equivalent to saying that, over the time scale of interest, the system is a reservoir of single molecules whose concentration differs little from its equilibrium value. At the other extreme, it is not necessary for $N(n \gg n^*)$ to be large in order for $f(n \gg n^*)/N(n \gg n^*)$ to vanish: all that is required is that $f(n \gg n^*)$ vanish (which it certainly does, otherwise the new phase would have already formed), while $N(n \gg n^*)$ stays finite [or, more rigorously, does not vanish as fast as $f(n \gg n^*)$].

The homogeneous nucleation rate is obtained by substituting the equilibrium embryo distribution into Equation (3.7). It is customary to replace the summation by an integral,

$$J = \beta N_{\text{tot}} \left[\int_{n \ll n*}^{n \gg n*} \exp \left(\frac{W(n)}{kT} \right) \frac{1}{F(n)} \, dn \right]^{-1}, \qquad (3.24)$$

---

[5]In addition to inconsistencies in the $n = 1$ limit (limiting consistency), the preexponential factor in (3.17) violates mass action, according to which $N(n) = [N(1)]^n K$, where K can only depend on temperature, but not on $N(1)$ or pressure if the vapor phase behaves ideally. In the case of incompressible liquid embryos in an ideal supersaturated vapor, for example, use of (3.10) in (3.17) results in an $N(1)$-dependent K, even after neglecting the pressure dependence of the liquid-phase chemical potential. Several authors have modified the preexponential term in (3.17) so as to recover mass action (Courtney, 1961; Dufour and Defay, 1963; Blander and Katz, 1972) or limiting consistency (Dufour and Defay, 1963; Girshik and Chiu, 1990; Girshik, 1991). The topic has been reviewed critically by Wilemski (1995). Ultimately, however, the way to improve classical nucleation theory is not by forcing consistency but by a critical reconsideration of its underlying assumptions, such as is described in Sections 3.1.2 and 3.1.3.

[6]The expression for the work of embryo formation [Figure 3.1; Equation (3.11)], when substituted into Equation (3.17), predicts an ever-increasing concentration of large $(n > n^*)$ embryos.

where n^* is the critical nucleus (Cohen, 1970). Since the exponential is sharply peaked at n^*, the overwhelming contribution to the integral comes from a narrow size range centered about n^*. Therefore the work of embryo formation can be expanded about the unstable equilibrium value, and $F(n)$ replaced by $F(n^*)$. For an incompressible embryo, the expansion has one independent variable,

$$W(n) \approx W(n^*) + \frac{1}{2}W''(n^*)(n - n^*)^2 = W(n^*) + \frac{1}{2}W''(n^*)(\delta n)^2, \quad (3.25)$$

where $W''(n^*)$ denotes the curvature of the $W(n)$ function, evaluated at n^*. Substituting into (3.24),

$$J = \beta N_{\text{tot}} F(n^*) \exp\left[\frac{-W(n^*)}{kT}\right]$$
$$\cdot \left[\int_{-\infty}^{\infty} \exp\left\{-\frac{1}{2kT}\left[-W''(n^*)\right](\delta n)^2\right\} d(\delta n)\right], \quad (3.26)$$

where we have changed integration variable from n to δn, and we have used the sharpness of the exponential to replace the limits of integration from $[1 - n^*, \Lambda - n^* \ (\Lambda \gg n^*)]$. Upon performing the integration, we obtain

$$J = \{\beta F(n^*)\}\left[\sqrt{\frac{-W''(n^*)}{2\pi kT}}\right] \cdot \left\{N_{\text{tot}} \exp\left[-\frac{W(n^*)}{kT}\right]\right\}$$
$$= j(n^*) Z N(n^*). \quad (3.27)$$

Note that the rate expression can be written as the product of three terms. The first one $[j(n^*)]$ is the product of β times the surface area of the critical nucleus, and represents the frequency of arrival of single molecules to the critical nucleus. The third term, $N(n^*)$, is the equilibrium concentration of critical nuclei. Therefore, the second term, Z, can be interpreted as a factor that corrects for the fact that the concentration of critical nuclei differs from the equilibrium value. This term is frequently referred to as the Zeldovich nonequilibrium factor (Zeldovich, 1942).

For incompressible embryos (for example, droplets in a supercooled vapor) use of (3.11)–(3.15) in (3.27) yields

$$J = 2\beta N_{\text{tot}} \sqrt{\frac{\sigma(v')^2}{kT}} \exp\left\{-\frac{16\pi}{3kT}\left[\frac{v'\sigma^{3/2}}{-\Delta\mu}\right]^2\right\}. \quad (3.28)$$

If we neglect the pressure dependence of the liquid chemical potential, and assume the vapor phase to behave ideally, we obtain for the homogeneous nucleation rate in a supercooled vapor

$$
\begin{aligned}
J &= 2\beta N_{\text{tot}} \sqrt{\frac{\sigma (v')^2}{kT}} \exp \left\{ -\frac{16\pi}{3} \left[\frac{\sigma (v')^{3/2}}{kT} \right]^3 \left[\frac{1}{\ln S} \right]^2 \right\} \\
&= \frac{\rho_v^2}{\rho_l} \sqrt{\frac{2\sigma}{\pi m}} \exp \left\{ -\frac{16\pi}{3kT} \left[\frac{v'\sigma^{3/2}}{-\Delta\mu} \right]^2 \right\},
\end{aligned}
\tag{3.29}
$$

where S is the supersaturation (P/P^e); P^e is the equilibrium vapor pressure at the given temperature, $\rho_v(= N_{\text{tot}})$ is the number density in the vapor, $\rho_l(= 1/v')$ is the number density in the liquid, and m is the molecular mass. The distinguishing feature of (3.28) and (3.29) is the extremely sharp dependence of the nucleation rate on the degree of penetration into the metastable region.

In Equation (3.29), use has been made of the ideal-gas expression for β,

$$
\beta = \frac{P}{\sqrt{2\pi m k T}}.
\tag{3.30}
$$

Cohen (1970) investigated the numerical implications involved in replacing the summation in Equation (3.7) by an integral, and in assuming that the integrand contributes significantly only near n^*. He found the errors introduced by these approximations to be negligible, even for very small critical nucleus sizes.

The development of a rigorous theory of nucleation in binary mixtures originates with the classic work of Reiss (1950). To formulate the problem, one starts with the analogue of (3.4),

$$
\frac{\partial f (n_a, n_b, t)}{\partial t} = -\left[\frac{\partial J_a}{\partial n_a} + \frac{\partial J_b}{\partial n_b} \right],
\tag{3.31}
$$

where n_a and n_b are the number of molecules of type a and b, respectively, in an embryo. The rate terms, in analogy with the continuous counterpart of (3.3), are given by

$$
\begin{aligned}
J_a &= -N(n_a, n_b) R_a \frac{\partial [f(n_a, n_b, t)/N(n_a, n_b)]}{\partial n_a}, \\
J_b &= -N(n_a, n_b) R_b \frac{\partial [f(n_a, n_b, t)/N(n_a, n_b)]}{\partial n_b},
\end{aligned}
\tag{3.32}
$$

where R_a and R_b are the rates (time^{-1}) at which single molecules of type a or b collide with embryos.[7] The equilibrium embryo distribution, in analogy with (3.17), is given by

$$
N(n_a, n_b) = N_{\text{tot}} \exp[-W_{\min}(n_a, n_b)/kT],
\tag{3.33}
$$

---

[7]As in the single-component theory, this rate is the product of a flux times a surface area. It is usually assumed that the former is independent of the size and composition of the embryo.

where $W_{\min}(n_a, n_b)$ is the free energy cost of forming an embryo containing n_a a-type and n_b b-type molecules. The steady-state nucleation rate (Stauffer, 1976; Shi and Seinfeld, 1990) is given by

$$J = \frac{kTu}{\sqrt{W_{aa}^* W_{bb}^* - \left(W_{ab}^*\right)^2}} \cdot N_{\text{tot}} \cdot \exp\left[-(W_{\min}^*/kT)\right], \qquad (3.34)$$

$$u = -\frac{1}{2}\left[D_{aa} + D_{bb} - \sqrt{(D_{aa} - D_{bb})^2 + 4D_{ab}^2}\right], \qquad (3.35)$$

$$kTD_{aa} = R_a^* W_{aa}^*, \quad kTD_{bb} = R_b^* W_{bb}^*, \quad kTD_{ab} = \sqrt{R_a^* R_b^*}\, W_{ab}^*, \qquad (3.36)$$

$$2W_{aa}^* = \left(\partial^2 W_{\min}/\partial n_a^2\right)_*,$$

$$2W_{bb}^* = \left(\partial^2 W_{\min}/\partial n_b^2\right)_*, \quad 2W_{ab}^* = \left(\partial^2 W_{\min}/\partial n_a \partial n_b\right)_*. \qquad (3.37)$$

In the above expressions, the asterisk denotes a saddle point in the $W_{\min}(n_a, n_b)$ three-dimensional surface, whose coordinates define the critical nucleus (Reiss, 1950). Thus, R_b^* is the rate at which b-type, molecules collide with an embryo containing n_a^* a-type and n_b^* b-type molecules.

The net number of b-type molecules that are incorporated into a critical nucleus between successive incorporations of a-type molecules defines the direction of the nucleation current at the saddle point (Stauffer, 1976). The angle α that this direction makes with the n_a axis on the (n_a, n_b) plane is given by

$$\tan \alpha = \frac{1}{2D_{ab}} \cdot \left[D_{bb} - D_{aa} + \sqrt{(D_{aa} - D_{bb})^2 + 4D_{ab}^2}\right] \qquad (3.38)$$

(Shi and Seinfeld, 1990). When $R_a^* = R_b^*$, this expression reduces to the form originally derived by Reiss (1950). When $R_a^* \gg R_b^*$, $\tan \alpha \to 0$, and the direction of nucleation is along the n_a axis. On the other hand, when $R_a^* \ll R_b^*$, $\tan \alpha \to \infty$, and the direction of nucleation is along the n_b axis.

The minimum, reversible work of formation of a spherical, incompressible embryo of radius r containing n_a molecules of type a and n_b molecules of type b can be written as

$$\begin{aligned}
W_{\min} &= n_a\left[\mu_a'(P, x) - \mu_a(P, y)\right] + n_b\left[\mu_b'(P, x) - \mu_b(P, y)\right] \\
&\quad + n_a^\sigma\left[\mu_a^\sigma - \mu_a(P, y)\right] \\
&\quad + n_b^\sigma\left[\mu_b^\sigma - \mu_b(P, y)\right] + 4\pi r^2 \sigma.
\end{aligned} \qquad (3.39)$$

This equation, like its single-component counterpart (3.10), is based on Gibbs' theory of capillarity (Gibbs, 1875–76, 1877–78, 1961), in which the interface is treated as a mathematical surface of discontinuity having no thickness. This approach is discussed in Appendix 2, where Equation (3.39) is derived. Primed chemical potentials in (3.39) refer to the embryo (mole fraction x), and un-primed chemical potentials to the bulk metastable phase (mole fraction y), both evaluated at the bulk pressure.[8] The quantities n_a^σ and n_b^σ denote the difference between the actual number of molecules of the respective species present and that calculated on the assumption that the bulk and embryo phases are both uniform up to the mathematical surface of discontinuity. μ_a^σ and μ_b^σ denote the chemical potentials of a and b in the nonuniform interfacial region. It is common in the nucleation literature (e.g., Reiss and Shugard, 1976) to leave out both these terms (i.e., the products of n_a^σ and n_b^σ times the respective chemical potential difference) from (3.39). As shown in Appendix 2, this is incorrect.

There is an important difference between (3.39) and its single-component counterpart, (3.10). Actual interfaces have finite thickness and their composition differs from that of either bulk phase. Hence the important question arises (Wilemski, 1984, 1987) as to what is the thermodynamically consistent way of calculating the effective composition, size, and work of formation of a binary critical nucleus in classical nucleation theory. This so-called surface enrichment problem and its solution (Wilemski, 1984, 1987) are discussed in Appendix 2.

To date, most work on the energetics of embryo formation in binary systems within the context of classical nucleation theory has addressed incompressible embryos, and is therefore applicable to condensation from supercooled vapors or to liquid-liquid phase separation (e.g., Doyle, 1961; Mirabel and Katz, 1974; Shugard et al., 1974; Wilemski, 1975; Reiss and Shugard, 1976; Mirabel and Clavelin, 1978; Flageollet-Daniel et al., 1983; Rasmussen, 1986a,b; Wilemski, 1987; Kreidenweis and Seinfeld, 1988; Wyslouzil et al., 1991a,b).

Progress in the development of improved diffusion (e.g., Hung et al., 1989) and expansion chambers (e.g., Schmitt, 1981; Miller et al., 1983; Wagner and Strey, 1984) during the last fifteen years has made it possible to measure rates of nucleation in supercooled vapors directly (Laaksonen et al., 1995). Previously, the quantity that was commonly measured was the supersaturation required to produce a certain measurable rate of nucleation, typically 1 cm^{-3} sec^{-1} (see McGraw, 1981, for a review). This quantity is referred to as the critical supersaturation. Using the density and surface tension of water, Equation (3.29) predicts an increase of seventeen orders of magnitude in the nucleation rate as a result of a 10% change in supersaturation, from 2 to 2.2. It follows that the critical supersaturation is not a sensitive probe of the accuracy of nucleation theories. Reliable experimental tests of nucleation theories, in other words, were not available prior to the early 1980s. Laaksonen et al. (1995), and Heist

---

[8]The temperature dependence of the chemical potentials has not been written explicitly in (3.39). It is understood that all quantities are evaluated at the system's temperature.

and He (1994) have reviewed recent progress in the experimental study of nucleation in supercooled vapors.

Although the subject of this book is metastable liquids, not vapors, a summary of the main conclusions from recent measurements of nucleation in vapors is appropriate here. This is so because these measurements provide a direct test of the validity of classical nucleation theory, whose key ideas and approximations underlie most quantitative treatments of nucleation in liquids. Two important conclusions can be drawn from the experiments:

(i) The isothermal dependence of the nucleation rate on supersaturation is well predicted by classical theory: experimental and theoretical slopes of ln J vs. S are in good agreement.

(ii) The temperature dependence of the nucleation rate is systematically different from that predicted by the classical theory: measured nucleation rates tend to be lower than theoretical predictions at high temperature, and higher at low temperatures. This effect is quite pronounced: the ratio $J_{\text{theoretical}}/J_{\text{experimental}}$ in n-nonane (Wagner and Strey, 1984; Adams et al., 1984; Hung et al., 1989) changes from a number of order 10^{-5} at 233 K to 10^{3} at 315 K (Hung et al., 1989).

Similar trends have also been reported for water (Viisanen et al., 1993) and short-chain alcohols (Schmitt et al., 1982; Kacker and Heist, 1985; Strey et al., 1986). It is clear that classical nucleation is not quantitatively accurate for the prediction of rates of nucleation. Microscopic and kinetic modifications of the theory are described in the following sections.

3.1.2 Energetics of Embryo Formation: Rigorous Approaches

Classical nucleation theory is based on two approximations: the need to invoke an equilibrium distribution of embryos as a means of calculating α, the dissociation or "evaporation" coefficient in Equation (3.1), and the treatment of small embryos as though they had bulk properties. The former approximation transforms what should really be a kinetic theory into a thermodynamic one; the latter, often referred to as the capillarity approximation, circumvents statistical mechanics in favor of heuristics. In this section we accept the validity of the equilibrium assumption, and discuss more fundamental, microscopic approaches to the calculation of the equilibrium embryo distribution. The kinetic approach to nucleation theory is discussed in Section 3.1.3.

Two important advances towards a first-principle treatment of energetics in nucleation are the idea of a physically consistent cluster, due to Reiss and co-workers (Reiss and Katz, 1967; Reiss et al., 1968; Reiss, 1970; Reiss et al., 1990; Ellerby et al., 1991; Ellerby and Reiss, 1992; Weakliem and Reiss, 1993, 1994), and the application of density functional theory to nucleation, due to Oxtoby and co-workers (Oxtoby and Evans, 1988; Oxtoby, 1991; Zeng and Oxtoby, 1991a,b; Oxtoby, 1992a,b; Talanquer and Oxtoby, 1993, 1994, 1995; Laaksonen and Oxtoby, 1995). They are the subject of this section. In its present form, the approach of Reiss is only applicable to nucleation in supercooled vapors, but the underlying method is of considerable general interest.

At sufficiently subcritical temperatures, supercooled vapors can be treated as ideal gases, and molecular clusters[9] are so rare that they do not interact with each other or with the surrounding ideal gas. The microscopic theory of Reiss hinges on the proper definition of a cluster. A cluster is composed of i interacting molecules that occupy a spherical volume v. At least one ideal-gas molecule lies in a spherical shell of volume dv concentric with v, thereby defining the cluster's volume. The i molecules assume only those configurations that keep the center of mass on the center of sphere v. A cluster of i molecules and volume v is called an i/v cluster. The probability $P_i(v)\,dv$ of observing one (and only one) i/v cluster in the given volume V is given by

$$P_i(v)dv = P(N-i) \cdot P(i+1 \mid N-i) \cdot P(i/v \mid N-i), \qquad (3.40)$$

(Ellerby and Reiss, 1992) where $P(N-i)$ is the probability that $(N-i)$ ideal-gas molecules are outside v; $P(i+1 \mid N-i)$ is the probability that any one of the $(N-i)$ molecules is in the shell dv around v, given that there are $(N-i)$ outside of v; $P(i/v \mid N-i)$ is the probability that a cluster has i molecules in v given that there are $(N-i)$ ideal-gas molecules outside of v. Clearly,

$$P(N-i) = (1 - v/V)^{N-i}. \qquad (3.41)$$

Therefore,

$$\lim_{N\to\infty} \ln P(N-i) \approx N \ln(1 - v/V) \approx -\rho v \qquad (3.42)$$

where $\rho = N/V$. In addition,

$$\lim_{N\to\infty} P(i+1 \mid N-i) = (N-i)\frac{dv}{V-v} \approx \rho dv. \qquad (3.43)$$

Finally, we have

$$P(i/v \mid N-i) = Q_{i/v}Q_i^{-1}, \qquad (3.44)$$

where $Q_{i/v}$ is the canonical partition function (Appendix 3) for i interacting particles in v, such that their center of mass coincides with that of v, and v is free to move throughout V; and Q_i is the partition function for i non-interacting,

---

[9]Reiss and co-workers used this term to denote embryos.

ideal-gas molecules in V,

$$Q_{i/v} = \Lambda^i \left(\frac{N^i}{i!}\right) \left(\frac{i^3}{i!}\right) V \int_0^{v^*} \cdots \int_0^{v^*} \exp(-\phi_i/kT)\, d^3r_1 \cdots d^3r_{i-1},$$

(3.45)

$$Q_i = \frac{\Lambda^i V^i}{i!},$$

(3.46)

where

$$\Lambda = \left(\frac{2\pi mkT}{h^2}\right)^{3/2}.$$

(3.47)

In the above equations, m is the molecular mass, h is Planck's constant, ϕ_i is the potential energy associated with a configuration of the i interacting molecules in v, and v^* is an upper limit on the cluster volume beyond which clusters become so large that they begin to interact. The factor $(N^i/i!)$ in (3.45) accounts for the number of ways of choosing i molecules out of N in the large-N limit; the second division by $i!$ corrects for the indistinguishability of the molecules; and the factor i^3 comes from writing

$$\mathbf{R} = \left(\sum_{j=1}^i \mathbf{r}'_j\right) i^{-1}, \quad i^3 d\mathbf{R} d\mathbf{r}_1 \cdots d\mathbf{r}_{i-1} = d\mathbf{r}'_1 \cdots d\mathbf{r}'_i,$$

(3.48)

where primes denote laboratory (fixed) coordinates, $\mathbf{R}$ is the location of the center of mass in laboratory coordinates, and unprimed vectors are relative to the center of mass. Note that the indistinguishability and kinetic energy factors $(i!, \Lambda^i)$ cancel, and have been included only to make the significance of the quantities clear.

Now, the probability $P_{i/v}$ of finding at least one i/v cluster in V is the sum of probabilities of finding exactly one, two, three, $\ldots$, clusters in V. For noninteracting clusters, the probability of finding exactly j clusters can be written as

$$P_{i/v}^j = \left(P_{i/v}^1\right)^j = [P_i(v)dv]^j.$$

(3.49)

Therefore the equilibrium number of i/v clusters in V is given by

$$n_i(v)dv = \sum_{j=1}^\infty j P_{i/v}^j = \sum_{j=1}^\infty j \left(P_{i/v}^1\right)^j \approx P_{i/v} = P_i(v)dv,$$

(3.50)

where $P_{i/v}^j \approx 0$ for $j \geq 2$ because such terms are proportional to $(dv)^j$. From (3.42)–(3.44),

$$\begin{aligned}
n_i(v)dv &= \rho dv \exp(-\rho v)\frac{Q_{i/v}}{Q_i} \\
&= \rho dv \exp\left\{-\beta\left[-\beta^{-1}\ln\left(\frac{Q_{i/v}}{Q_i}\right) + \beta^{-1}\rho v\right]\right\}
\end{aligned}$$

(3.51)

where $\beta = 1/kT$. Equation (3.51) can be written more concisely as

$$
\begin{aligned}
n_i(v)dv &= \rho dv \exp(-\rho v)\frac{Q_{i/v}}{Q_i} \\
&= \rho dv \exp\left\{-\beta\left[W'_{min} + Pv\right]\right\} \\
&= \rho dv \exp\left(-\beta W_{min}\right),
\end{aligned}
\tag{3.52}
$$

$$
W'_{min} = -kT \ln\left(\frac{Q_{i/v}}{Q_i}\right),
\tag{3.53}
$$

where W'_{min} is the reversible work needed to put i molecules in a preexisting hole of volume v, and Pv is the reversible work of making a cavity of volume v when the supercooled vapor is at a pressure P (the ideal-gas law, $P = \rho kT$, has been used). Therefore $W_{min} \left(= W'_{min} + Pv\right)$ is the minimum (reversible) work of embryo formation. After rearrangement, this quantity can be written as

$$
W_{min} = A^*_{i/v} - kT \ln \Lambda i^{3/2} V - i\mu_1 + Pv,
\tag{3.54}
$$

where the Helmholtz energy of the i/v cluster is given by

$$
A^*_{i/v} = -kT \ln\left\{\frac{\Lambda^{i-1}i^{3/2}}{i!}\int_0^{v^*}\cdots\int_0^{v^*}\exp\left(-\phi_i/kT\right)d^3r_1\cdots d^3r_{i-1}\right\}
\tag{3.55}
$$

and where $\mu_1 = kT \ln(\rho/\Lambda)$ is the chemical potential of the ideal supercooled vapor. The second term in the right-hand side of (3.54) is the contribution to the free energy due to the translation of the cluster's center of mass.

Equation (3.52) gives the equilibrium cluster distribution. It is identical in form to (3.17), but the reversible work of cluster formation is here written exactly, rather than heuristically. However, the calculation of the work of embryo formation now requires computer simulations. Thus the price that has been paid for dropping the assumption that continuum thermodynamics applies to small embryos, and for not having to contend with missing degrees of freedom (the price, in short, of rigor) is loss of analyticity. Lee et al. (1973) did a classic computational study of embryo free energies. They evaluated $Q_{i/v}$ by integration of the Helmholtz energy derivatives with respect to the embryo-constraining volume[10] and temperature over several simulations. Some of their results are shown in Figure 3.2, in which the Helmholtz energy is simply $-kT \ln Q_{i/v}$. Figure 3.3 shows more recent results due to Weakliem and Reiss

---

[10]The constraining volume in Lee et al.'s (1973) work was taken to be a spherical volume of specified radius centered at the center of mass of the embryo. During a simulation, configurations in which one or more atoms were outside the constraining volume were rejected.

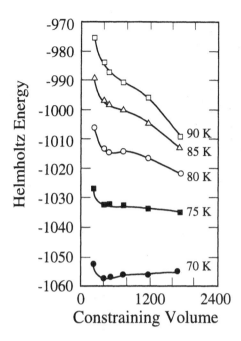

Figure 3.2. Dependence of the embryo Helmholtz energy on the constraining volume used to define an 87-atom Lennard-Jones cluster at several temperatures. The Helmholtz energy is in units of kT, and the volume in units of σ^3, where σ is the Lennard-Jones size parameter. (Adapted from Lee et al., 1973)

(1993). These authors used the method of Lee et al. (1973) to calculate the cluster's internal Helmholtz energy $A^*_{i/v}$ and hence, via (3.54), W_{min}. The surfaces shown in Figure 3.3 correspond to 1 cm^3 of supercooled argon vapor, modeled with the Lennard-Jones potential, at 70 K and various bulk pressures. The ridge that runs not quite parallel to the v axis is the barrier that clusters must overcome as they grow in i and become drops. Not visible in the figure is a valley on the far side of the ridge, whose axis runs not quite parallel to the i axis. This valley channels the growing clusters that become drops. Note the decrease in W_{min} and the movement of the ridge towards lower i values as the pressure is increased.

The work of Reiss and co-workers puts the question of the equilibrium distribution of liquidlike embryos in supercooled vapors on sound conceptual ground. However, having to calculate free energies by simulation rules out the use of this approach in practical applications. To overcome this limitation, Weakliem and Reiss (1993) defined a modified liquid drop theory that combines

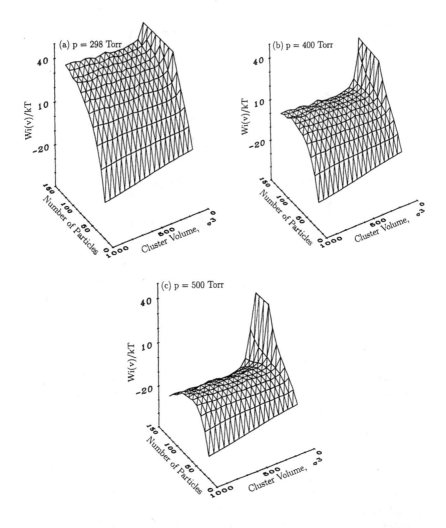

Figure 3.3. Reversible work of formation W_{min} of a physically consistent cluster, as a function of the number of molecules in the cluster, i, and the cluster volume v. Monte Carlo simulation for supercooled Ar vapor (Lennard-Jones potential) at 70 K and various pressures. [Reprinted, with permission, from Weakliem and Reiss, *J. Chem. Phys.* 99: 5374 (1993)]

elements of the physically consistent cluster with the conventional capillarity approximation. The i molecules of a cluster are partitioned between a liquid drop of radius r containing n molecules and a surrounding ideal vapor containing s molecules ($i = s + n$). The cluster is contained inside a spherical volume v. The droplet has the density and interfacial tension of the bulk liquid, and equilibrium between the vapor and the droplet is imposed. This yields two equations,

$$P^* = \frac{skT}{v - 4\pi r^3/3} = P_{vp}(T) \exp\left(\frac{2\sigma v_l}{rkT}\right),$$
(3.56)

$$n = \frac{4\pi r^3}{3v_l} = i - s,$$
(3.57)

where P^* is the pressure of the vapor in the spherical container, σ is the interfacial tension, v_l is the liquid-phase bulk molecular volume, and $P_{vp}(T)$ is the equilibrium vapor pressure at the given temperature. Equation (3.56) is the Gibbs-Thomson condition of equilibrium between the ideal vapor and the droplet. Equations (3.56) and (3.57) map every (i, v) pair uniquely into a corresponding (i, s) or (r, P^*) pair. Thus, to each physically consistent cluster there corresponds a modified liquid drop. Calculation of the reversible work of forming an (i, s) drop, however, does not require simulations: Weakliem and Reiss (1994) obtained the $W_{min}(i, s)$ surface analytically.

In addition to addressing the equilibrium distribution function for modified drops, Weakliem and Reiss developed a rate theory. The starting point is expressions analogous to (3.1) for the net rates of transition between points (n, s) and $(n - 1, s + 1)$, $J_s(n, s)$; and between points (n, s) and $(n + 1, s)$, $J_i(n, s)$. A transition from (n, s) to $(n - 1, s + 1)$, for example, corresponds to an increase in the size of an i cluster due to evaporation of a molecule from the liquid droplet into the surrounding ideal vapor; a transition from (n, s) to $(n + 1, s)$ corresponds to condensation of a molecule from the external supercooled vapor onto the droplet. The fluxes are written in terms of equations analogous to (3.1), and microscopic reversibility is invoked to calculate evaporation coefficients, in analogy to (3.2). Thus the fluxes are a function of ratios of actual to equilibrium distribution functions, in analogy to Equation (3.3). Since the fluxes are related by the conservation equation

$$J_s(n + 1, s - 1) + J_i(n - 1, s) = J_s(n, s) + J_i(n, s),$$
(3.58)

application of this equation to each interior (n, s) point gives rise to a set of linear equations in the ratio of nonequilibrium to equilibrium cluster distribution functions [f/N at each node, in the notation of (3.1)–(3.3)]. Solution of these equations subject to appropriate flux conditions at the boundaries (Weakliem and Reiss, 1994) yields the nonequilibrium to equilibrium distribution function ratios at each point. This, coupled with the analytically calculated equilibrium distribution functions, yields the fluxes at each point, in analogy to (3.3). The nucleation rate is then obtained by summing the i components of flux across the valley.Figure 3.4 (Weakliem and Reiss, 1994) shows the nucleation rates calculated as explained above, compared to the classical nucleation theory predictions. Each isotherm is parallel to the dashed line along which the classical and modified liquid droplet rates agree. This means that the modified liquid

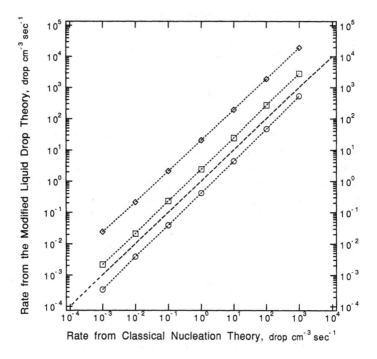

Figure 3.4. Comparison of nucleation rates calculated from the modified liquid drop theory and from classical nucleation theory for Ar. $\diamond = 85\,\mathrm{K}$, $\square = 90\,\mathrm{K}$, and $\bigcirc = 95\,\mathrm{K}$. The theories agree along the dashed line with no symbols. [Reprinted, with permission, from Weakliem and Reiss, *J. Chem. Phys.* 101: 2398 (1994)]

droplet theory yields nucleation rates whose isothermal dependence upon supersaturation agrees with classical nucleation theory, although the temperature dependence differs.

    The molecular theory of vapor-phase nucleation of Reiss and co-workers is based on a careful definition of what constitutes a cluster. An important advantage of adopting a rigorous molecular viewpoint is conceptual clarity. Thus, invoking the statistical mechanics of a low-density supercooled vapor composed of noninteracting physically consistent clusters, Reiss and Katz (1967) and Reiss et al. (1968) were able to clarify the so-called rotation-translation paradox (Lothe and Pound, 1962, 1966, 1968; Abraham and Pound, 1968; Feder et al., 1966), according to which the improper neglect of a cluster's translational and rotational degrees of freedom in the classical theory gives rise to nucleation rates that are twelve to eighteen orders of magnitude smaller than they should be. Reiss et al. (1968) showed that this is incorrect, and thereby resolved a lively controversy (Reiss and Katz, 1967; Abraham and Pound, 1968; Stillinger, 1968). Note, however, that the molecular theory of Reiss and co-workers applies only to nucleation in low-density supercooled vapors. Extending this approach to situations where embryos interact with each other and with the bulk metastable phase, that is to say, extending the theory to nucleation in metastable liquids, appears all but impossible at present.

Oxtoby and co-workers (Harrowell and Oxtoby, 1984; Oxtoby and Evans, 1988; Zeng and Oxtoby, 1991a,b; Oxtoby, 1991, 1992a,b; Talanquer and Oxtoby, 1993, 1994, 1995; Laaksonen and Oxtoby, 1995) have tackled the problem of improving upon the capillarity approximation by using density functional theory. In this approach, first applied to clusters by Abraham et al. (1974), the free energy is taken to be not a function of the embryo's radius but a functional of the density profile across the interface separating the embryo from the bulk metastable fluid. Functional differentiation of the free energy then yields the density profile, and thence the free energy barrier to nucleation. Density functional theory is intermediate between a fully microscopic treatment and the capillarity approximation, in that the problem is posed in terms of a density profile, rather than atomic coordinates; nevertheless, the distance over which the density profile changes is characteristic of molecular length scales (Oxtoby, 1992b). In contrast to the molecular theory of Reiss and co-workers, density functional theory is not limited to vapor-phase nucleation.

In its simplest form, density functional theory involves writing the Helmholtz energy of an inhomogeneous fluid as a sum of two terms,

$$A\{\rho(\mathbf{r})\} = \int d\mathbf{r} a_h [\rho(\mathbf{r})] + \frac{1}{2} \int \int d\mathbf{r} \, d\mathbf{r}' \rho(\mathbf{r}) \rho(\mathbf{r}') \phi(|\mathbf{r} - \mathbf{r}'|), \quad (3.59)$$

where A, the Helmholtz energy, depends on the density profile $\rho(\mathbf{r})$ across the embryo-bulk interface, a_h is the Helmholtz energy density of a uniform hard sphere fluid (Appendix 3) at the local density ρ, and ϕ is the attractive part of the pairwise-additive intermolecular potential. According to (3.59), the Helmholtz energy of an inhomogeneous fluid is written as the sum of repulsive (hard sphere) and nonlocal perturbation terms. Correlations between molecules are ignored in the perturbation terms. Consider now an open control volume V within the bulk metastable fluid, containing a smooth density inhomogeneity, the embryo. The grand potential (Hill, 1956; see also Appendix 3) can then be written as

$$-kT \ln \Xi = -PV \equiv \Xi' = A - G = A\{\rho(\mathbf{r})\} - \mu \int d\mathbf{r} \rho(\mathbf{r}), \quad (3.60)$$

where μ is the chemical potential and G, the Gibbs energy. The Gibbs energy of this same system is then given by

$$G = \Xi'\{\rho(\mathbf{r})\} + PV + \mu N, \quad (3.61)$$

where P and μ are the bulk pressure and chemical potential of the metastable fluid. Rearranging,

$$G - \mu N = \Xi'\{\rho(\mathbf{r})\} + PV. \quad (3.62)$$

The left-hand side is the difference between the Gibbs energy of the inhomogeneous system and that of a homogeneous system with the same number of particles at bulk conditions, ΔG. The right-hand side is the difference between the grand potential of the inhomogeneous system and that of the homogeneous system with the same number of particles at bulk conditions, $\Delta\Xi'$. Hence the free energy barrier to the formation of a nonuniform embryo is also the grand potential barrier (Oxtoby and Evans, 1988). Equating to zero the functional derivative of the grand potential with respect to the density yields an integral equation for the density profile of the critical embryo,

$$\mu_h\{\rho(\mathbf{r})\} = \mu - \int d\mathbf{r}'\rho(\mathbf{r}')\phi(|\mathbf{r}-\mathbf{r}'|), \qquad (3.63)$$

where μ_h is the chemical potential of the hard sphere reference fluid. Equation (3.63) must be solved iteratively: a density profile is guessed and substituted into the right-hand side; the hard sphere chemical potential is then solved for the density at each point, which yields a new density profile. Because the critical embryo is unstable, it is a saddle point, not a minimum, of the grand potential. Hence the iteration process is unstable; the numerical technique needed to converge on the saddle point is discussed by Oxtoby and Evans (1988) and Zeng and Oxtoby (1991a).

Figure 3.5 (Oxtoby and Evans, 1988) shows the calculated density profiles for the critical droplet and bubble at $0.6T_c$, for a Yukawa attractive potential,

$$\phi(r) = -\alpha\lambda^3\exp(-\lambda r)/4\pi\lambda r. \qquad (3.64)$$

According to the classical theory, the critical droplet (bubble) should have the same density as the liquid (vapor) at coexistence. The density functional calculations show significant nonclassical effects: the density at the center of the critical droplet is lower than the liquid density at coexistence, and lower than the density of the liquid at the same bulk chemical potential; and the density at the center of the critical bubble is twice the density of the saturated vapor.

For droplet condensation in supercooled vapors or bubble formation in superheated liquids, density functional theory predicts that the free energy barrier to nucleation vanishes at the spinodal curve. This is an important improvement on classical nucleation theory, which predicts finite barriers irrespective of the depth of penetration into the two-phase region.[11] It will be shown in Section

---

[11] Oxtoby and Evans (1988) illustrated this point by comparing the dependence of the free energy cost of forming a critical droplet upon the bulk density of the supercooled vapor, for classical nucleation and density functional theories. The classical calculation was done using (3.14) and (3.15), with σ calculated for a planar interface using the same free energy functional as in the density functional calculations. The free energy barrier calculated by density functional theory vanished at the spinodal, whereas classical theory predicted a finite free energy barrier even when the bulk fluid (hard sphere with Yukawa tail) was unstable.

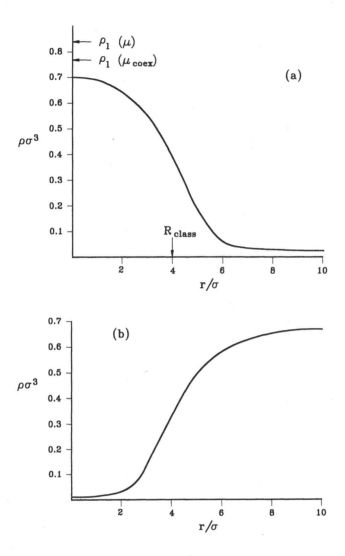

Figure 3.5. Density profile of the critical droplet (a) and bubble (b) for a metastable hard sphere fluid with Yukawa attractive tail [Equation (3.64)] at $0.6T_c$. $\Delta\mu$, the difference in chemical potential of the bulk supercooled vapor (a) and superheated liquid (b) with respect to saturation, is $0.62kT_c$ (a) and $-0.52kT_c$ (b). σ is the hard sphere diameter. R_{class} is the radius of the critical droplet predicted by classical nucleation theory, Equation (3.14). $\lambda\sigma = 1$. $\rho_l(\mu)$ and $\rho_l(\mu_{coex})$ are the liquid density at the given μ and T, and at coexistence. Density functional theory calculations. (Adapted from Oxtoby and Evans, 1988)

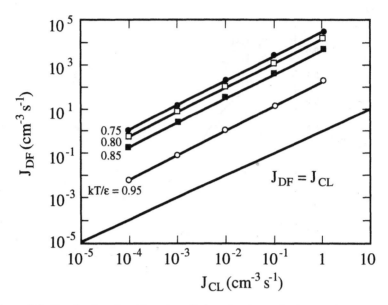

Figure 3.6. Density-functional-based predictions for the nucleation rate in the super-cooled Lennard-Jones vapor at several temperatures. Classical and density functional predictions agree along the $J_{DF} = J_{CL}$ line. (Adapted from Zeng and Oxtoby, 1991b)

3.2 that the vanishing of the nucleation barrier in an unstable fluid follows quite generally from allowing for the finiteness of the bulk-embryo interface. This is in fact the starting point of the van der Waals–Cahn–Hilliard theory of in-homogeneous fluids, which forms the basis of the Cahn-Hilliard treatment of spinodal decomposition (see Section 3.2).

Oxtoby and co-workers have applied density functional theory to the pre-diction of nucleation rates, by combining the prefactors given by the classical theory with the critical free energy barriers obtained from free energy func-tionals as explained above. Figure 3.6 shows a comparison between rates of droplet nucleation from the supercooled Lennard-Jones vapor as predicted by the classical theory and the density functional approach. The classical rates were computed with (3.29), using a flat interface and the same free energy functional as in the density functional calculations. The fact that the density functional predictions are almost parallel to the line along which both theo-ries agree means that both approaches predict essentially the same dependence of nucleation rate upon supersaturation. The fact that the density functional isotherms lie above the agreement line means that the respective predictions on the temperature dependence of the nucleation rate disagree by several orders of magnitude.

Density functional theory is an extremely powerful technique for the rigorous calculation of free energy barriers to nucleation. It has the important advantage with respect to the physically consistent cluster treatment of Reiss and co-workers that it is not limited to nucleation of liquids in ideal vapors. Examples of calculations in nonideal systems include bubble nucleation (see Section 3.1.4) in the superheated Yukawa and Lennard-Jones liquids (Oxtoby and Evans, 1988; Zeng and Oxtoby, 1991a); liquid nucleation in dipolar vapors (Talanquer and Oxtoby, 1993); binary nucleation of liquids from vapors (Zeng and Oxtoby, 1991b; Laaksonen and Oxtoby, 1995) and of bubbles from liquids (Talanquer and Oxtoby, 1995); and crystal nucleation (Harrowell and Oxtoby, 1984; see Section 3.1.5).

3.1.3 Kinetic Nucleation Theories

The goal of nucleation theory is to predict the rate of formation of embryos of a new phase that exceed a threshold size and grow spontaneously. Clearly, a truly rigorous theory should be kinetic rather than thermodynamic. Nevertheless, much of the discussion in Sections 3.1.1 and 3.1.2 has been thermodynamic. The reason for this is that the dissociation rate α [Equation (3.1)] is not known in general, whereas the arrival rate β is easier to obtain: it can, for example, be calculated exactly when the metastable phase is an ideal gas. Therefore equilibrium is usually invoked, and detailed balancing allows α to be expressed in terms of β [see Equation (3.2)]. As a result, the key question in the theory becomes the calculation of the equilibrium embryo distribution or, equivalently, of the energetics of embryo formation.

Katz and co-workers (Katz and Wiederisch, 1977; Wiederisch and Katz, 1979; Katz and Donohue, 1979; Chiang et al., 1988) questioned the nature of the equilibrium state invoked in the classical theory. They argued that the equilibrium distribution $N(n)$ [Equation (3.2)] is hypothetical and unattainable, because it describes the population of embryos that would exist if the system were constrained to remain at equilibrium at the given temperature and supersaturation. Instead, they applied Equation (3.2) to a stable equilibrium state, namely, unit supersaturation at the given temperature. Examples of embryo distributions at stable equilibrium include the distribution of droplets in a saturated vapor and the distribution of bubbles in a saturated liquid. Assuming only that $\alpha(n)$, the rate at which molecules leave the n-molecule embryo, is not a function of supersaturation, and repeating the argument that leads from (3.1) to (3.7), they obtained the following expression for the nucleation rate:

$$ J = \frac{[f(1)/N(1)][\beta_e/\beta]}{\displaystyle\sum_{n=1}^{\Lambda-1} \frac{1}{\beta F(n)N(n)\left(\frac{\beta}{\beta_e}\right)^n}}, \qquad (3.65) $$

where β and β_e are the arrival rates of single molecules at the given super-saturation and at equilibrium, respectively, both calculated at the same temperature. The other symbols have already been defined. The assumption of a supersaturation-independent α should be valid in dilute systems, where the nucleating species interacts negligibly with other bulk molecules upon leaving the embryo. For pure substances, this situation would correspond to droplet condensation from a supercooled vapor at low pressures.

The kinetic quantities required for the calculation of the nucleation rate are the arrival rate β and the arrival rate ratio β/β_e; the required thermodynamic quantities are the equilibrium embryo distribution at unit supersaturation and the ratio of single-molecule concentration at the given supersaturation to that at equilibrium. Since the bulk phase consists overwhelmingly of single molecules, the quantity $f(1)/N(1)$ is the ratio of the bulk-phase concentration at the given pressure to that at the equilibrium pressure (unit supersaturation). For condensation from a supercooled ideal vapor, Equation (3.65) yields the same nucleation rate as the classical theory if macroscopic arguments are used to calculate free energies of embryo formation. Katz and Donohue (1979) also discussed extensions of their approach to cases where the assumption of a supersaturation-independent α does not apply, such as crystal nucleation from the melt, or bubble formation in superheated liquids.

Because the driving force for nucleation is a kinetic quantity (i.e., the ratio of arrival rates), this theory has been called kinetic. However, the theory of Katz and co-workers is really an equilibrium one. It differs from classical nucleation theory in that the equilibrium condition invoked to calculate the evaporation rate α is stable rather than metastable. In arriving at (3.65), Katz and co-workers argued that the equilibrium condition invoked in the classical theory is unattainable. However, the classical theory does not assume the existence of an equilibrium embryo distribution beyond the critical size. Rather, it simply requires a mathematical closure condition in the large-embryo limit. It is not necessary to assign a numerical value to $N(\Lambda + 1)$ [see (3.6)] in order to accept the existence of some Λ beyond which $f(\Lambda + 1)/N(\Lambda + 1)$ vanishes. For subcritical embryos, the existence of an equilibrium distribution at the given supersaturation is not unreasonable: it merely requires that the characteristic time for density fluctuations be much shorter than the time of formation of critical nuclei (Lifshitz and Pitaevskii, 1981). In other words, the equilibrium distribution invoked in the classical theory is unphysical only for supercritical embryos and long times. Since the classical theory is pretransitional (that is to say, it describes events taking place before the catastrophic growth of the new phase), it does not apply to either one of the above conditions. Though not strictly a kinetic theory (detailed balance is invoked at the outset), the approach of Katz and co-workers is quite interesting because it leads to a kinetic driving force (β/β_e) by making reference to a condition of stable equilibrium. The practical applications of this work are limited by the fact that in cases other than droplet formation in a supercooled vapor that behaves ideally, the quantities β, β_e, and $N(n)$ cannot be calculated exactly.[12]

Ruckenstein and co-workers proposed a kinetic approach to nucleation in liquids (Narsimhan and Ruckenstein, 1989; Nowakowski and Ruckenstein, 1990; Ruckenstein and Nowakowski, 1990) and gases (Nowakowski and Ruckenstein, 1991a,b; Ruckenstein and Nowakowski, 1991). The theory avoids the assumption of equilibrium, and is based on the calculation of the rate $\alpha(n)$ at which single molecules abandon a cluster of n molecules. The physical picture is one of molecules moving within a thin layer surrounding a cluster; dissociation occurs when molecules acquire enough energy to overcome the effective potential energy barrier due to the cluster and the surrounding fluid. We illustrate Ruckenstein's approach as it applies to nucleation in a pure supercooled liquid. In this case, because of the high density, molecules experience many collisions before escaping the potential well, and hence their kinetic energy can be thought of as being in equilibrium, that is to say, the velocity distribution is Maxwellian. The diffusive motion of molecules under the influence of a potential field can then be described by the Smoluchowski equation (see, for example, Chandrasekhar, 1943)

$$\frac{\partial \rho(\mathbf{r}, t)}{\partial t} = \nabla \cdot \left(D \nabla \rho + \frac{D}{kT} \rho \nabla \phi \right), \qquad (3.66)$$

where $\rho(\mathbf{r}, t)$ is the number density, at time t and location $\mathbf{r}$, of molecules diffusing in a potential field $\phi(\mathbf{r})$; $\mathbf{r}$ is the location of a molecule measured from the center of the cluster; $\beta = 1/kT$, with k Boltzmann's constant; and D is the diffusion coefficient. If the diameter of the effective rigid core of a molecule is η, then a surface molecule is considered bound to a spherical cluster of radius R so long as it lies within a distance r of its center, where $R - \eta \leq r \leq R + \eta$. If $r > R + \eta$, the molecule dissociates from the cluster. The probability that a molecule initially at ξ will still be in the surface layer at time t can be written as

$$Q(t/\xi) = 4\pi \int_R^{R+\eta} r^2 p(r, t/\xi) \, dr, \qquad (3.67)$$

where $p(r, t/\xi)$ is the probability that a molecule initially at ξ will be at r after an elapsed time t. The function $p(r, t/\xi)$ is the solution to the Smoluchowski equation with initial condition $\rho(r, 0) = \delta(r - \xi)$, where δ is the Dirac delta function. The probability that the escape will occur in a time shorter than t is $1 - Q$, and the mean dissociation time for molecules initially at ξ is therefore given by

$$\tau(\xi) = -\int_0^\infty t \frac{\partial Q(t/\xi)}{\partial t} \, dt. \qquad (3.68)$$

---

[12]A rigorous calculation of $N(n)$ would require computations such as those outlined in Section 3.1.2, but applied to a saturated vapor.

The mean dissociation time averaged over all initial positions is given by

$$\langle \tau \rangle = \int_R^{R+\eta} \tau(\xi) p(\xi) \, d\xi, \quad p(\xi) = \frac{\xi^2 e^{-\phi(\xi)/kT}}{\int_R^{R+\eta} \xi^2 e^{-\phi(\xi)/kT} \, d\xi}, \tag{3.69}$$

and thus the rate at which molecules abandon the cluster can be written as

$$\alpha(n) = \frac{N_s(n)}{\langle \tau(n) \rangle}, \tag{3.70}$$

where N_s is the number of molecules in the surface layer. For example, if the cluster is an amorphous, uniform spherical solid of density ρ_s,

$$n = \frac{4\pi \rho_s R^3}{3}, \tag{3.71}$$

$$N_s = \frac{4\pi \rho_s R^3}{3} \left[1 - \left(1 - \frac{\eta}{R} \right)^3 \right]. \tag{3.72}$$

The expression for $\alpha(n)$, valid for a spherical amorphous cluster of uniform density, is given in the paper by Ruckenstein and Nowakowski (1990). Coupled with an equation for $\beta(n)$, the arrival rate, it can be used to calculate the nucleation rate without invoking equilibrium. The arrival rate follows also from solving the Smoluchowski equation, but with different boundary conditions. The result is

$$\beta(n) = \frac{4\pi D\rho_{\text{bulk}}}{\int_{R+\eta}^{\infty} \xi^{-2} e^{-\phi(\xi)/kT} \, d\xi} \tag{3.73}$$

where ρ_{bulk} is the density of molecules in the bulk liquid (Ruckenstein and Nowakowski, 1990). The kinetic equations now read[13]

$$J(n) = \beta(n-1)f(n-1) - \alpha(n)f(n), \tag{3.74}$$

$$\frac{\partial f(n,t)}{\partial t} = J(n) - J(n+1). \tag{3.75}$$

Assuming steady state and using the boundary condition $f(n) \to 0$ for $n \to \infty$ leads, after rewriting the difference equations (3.74) and (3.75) in continuous form, to the following equation for the homogeneous nucleation rate:

$$J = \frac{\beta(1)n_{\text{bulk}}}{2 \int_0^{\infty} \exp[-2w(g)] \, dg}, \quad w(g) = \int_0^g \frac{\beta(g) - \alpha(g)}{\beta(g) + \alpha(g)} \, dg, \tag{3.76}$$

where $\beta(1)$ is the rate of single-molecule coagulation. The number of molecules in the critical nucleus (n^*) is obtained from the condition $\alpha(n^*) = \beta(n^*)$. The supersaturation is defined as $\rho_{\text{bulk}}/\rho_{\text{eq}}$, where ρ_{eq} is the bulk density that

---

[13]Note that α and β in (3.74) include the surface area of the cluster.

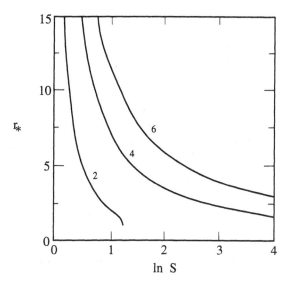

Figure 3.7. Size of the critical nucleus (in units of the hard core molecular diameter) as a function of supersaturation according to the Ruckenstein-Nowakowski kinetic theory of nucleation in supercooled liquids. Labels on the curves are reciprocal temperatures, ε/kT. (Adapted from Ruckenstein and Nowakowski, 1990)

coexists with an infinite critical nucleus; ρ_{eq} is obtained from the condition $\alpha(n^*) = \beta(n^*)$, in the limit $R/\eta \to \infty$. Thus, the theory allows the calculation of the nucleation rate and of the size of the critical nucleus as a function of supersaturation, entirely from kinetic considerations. Figures 3.7 and 3.8 show typical results of this kinetic theory.[14]

The work of Ruckenstein and co-workers is an important step in the direction of developing a kinetic theory of homogeneous nucleation. Much remains to be done: the bulk medium's contribution to the effective potential needs to be incorporated rigorously; calculations must be done for realistic intermolecular potentials; comparisons with good experimental data for supercooled liquids have yet to be made. Nevertheless, the work provides a valuable point of departure for future treatments of nucleation that do not rely on equilibrium assumptions or on the application of continuum thermodynamics to small objects.

---

[14]Ruckenstein and Nowakowski (1990) present a comparison between their calculated nucleation rates and classical nucleation predictions. These are based on comparing their kinetic expressions for the relation between supersaturation and critical nucleus size with the Kelvin equation. The latter follows from (3.14) upon writing $\Delta\mu = kT \ln S$ (S = supersaturation ratio). However, this relationship applies to bubble nucleation in liquids ($S = P_{eq}/P$, with P_{eq} the saturation pressure at the given temperature), and to droplet nucleation in vapors ($S = P/P_{eq}$), but not to crystallization.

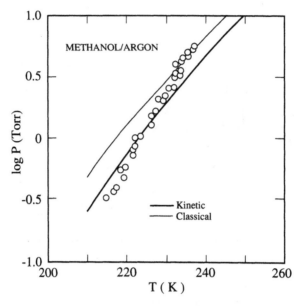

Figure 3.8. Partial pressure of methanol in argon that will give rise to a methanol nucleation rate of 10^8 cm^{-3} sec^{-1}. The lower curve is the prediction according to the Ruckenstein-Nowakowski kinetic nucleation theory (extended to vapor condensation). The upper line is the prediction according to classical nucleation theory. The symbols are experimental results (Peters and Paikert, 1989). (Adapted from Nowakowski and Ruckenstein, 1991a)

3.1.4 Homogeneous Nucleation in Superheated Liquids

By reducing the concentration of dissolved gases and suspended impurities, and by eliminating contact with solid interfaces, it is possible to superheat liquids far above their boiling point: up to roughly 90% of their critical temperature at atmospheric pressure, for example (Blander and Katz, 1975). Under these conditions, the attainable extent of superheating is largely controlled by homogeneous nucleation. An article by Reid (1976) is a very good introduction to the general topic of superheated liquids. The measurement of attainable limits of superheating is among the most carefully studied aspects of the physics of metastable liquids. A large body of information on this topic has been critically compiled in an excellent review by Avedisian (1985). It tabulates experimental data for ninety substances and twenty-eight mixtures. The two books by Skripov and co-workers (Skripov, 1974; Skripov et al., 1988), and an earlier review article by Blander and Katz (1975) also contain extensive tabulations of superheat limits. Readers interested in data on the limits of superheat of specific liquids or mixtures should consult these compilations, and the many references therein. In this section we discuss the basic principles of homogeneous nucleation in superheated liquids, and how these relate to the collection and critical interpretation of data on attainable limits of superheating.

The embryos that trigger vapor formation in a superheated liquid are microscopic bubbles: small regions where the density is smaller than in the bulk. To calculate the rate of homogeneous nucleation in a superheated liquid according to the classical theory, one must therefore consider the energetics of bubble formation. The contents of vapor embryos can be treated as an ideal gas except near the critical point. Let P^* be the pressure inside the critical nucleus. Then, P being the bulk pressure in the superheated liquid, and denoting embryo properties with primes as in Section 3.1.1, we must have $\mu'(T, P^*) = \mu(T, P)$, since the critical nucleus is in unstable equilibrium with the bulk. For an ideal gas,

$$\mu'(T, P') - \mu'(T, P^*) = kT \ln \frac{P'}{P^*}, \tag{3.77}$$

and for a spherical ideal-gas embryo of radius r containing n molecules,

$$\frac{4\pi r^3}{3} P' = nkT, \tag{3.78}$$

whereupon the reversible work of embryo formation becomes [see (3.8)]

$$W_{\min}(r, P') = 4\pi r^2 \sigma - \frac{4\pi r^3}{3}(P' - P) + \frac{4\pi r^3 P'}{3} \ln \frac{P'}{P^*}. \tag{3.79}$$

The work of embryo formation is a function of two variables (embryo radius and pressure). In contrast, for incompressible embryos, $W_{\min}$ is a function of r alone. At equilibrium, we must have

$$P' = P^* \tag{3.80}$$

and also

$$P' - P = \frac{2\sigma}{r^*} \tag{3.81}$$

(Appendix 2). In what follows, we use the superscript $*$ to denote any quantity pertaining to the critical nucleus. When (3.80) and (3.81) are satisfied, substitution into (3.79) yields the free energy of formation of the critical nucleus

$$W^*_{\min} = \frac{4\pi \sigma r^{*2}}{3} \tag{3.82}$$

(Gibbs, 1875–76, 1877–78, 1961). As with the incompressible-nucleus case, this condition is one of unstable equilibrium. For small departures from equilibrium, we can expand the free energy in Taylor series,

$$
\begin{aligned}
W_{\min}(r, P') \approx\ & W_{\min}^* + \frac{1}{2}W_{rr}\left(r - r^*\right)^2 \\
& + \frac{1}{2}W_{PP}\left(P' - P^*\right)^2 \\
& + W_{rP}\left(r - r^*\right)\left(P' - P^*\right),
\end{aligned}
\tag{3.83}
$$

where first-order derivatives vanish at equilibrium, and where W_{rr}, W_{PP}, and W_{rP} are second-order partial derivatives, evaluated at unstable equilibrium. Evaluating these partial derivatives from (3.79), Equation (3.83) becomes

$$
W_{\min}(r, P') \approx \frac{4\pi \sigma r^{*2}}{3} - 4\pi \sigma \left(r - r^*\right)^2 + \frac{2\pi r^{*3}}{3P^*}\left(P' - P^*\right)^2.
\tag{3.84}
$$

In order to calculate the relationship between r and P' away from equilibrium one must take into account the dynamics of bubble growth and decay. Such a hydrodynamic approach was developed by Zeldovich (1942) and Kagan (1960), and is discussed below. There are, however, two obvious limiting cases (Blander and Katz, 1975). For small departures from equilibrium, the pressure inside the embryo can be plausibly assumed to remain constant at its equilibrium value. In this case, (3.84) becomes

$$
W_{\min} \approx \frac{4\pi \sigma r^{*2}}{3} - 4\pi \sigma \left(r - r^*\right)^2.
\tag{3.85}
$$

On the other hand, if the embryo remains in mechanical equilibrium with its surroundings (i.e., $P' - P = 2\sigma/r$), (3.84) reads

$$
W_{\min} \approx \frac{4\pi \sigma r^{*2}}{3} - 4\pi \sigma \left(r - r^*\right)^2\left[1 - \frac{1}{3}\left(1 - \frac{P}{P^*}\right)\right].
\tag{3.86}
$$

In deriving Equation (3.86), the additional assumption is made that $(r/r^*)^2$ is not too different from unity.

At unstable equilibrium, the chemical potentials in the critical nucleus and in the metastable bulk phase are equal. Since the vapor in the nucleus is assumed to behave ideally, we must have

$$
P^* = z(T, P),
\tag{3.87}
$$

where z denotes the fugacity of the liquid at the given temperature and pressure. For an incompressible liquid,

$$
\ln \frac{z(T, P)}{z[T, P^e(T)]} = \frac{(P - P^e)v}{kT},
\tag{3.88}
$$

where P^e is the equilibrium vapor pressure of the liquid at the given temperature, and v its molecular volume. The liquid and vapor fugacities are equal at the equilibrium vapor pressure. If the vapor behaves ideally, this implies equality between liquid fugacity and vapor pressure,

$$z[T, P^e(T)] = P^e(T), \tag{3.89}$$

and, therefore,

$$P^* = P^e(T) \exp\left[(P - P^e) v/kT\right]. \tag{3.90}$$

The radius of the critical bubble then follows from (3.81) with $P' = P^*$.

The argument of the exponential in (3.90) is normally small because the specific volume of liquids is small compared to kT/P^e. Therefore, in general,

$$P^* = O[P^e(T)], \tag{3.91}$$

that is to say, the pressure inside the critical nucleus is not too different from the equilibrium vapor pressure at the given temperature. For example, if the liquid's specific volume is $1 \text{ cm}^3 \text{ g}^{-1}$, and $T = 300$ K, then for $P - P^e = -100$ bar, $P^* = 0.67 P^e(T)$ for a substance whose molecular weight is 100 g mol^{-1}. Thus, for moderate superheats, the quantity in brackets in (3.86) is close to 1; it approaches 2/3 when P becomes small. Reality will be intermediate between the two idealized situations (thermodynamic and mechanical equilibrium) considered here.

To calculate nucleation rates for the limiting cases in which near-critical vapor embryos are assumed to be either in mechanical or in thermodynamic equilibrium with the bulk, we use (3.86). Substituting into Equation (3.7), replacing the summation by an integral, changing the integration variable from n to r and from r to δr, and the integration limits to $[-\infty, +\infty]$ we obtain[15]

$$\begin{aligned} J &= N_{\text{tot}} \sqrt{\frac{2\sigma}{\pi m B}} \exp\left[-\frac{16\pi}{3kT} \cdot \frac{\sigma^3}{\delta^2 (P^e - P)^2}\right] \\ &\approx \rho_l \sqrt{\frac{2\sigma}{\pi m B}} \exp\left[-\frac{16\pi}{3kT} \cdot \frac{\sigma^3}{\delta^2 (P^e - P)^2}\right], \end{aligned} \tag{3.92}$$

(Blander and Katz, 1975), where m is the molecular mass, ρ_l is the number density in the bulk liquid phase, and δ and B are defined as follows:

$$\delta = 1 - \frac{P^e v}{kT}, \tag{3.93}$$

$$B \equiv 1 - \frac{1}{3}\left(1 - \frac{P}{P^*}\right) \approx 1 - \frac{1}{3}\left(1 - \frac{P}{P^e}\right). \tag{3.94}$$

---

[15]Note that the preexponential factors for droplet and bubble nucleation [Equations (3.29) and (3.92)] differ by the factor $(\rho_l/\rho_v)^2$.

The quantity B is the numerical factor that differentiates Equation (3.85) from Equation (3.86). Thus, if bubbles are assumed to be in thermodynamic equilibrium with the superheated liquid, $B = 1$, while $B \neq 1$ for mechanical equilibrium. In arriving at (3.92), it is assumed that the flux of single molecules arriving at the bubble-liquid interface from within a bubble inside which the pressure is P is given by

$$\beta = \frac{P}{\sqrt{2\pi mkT}}. \tag{3.95}$$

The problem of relating the pressure inside noncritical embryos to their size was addressed by Zeldovich (1942) and Kagan (1960). This approach leads to expressions for the nucleation rate that are not based on the assumption of mechanical or thermodynamic equilibrium between noncritical bubbles and the superheated liquid. Kagan (1960) considered nucleation as a flux in embryo size space, and wrote

$$J(n, t) = \dot{n} f(n, t) - D \frac{\partial f(n, t)}{\partial n}, \tag{3.96}$$

where $\dot{n}$ denotes dn/dt. The first term accounts for the hydrodynamic growth or decay of bubbles, and the diffusive term accounts for fluctuational growth. The steady-state solution of (3.96) reads

$$f(n) = J \exp\left(\int \frac{\dot{n}}{D} \, dn\right) \int_0^\infty D^{-1} \exp\left(-\int \frac{\dot{n}}{D} \, dn\right) dn, \tag{3.97}$$

where the boundary condition $f(n) \to 0$ for $n \to \infty$ has been used. At equilibrium, the nucleation rate vanishes, and thus we have

$$\dot{n} = D \frac{\partial \ln N(n)}{\partial n} = -\frac{D}{kT} \frac{dW_{\min}}{dn}, \tag{3.98}$$

where $N(n)$ is the equilibrium distribution of bubbles in the saturated liquid at the given temperature. Combining (3.97) and (3.98) leads to

$$J = N_{\text{tot}} D^* \left(\frac{dr}{dn}\right)_* \left[\int_{r \ll r^*}^\infty \exp\left(\frac{W_{\min}(r)}{kT}\right) dr\right]^{-1}, \tag{3.99}$$

where $*$ denotes quantities evaluated at the critical embryo condition, N_{tot} is the concentration of molecules in the bulk superheated liquid, and r is the radius of a bubble. In arriving at (3.99), the sharpness of the exponential term in the integrand at $r = r^*$ has been used to remove preexponential factors outside the integral, as constants evaluated at r^*. The key quantity required for the calculation of J is D^*. From (3.98), we have

$$D = -kT \frac{\dot{n}(dn/dr)}{dW_{\min}/dr}. \tag{3.100}$$

At r^*, both dn/dt and dW_{min}/dr vanish. Removing the indetermination,

$$D^* = -kT \frac{(d\dot{n}/dr)_* \, (dn/dr)_*}{(d^2 W_{min}/dr^2)_*}. \tag{3.101}$$

The derivatives in the numerator of (3.101) are obtained from the system of equations

$$P' = P + 2\sigma/r + \rho r\ddot{r} + 3\rho\dot{r}^2/r/2 + 4\eta\dot{r}/r, \tag{3.102}$$

$$\dot{n} = 4\pi r^2 \xi \left(\frac{P^* - P'}{\sqrt{2\pi mkT}} \right), \tag{3.103}$$

$$\dot{n} = \frac{4\pi}{3kT} \frac{d}{dt}(P'r^3). \tag{3.104}$$

Using (3.93) for P' in (3.103) and (3.104), differentiating with respect to r, and equating $d\dot{n}/dr$ obtained from (3.103) and (3.104) yields a cubic equation for $d\dot{r}/dr$. This allows the calculation of $d\dot{n}/dr$ and hence of D^*.

The hydrodynamic equation (3.102) gives the instantaneous relation between the pressure inside a bubble (P') and its radius, with P, η, and ρ denoting the bulk pressure, viscosity, and mass density of the superheated liquid, σ the interfacial tension, and dots and double dots first and second time derivatives. Equation (3.103) expresses the fact that bubbles grow by evaporation of molecules from the liquid-bubble interface into the bubble, and shrink by condensation of bubble molecules into the bubble-liquid interface. The driving force for this flux is then approximately equal to the difference between the vapor pressure at the given temperature and P' [see (3.91)]. ξ is a dimensionless condensation coefficient, and m is the molecular mass. Kagan also included thermal effects by considering the balance between heat flux at the interface and latent heat released (or absorbed) when molecules evaporate or condense at the bubble-liquid interface. This analysis leads to

$$\frac{P^*|_{T_i}}{P^*|_T} = 1 - \frac{\dot{n} \, (\Delta h)^2}{4\pi kT^2 \lambda r}, \tag{3.105}$$

where T_i is the temperature at the interface, T is the bulk temperature, Δh is the latent heat of vaporization, and λ is the thermal conductivity of the superheated liquid.

The nucleation rate expressions in Kagan's theory have the general form

$$J = N_{tot} \sqrt{\frac{\sigma}{kT}} A \exp \left\{ -\frac{16\pi\sigma^3}{3kT\delta^2 [P^e - P]^2} \right\}, \tag{3.106}$$

where the prefactor A, a characteristic velocity, depends on the controlling mechanism for bubble growth,[16] and δ is defined in (3.93). When the inertial

---

[16] In the classical theory, $A = (2kT/\pi mB)^{1/2}$.

Table 3.1. Hydrodynamic Regimes and Prefactors A in Kagan's Theory

Criterion	Controlling Mechanism or Range of Validity	Prefactor A
$\omega' \ll \omega^2$; $\omega \gg 1; \omega \gg 3/b$	Viscosity control; Large superheats and/or $P \ll 0$	$\sigma/\eta b$
$(3/b) - 1 \gg \omega$	$P > 0$; also when $P \le 0$ (if $|P|$ is small)	$\dfrac{\sqrt{2kT/\pi m}}{1+\delta_c}$
$\delta_c \gg 1$	Heat transfer to bubble	$\dfrac{\lambda bT}{\sigma}\left(\dfrac{kT}{\Delta h}\right)^2$

term in (3.102), $3\rho\dot{r}^2/2$, can be neglected, the physics is described by four dimensionless groups,

$$\delta_c = \frac{P^*}{P^* - P} \cdot \left(\frac{\Delta h}{kT}\right)^2 \cdot \sqrt{\frac{2k\sigma^2}{\pi m\lambda^2 T}}, \tag{3.107}$$

$$\omega = \sqrt{\frac{18kT}{\pi m}} \cdot \frac{\eta}{\sigma\,(1 + \delta_c)}, \tag{3.108}$$

$$\omega' = \frac{16\rho kT}{3\pi m\,(1 + \delta_c)^2\,(P^* - P)}, \tag{3.109}$$

$$b = \frac{P^* - P}{P^*} \approx \frac{P^e - P}{P^e}, \tag{3.110}$$

where P^* denotes the equilibrium pressure in critical-sized bubbles (or, to a good approximation, the vapor pressure at bulk temperature). The various regimes and the corresponding values of A are given in Table 3.1.

Figure 3.9 shows the nucleation rate in superheated diethyl ether [$T_c = 193.5°C$ (Reid et al., 1987)] calculated using Kagan's theory, but without neglecting inertial effects (Skripov, 1974). The temperature range in Figure 3.9 is $0.88 < T/T_c < 0.97$. The curves saturate at a nucleation rate of approximately 10^{30} cm^{-3} sec^{-1}, corresponding to a vanishing free energy barrier to nucleus formation. Mathematically, such rates are given by the preexponential terms in (3.92) or (3.106). The average frequency of formation of critical nuclei in 1 cm^3 of diethyl ether at atmospheric pressure (boiling point = 34.4°C) is once every 317 years (10^{10} seconds) at 141°C, once every second at 145°C, and once every 10^{-10} seconds at 149°C, which amounts to a change of twenty orders of magnitude in the nucleation rate over an 8°C temperature range. This extraordinarily sensitive dependence of the nucleation rate upon the extent of superheating causes a very sudden change from apparent stability to catastrophic boiling.

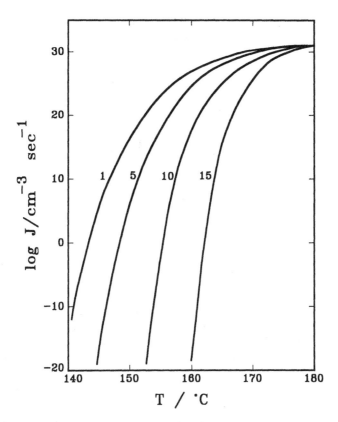

Figure 3.9. Homogeneous nucleation rate (J, in cm^{-3} sec^{-1}) in superheated diethyl ether ($T_c = 193.5°C$). Calculations are according to Kagan's hydrodynamic theory. The labels on each curve are pressures in bar. (Adapted from Skripov, 1974)

Such increases in nucleation rates upon isobaric superheating are due to the rapid decrease in surface tension close to the critical point, and to the increase in vapor pressure with temperature [see Equation (3.92)]. This behavior is typical of superheated liquids at atmospheric or higher pressure.[17]

The sharp transition to explosive boiling defines an attainable limit of superheating that is practically independent of the experimental technique used to measure it. This is because different methods of superheating give rise to different nucleation rates and the transition to catastrophic boiling is caused

---

[17]Nucleation rates in superheated liquids under high tension at temperatures considerably lower than T_c, on the other hand, are much less sensitive to temperature. This is because the temperature dependence of the surface tension is less pronounced away from the critical point. Furthermore, for $(-P) \gg P^e$, the temperature dependence of the vapor pressure is a less important contribution to the exponential in (3.92) (Blander and Katz, 1975).

Table 3.2. Typical Nucleation Rates Attained with Different Superheat-Measuring Techniques

Technique	J $(cm^{-3} sec^{-1})$
Bubble chamber	$10^2 - 10^3$
Capillary tube	$10^5 - 10^6$
Droplet superheat	$10^6 - 10^7$
Pulse heating	$10^{18} - 10^{23}$

by the nucleation rate changing by many orders of magnitude across a very small temperature range. Thus the kinetically controlled limit of attainable superheating is, in practice, a well-defined property of a given liquid. Table 3.2 gives nucleation rates attainable with some common experimental techniques.

In bubble chambers, the test liquid is placed in a sealed capillary, with one end connected to a pressurizing cylinder. Superheating is triggered by rapid release of the cylinder's pressure. The sudden formation of bubbles is detected audibly. Sources of uncertainty include the temperature drop during the rapid decompression, nucleation at the capillary walls, and nonuniformities in the temperature distribution in the test liquid (Avedisian, 1985). In capillary tube experiments, the test liquid is heated isobarically in an open capillary. Boiling is detected audibly or visually. The limitations of this technique are similar to those of bubble chambers. In pulse heating, a short pulse of current (25–1000 μsec) is passed through a small-diameter wire immersed in the test liquid. When bubbles are formed, the heat transfer rate between the wire and the adjacent liquid is altered; this causes the temperature and hence the resistance of the wire to change. Therefore measuring the resistance allows the determination of the temperature at which bubbles are formed. Sources of uncertainty include spatial nonuniformities in the wire's temperature and poor reproducibility of the calibration that relates temperature and resistance (Avedisian, 1985). Thorough discussions of experimental methods can be found in Skripov (1974), Avedisian (1985), and Skripov et al. (1988).

The insensitivity of measured superheat limits to experimental conditions is well illustrated by the droplet superheating technique (e.g., Apfel, 1972; see also Figure 1.4), in which small droplets of a liquid rise in an immiscible and denser host liquid, in the presence of a vertical temperature gradient. Consider a rising droplet of volume V being heated at a rate dT/dt (Skripov, 1974). Then Z, the number of critical nuclei that can form in the droplet during an interval

τ, is given by

$$Z = \dot{V} \int_0^\tau J \, dt = V \int_{T_1}^{T_2} \frac{J}{dT/dt} \, dT, \qquad (3.111)$$

where J is the nucleation rate. Since we are interested in events occurring within a narrow temperature interval around the point at which the droplet boils suddenly, we can linearize the reversible work of bubble formation about a reference temperature T_0,

$$J(T) \approx C \exp(bT), \qquad (3.112)$$

$$C = B \exp \left\{ - W_{\min}^*/kT \big|_0 + T_0 \left. \frac{d W_{\min}^*/kT}{dT} \right|_0 \right\}, \qquad (3.113)$$

$$b = - \left. \frac{d W_{\min}^*/kT}{dT} \right|_0, \qquad (3.114)$$

where B is the preexponential term in (3.92) or (3.106). It follows that we can write

$$Z \approx \frac{V J(T_2)}{b(dT/dt)} \qquad (3.115)$$

where we have assumed $J(T_2) \gg J(T_1)$. In order for Equation (3.115) to be relevant to the explosive boiling of a droplet, Z must be at least 1, and T_2 must be the measured superheat limit temperature T_{sh} at which the rising droplet boils explosively. Then,

$$J(T_{sh}) \approx \frac{b(dT/dt)}{V}. \qquad (3.116)$$

Substituting typical values for b [10 per degree (Skripov, 1974)], V (10^{-5} cm^3, corresponding to ca. 100 μm droplets), and dT/dt (10 degrees per second), we obtain nucleation rates of 10^6–10^7 cm^{-3} sec^{-1} for a typical droplet superheat experiment (see Table 3.2). It follows from (3.112) that $\Delta T = b^{-1} \Delta \ln J$. Since dT/dt is proportional to the droplet rise velocity, and hence to $V^{1/3}$, it follows that $\Delta T = (-2b/3) \Delta \ln V$. Thus, three-order-of-magnitude changes in either the nucleation rate or the droplet size cause the observed superheat limit to change by less than one degree Celsius.

Figure 3.10 shows the attainable limits of superheating for argon in capillary tubes (Skripov et al., 1973). Also shown are the locus of superheat temperatures corresponding to a nucleation rate of 10^3 cm^{-3} sec^{-1} (see Table 3.2), as predicted by classical nucleation theory [Equation (3.92)]; the spinodal curve calculated with the Peng-Robinson equation of state (Peng and Robinson, 1976); and the saturation line. Similarly, Figure 3.11 shows attainable limits of superheating for hexane in bubble column (Skripov and Ermakov, 1964) and pulse heating (Pavlov and Skripov, 1970) experiments; the locus of

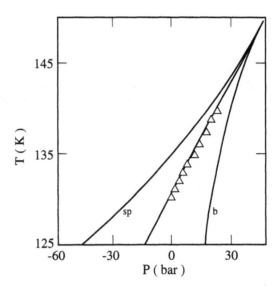

Figure 3.10. Attainable limits of superheat for argon, measured in capillary tubes ($\triangle$). The line through the data points is the classical nucleation theory prediction, using a nucleation rate of 10^3 cm^{-3} sec^{-1}. Curves b and sp are the binodal and the spinodal, the latter according to the Peng-Robinson equation. (Adapted from Skripov et al., 1973)

superheat temperatures corresponding to a nucleation rate of 10^6 cm^{-3} sec^{-1} (see Table 3.2); the Peng-Robinson spinodal; and the saturation line. Figure 3.12 shows the saturation line, attainable limits of superheat in droplet superheat experiments (Ermakov and Skripov, 1967), and the Redlich-Kwong spinodal (Redlich and Kwong, 1949) for six perfluoroparaffins.

Some general conclusions follow from the above discussion. The limit of superheat of liquids is a well-defined experimental quantity. At atmospheric pressure, liquids can be superheated to roughly 90% of their critical temperature (Blander and Katz, 1975). The droplet superheat technique, which corresponds to nucleation rates of ca. 10^6 cm^{-3} sec^{-1}, allows deep penetration into the metastable region, eliminates solid-liquid interfaces, and yields reproducible results in excellent agreement with the predictions of classical nucleation theory. The pulse heating technique corresponds to nucleation rates of ca. 10^{20} cm^{-3} sec^{-1}. It allows deeper penetration into the metastable region. However, pulse heating has been less widely used than droplet superheating, and its reliability is more difficult to assess (Avedisian, 1985).

The superheat limits of binary mixtures have been extensively investigated. Compilations of data for numerous systems can be found in the review of

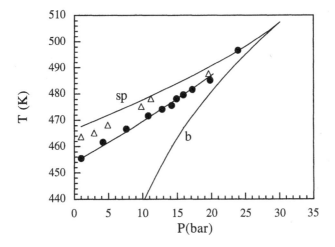

Figure 3.11. Attainable limits of superheat for n-hexane, measured in droplet superheat (●; Skripov and Ermakov, 1964), and pulse heating (△; Pavlov and Skripov, 1970) experiments. The line through the droplet superheat data is the classical nucleation theory prediction, using a nucleation rate of 10^6 cm^{-3} sec^{-1}. Curves b and sp are the binodal and the spinodal, the latter according to the Peng-Robinson equation.

Avedisian (1985), and in the monograph by Skripov et al. (1988). Table 3.3 lists literature references where data for mixtures can be found.

In many cases (e.g., Holden and Katz, 1978), results for mixtures are well correlated by modifying (3.92) in such a way that N_{tot}, the total number density, is written as the sum of component number densities in the liquid, and $m^{-1/2} = [y_1 (m_2)^{1/2} + (1-y_1) (m_1)^{1/2}]/ (m_1 m_2)^{1/2}$, where y_1 is component 1's vapor-phase mole fraction. This is equivalent to assuming that nuclei close to the critical size have the bulk-liquid composition (Holden and Katz, 1978). For mixtures that do not deviate greatly from ideality, the measured superheat limits are closely approximated by mole fraction averages of the respective pure-component values. The composition dependence of measured and calculated superheat limits of a nonideal mixture at atmospheric pressure are shown in Figure 3.13 (calculations as per the above-described procedure; droplet superheat experiments).

3.1.5 Homogeneous Nucleation in Supercooled Liquids

Homogeneous nucleation in supercooled liquids (Turnbull and Fisher, 1949; Frenkel, 1955; Buckle, 1961; Jackson, 1965; Turnbull, 1969; Woodruff, 1973;

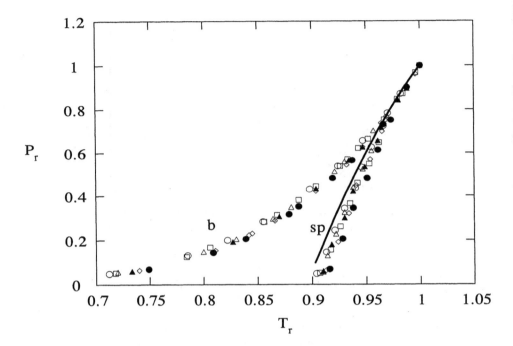

Figure 3.12. Saturation line b and superheat limits in rising droplet experiments for perfluoropentane ($\bigcirc$), perfluorohexane ($\square$), perfluoroheptane ($\triangle$), perfluoro-octane ($\blacktriangle$), perfluorononane ($\diamond$), and perfluorodecane ($\bullet$). Curve sp is the Redlich-Kwong spinodal. (Adapted from Ermakov and Skripov, 1967)

Oxtoby, 1988, 1991; Tiller, 1991; Myerson and Izmailov, 1993) involves the formation of embryos with different symmetry from that of the bulk metastable phase. In order for such ordered embryos to grow, molecules must cross an interface, the two sides of which differ not only in density but also in degree of order. The presence of this additional barrier is a distinguishing feature of nucleation in supercooled liquids.

Consider a supercooled liquid at a pressure P and temperature T, and let the equilibrium crystallization temperature corresponding to P be $T_m(> T)$. Expanding the chemical potentials in the solid and liquid phases about their equilibrium values at (T_m, P), differentiation of Equation (3.10) yields, for the radius of the critical nucleus,

$$r^* = \frac{2\sigma v^s T_m}{\Delta T \Delta h}, \tag{3.117}$$

where Δh and v^s are the heat of fusion and solid volume per molecule, and $\Delta T = T_m - T$. In discussing nucleation in supercooled liquids, it is customarily assumed that a single quantity, σ, can characterize what in reality are

Table 3.3. Literature References for Superheat Limits of Binary Mixtures[a]

Mixture	Reference
Chloroform, n-pentane	Buivid and Sussman (1978)
Carbon dioxide, chlorodifluoromethane	Mori et al. (1976)
Carbon dioxide, n-propane	Mori el al. (1976)
Carbon dioxide, isobutane	Mori et al. (1976)
Acetonitrile, acrylonitrile	Patrick-Yeboah and Reid (1981)
Ethane, n-propane	Porteous and Blander (1975)
Ethane, n-butane	Porteous and Blander (1975)
Ethanol, n-propanol	Avedisian and Sullivan (1984)
Ethanol, benzene	Danilov and Sinitsyn (1980)
Acetone, n-butanol	Danilov and Sinitsyn (1980)
Acetone, benzene	Danilov and Sinitsyn (1980)
n-Propane, 2-methyl propane	Porteous and Blander (1975)
n-Propane, n-butane	Renner et al. (1975)
n-Propanol, n-butanol	Avedisian and Sullivan (1984)
Diethyl ether, nitrogen	Forest and Ward (1977, 1978)
n-Butanol, n-pentanol	Avedisian and Sullivan (1984)
n-Pentane, cyclohexane	Eberhart et al. (1975)
n-Pentane, n-hexane	Holden and Katz (1978)
	Skripov (1974)
n-Pentane, n-heptane	Avedisian and Glassman (1981)
n-Pentane, n-octane	Avedisian and Glassman (1981)
	Renner el al. (1975)
n-Pentane, n-dodecane	Eberhart et al. (1975)
n-Pentane, n-hexadecane	Blander et al. (1971)
	Avedisian and Glassman (1981)
	Eberhart et al. (1975)
Benzene, cyclohexane	Holden and Katz (1978)
	Sinitsyn et al. (1971)
Benzene, n-hexane	Holden and Katz (1978)
Cyclohexane, n-hexane	Holden and Katz (1978)
n-Hexane, n-heptane	Skripov and Kukushkin (1961)

[a] Adapted from Avedisian (1985).

different surface energies associated with the various crystallographic orientations of a nucleus (Woodruff, 1973). The anisotropic embryo is thus replaced by an equivalent spherical object with a unique surface energy averaged (in an unspecified way) over all faces of the true polyhedral nucleus. With this assumption, the minimum work associated with the formation of a critical nucleus

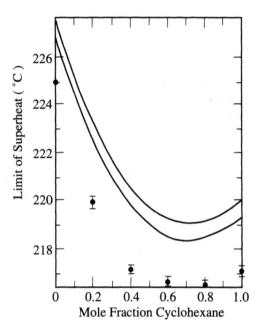

Figure 3.13. Composition dependence of the experimental (points) and theoretical (lines) superheat limits for a binary liquid mixture composed of benzene and cyclohexane at atmospheric pressure. The upper line is for a nucleation rate of 10^4 cm^{-3} sec^{-1}; the lower line for 10^2 cm^{-3} sec^{-1}. (Adapted from Holden and Katz, 1978)

[see Equations (3.10–3.15)] is given by

$$\frac{W^*_{min}}{kT} = \frac{16\pi}{3} \cdot \frac{\Delta s}{k} \cdot \left[\frac{\sigma \, (v^s)^{2/3}}{\Delta h} \right]^3 \cdot \frac{T_m^3}{T(T_m - T)^2}, \qquad (3.118)$$

where Δs is the entropy of fusion. The rate of homogeneous nucleation can be written as

$$J = \text{const} \cdot \exp \left\{ -\frac{16\pi}{3} \cdot \frac{\Delta s}{k} \cdot \left[\frac{\sigma \, (v^s)^{2/3}}{\Delta h} \right]^3 \cdot \frac{T_m^3}{T(T_m - T)^2} \right\} \cdot \exp \left[-\frac{E}{kT} \right], \qquad (3.119)$$

where E is the activation energy associated with the crossing of the liquid-solid interface. This exponential term can be identified with a diffusivity, or, equivalently, with a reciprocal viscosity (Frenkel, 1955; Turnbull, 1969). Differentiating (3.119) with respect to temperature (neglecting any temperature dependence in the preexponential factor and in the physical properties inside the exponentials) yields an implicit expression for the extent of supercooling that maximizes the nucleation rate:

$$\frac{[1 - \theta(J_{max})]^3}{5 - 3\theta(J_{max})} = \frac{16\pi}{3} \cdot \frac{\Delta h}{E} \cdot \left[\frac{\sigma (v^s)^{2/3}}{\Delta h}\right]^3, \quad \theta (J_{max}) = T (J_{max})/T_m.$$

(3.120)

An expression for the preexponential factor in (3.119) was obtained by Turnbull and Fisher (1949) using absolute reaction rate theory. Invoking their result, the rate expression becomes

$$J = N_{tot}\frac{kT}{h} \exp\left\{-\frac{16\pi}{3} \cdot \frac{\Delta s}{k} \cdot \left[\frac{\sigma (v^s)^{2/3}}{\Delta h}\right]^3 \cdot \frac{T_m^3}{T(T_m - T)^2}\right\} \exp\left[-\frac{E}{kT}\right],$$

(3.121)

where N_{tot} is the total concentration of molecules in the supercooled liquid, and h is Planck's constant. The exponentials in (3.121) can be written in dimensionless form,

$$J = N_{tot}\frac{kT}{h} \exp\left\{-\left[\frac{E'}{\theta} + \frac{\alpha^3\Gamma}{\theta (1 - \theta)^2}\right]\right\},$$

(3.122)

where

$$E' = \frac{E}{kT_m}, \quad \theta = \frac{T}{T_m}, \quad \alpha = \frac{\sigma (v^s)^{2/3}}{\Delta h}, \quad \Gamma = \frac{16\pi}{3} \cdot \frac{\Delta h}{kT_m}$$

(3.123)

(Turnbull, 1969). Figure 3.14 shows the dependence of the homogeneous nucleation rate upon the extent of supercooling, for different values of $\alpha^3\Gamma$ and E' (for comparison, σ, v^s, Δh, E, and T_m values of 0.03 N m^{-2}, 60 cm^3 mol^{-1}, 10 kJ mol^{-1}, 17 J mol^{-1}, and 300 K yield $\alpha^3\Gamma = 3.95$ and $E'=7$; a preexponential factor of 10^{35} cm^{-3} sec^{-1} was used in the calculations, corresponding to $N_{tot} = 0.055$ mol cm^{-3} and $T = 300$ K). Increasing E lowers the nucleation rate and raises the temperature at which the nucleation rate becomes maximum. Increasing the interfacial energy suppresses nucleation and lowers the temperature at which the nucleation rate is maximized. At high enough values of α, nucleation is suppressed. Under these conditions, supercooling leads to glass formation. This topic is discussed in Chapter 4.

Because σ in Equation (3.121) pertains to a hypothetical isotropic nucleus, it cannot be measured. Furthermore, true (anisotropic) solid-liquid interfacial energies are measured near the melting temperature (Franks, 1982), whereas what would be needed for an independent confirmation of the theory is $\sigma (T)$ in the supercooled region. Consequently, the theory of homogeneous nucleation, as it applies to supercooled liquids, has been used mainly to calculate effective interfacial tensions from measurements of nucleation rates (Woodruff, 1973; Franks, 1982).

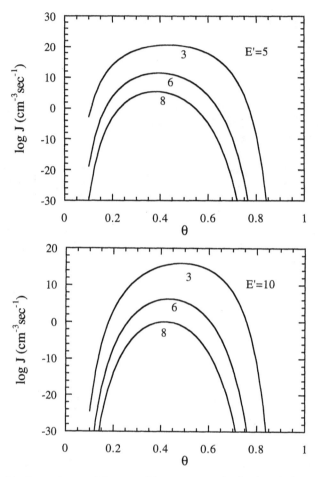

Figure 3.14. Dependence of the rate of homogeneous nucleation J (in cm^{-3} sec^{-1}) upon the extent of supercooling, $\theta = T/T_m$, where T_m is the equilibrium crystallization temperature. Numbers on the curves are values of $\alpha^3\Gamma$. α is the ratio of interfacial energy to enthalpy of fusion per unit molecular equivalent area, and Γ, up to a geometric factor, is the entropy of fusion in units of Boltzmann's constant. E' is the activation energy for viscosity normalized by kT_m. α, Γ, and E' are defined in (3.123). $N_{tot}kT/h = 10^{35}$ cm^{-3} sec^{-1}.

An effective way of supercooling a liquid is to subdivide the sample into small droplets. If the number of droplets is large compared to the number of nucleation-triggering impurities, a large fraction of the droplets can be extensively supercooled (Jackson, 1965; Perepezko, 1984). Cloud chambers (Cwilong, 1945, 1947; Thomas and Staveley, 1952; Coriell et al., 1971), and

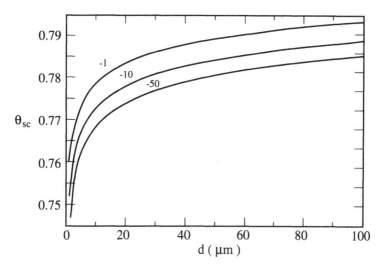

Figure 3.15. Attainable limits of supercooling, $\theta_{sc} = T_{sc}/T_m$, as a function of droplet diameter, calculated with Equation (3.124). Numbers on the curves are cooling rates (K sec^{-1}). $\alpha^3\Gamma = 2$, $E' = 2$, $T_m = 300$ K, and $N_{tot}kT/h = 10^{35}$ cm^{-3} sec^{-1}. α, Γ, and E' are defined in (3.123).

emulsions in an immiscible host liquid with a lower freezing point (Rasmussen and MacKenzie, 1972; Rasmussen et al., 1973) utilize this principle, and have been extensively applied to the study of supercooled liquids. Typical droplet sizes are 1 μm in cloud chambers, and 10 μm in emulsions (Angell, 1982). Assuming that nucleation is truly homogeneous, the number of critical nuclei that are formed in a droplet of volume V during a time interval δt is $VJ\delta t$. In order for crystallization to occur in a droplet, at least one critical nucleus must be formed. Then, the attainable limit of supercooling, the homogeneous nucleation temperature θ_{sc}, is given by

$$1 = \frac{VT_m}{dT/dt} \int_1^{\theta_{sc}} J(\theta) \, d\theta. \tag{3.124}$$

Figure 3.15 shows calculations of the attainable limits of supercooling as a function of droplet diameter for different values of the cooling rate. For a liquid with an equilibrium freezing temperature (T_m) of 300 K, the difference between the limits of supercooling attainable with 100 μm and 1 μm droplets is 10 K.

Table 3.4 lists measured attainable supercooling temperatures for some metals, miscellaneous inorganic and organic compounds, and alkali halides.

The extent of supercooling given in Table 3.4 for Al, Sb, Pb, Hg, and Sn (Perepezko, 1984) is considerably larger than the values measured by Turnbull

Table 3.4. Some Experimentally Reported Values of Limits of Supercooling[a]

Substance	T_m (K)	T_{sc} (K)	θ_{sc}	Reference
Aluminum	931.7	771.7	0.828	b
Antimony	903	693	0.767	b
Bismuth	544	317	0.583	b
Cadmium	594	484	0.815	b
Cobalt	1763	1433	0.813	c, d
Copper	1356	1120	0.826	c, d
Gallium	303	129	0.426	b
Germanium	1231.7	1004.7	0.816	c, d
Gold	1336	1106	0.828	c, d
Indium	430	320	0.744	b
Iron	1803	1508	0.836	c, d
Lead	600.7	447.7	0.745	b
Manganese	1493	1185	0.794	c, d
Mercury	234.3	146.3	0.624	b
Nickel	1725	1406	0.815	c, d
Palladium	1828	1496	0.818	c, d
Platinum	2043	1673	0.819	c, d
Silver	1233.7	1006.7	0.816	c, d
Tellurium	723	487	0.673	b
Tin	505.7	318.7	0.630	b
Ammonia	195.5	155.2 ± 1.5	0.794	e, d
Benzene	278.4	208.2 ± 2	0.748	e, d
Benzoic acid	395	275 ± 2	0.696	e, d
Boron trifluoride	144.5	126.7	0.877	f, d
1,2-Dibromoethane	282.7	216.2 ± 2	0.765	e, d
Bromomethane	179.4	155	0.863	f, d
Carbon tetrachloride	250.2	200.2 ± 2	0.800	e, d
Carbon tetrabromide	363.3	281 ± 5	0.773	e, d
Chloroform	209.7	157.2	0.750	f, d
Cyclopropane	145.8	128	0.878	f, d
Methylamine	179.7	144	0.801	f, d
Naphthalene	353.1	258.7 ± 1	0.733	e, d
Sulfur dioxide	197.6	164.6	0.833	f, d
Thiophene	234.9	184.2	0.784	f, d
Cesium bromide	908	747	0.823	g, d
Cesium chloride	918	766	0.834	g, d
Cesium fluoride	955	823	0.862	g, d
Cesium iodide	894	701	0.784	g, d

Table 3.4. *Continued.*

Substance	T_m (K)	T_{sc} (K)	θ_{sc}	Reference
Lithium bromide	823	629	0.764	g, d
Lithium chloride	887	697	0.786	g, d
Lithium fluoride	1121	889	0.793	g, d
Potassium bromide	1013	845	0.834	g, d
Potassium chloride	1045	874	0.836	g, d
Potassium iodide	958	799	0.834	g, d
Rubidium chloride	988	832	0.842	g, d
Sodium bromide	1023	857	0.838	g, d
Sodium chloride	1074	905	0.843	g, d
Sodium fluoride	126	984	0.778	g, d

[a] Adapted in part from Jackson (1965).
[b] Perepezko (1984).
[c] Turnbull and Cech (1950).
[d] Jackson (1965).
[e] Thomas and Staveley (1952).
[f] DeNordwall and Staveley (1954).
[g] Buckle and Ubbelohde (1960).

and Cech (1950). This shows that reported limits of supercooling are quite sensitive to the experimental technique [droplet dispersion on a glass substrate (Turnbull and Cech, 1950); droplet suspension in an immiscible host solvent (Perepezko, 1984)], and that in many cases what is observed is not homogeneous nucleation.

Deeply supercooled systems can exhibit the interesting phenomenon of phase selection, in which a metastable crystalline phase is formed instead of the stable structure (Follstaedt et al., 1986; Yoon et al., 1986). This has been observed in Bi, Ga, and Sn (Bosio et al., 1971; Akhtar et al., 1979; Perepezko and Anderson, 1980; Yoon et al., 1986). Perepezko and co-workers have also studied a related form of phase selection: they used a pulsed laser to rapidly anneal a low-temperature metastable allotrope of Mn (the α form), thereby producing a melt that is highly supercooled with respect to the stable δ form (Follstaedt et al., 1986). This technique allows the study of nucleation at very high supercoolings, and of the formation of different phases than the one that coexists with the liquid at high temperature.

The application of density functional theory to nucleation in supercooled liquids is a promising development (Harrowell and Oxtoby, 1984; Oxtoby, 1984, 1991). The theory yields the free energy barrier to nucleation, though not the preexponential factor that allows the calculation of a rate. Assuming the critical nucleus to be spherical, the local density $\rho(\mathbf{r})$ of the nonuniform system is written as

$$\rho\left(\mathbf{r}\right) = \rho_0 \left[1 + \eta\left(\mathbf{r}\right) + \sum_n \mu_n\left(\mathbf{r}\right) \exp\left(i\mathbf{k}_n \cdot \mathbf{r}\right) \right], \qquad (3.125)$$

where ρ_0 is the bulk density of the liquid, and the μ_n are structural order parameters corresponding to the reciprocal lattice vectors $\mathbf{k}_n$. $\eta\left(\mathbf{r}\right)$ is the generalization to nonuniform systems of the constant η, the fractional density change on freezing (Ramakrishnan and Yussouff, 1979; Haymet and Oxtoby, 1981; Oxtoby and Haymet, 1982). The order parameters $\eta(\mathbf{r})$ and $\{\mu_n\left(\mathbf{r}\right)\}$ change smoothly from zero in the liquid to their bulk solid values.

The excess free energy of the nonuniform liquid + nucleus system with respect to that of the uniform liquid is written as a perturbation expansion in $\Delta\rho\left(\mathbf{r}\right) = \rho\left(\mathbf{r}\right) - \rho_0$. Assuming that the order parameters vary slowly over interatomic distances, they can be expanded in Taylor series whose coefficients are order parameter gradients. Keeping only second order terms and performing functional minimization of the excess free energy with respect to the order parameter profiles leads to a coupled set of second-order, nonlinear differential equations. Their numerical solution yields the order parameter profile across the interface between the critical nucleus and the liquid, as well as the free energy excess with respect to the liquid. The liquid's structure factor (Appendix 3) is required as input to the theory.

Calculations for nucleation in supercooled sodium (Harrowell and Oxtoby, 1984) showed reasonable agreement with the classical theory predictions for the height of the free energy barrier. Interestingly, the density functional calculations also predicted that the order parameter μ_1 at the center of the critical nucleus was independent of the extent of supercooling, and corresponded to the bulk-solid value. This picture of the critical nucleus as having essentially the properties of the bulk phase is also in agreement with the classical theory.

3.1.6 The Approach to Steady State

A finite time is required in order for the concentration of embryos and the nucleation rate to attain their steady-state values. This time lag can be important (that is to say, comparable to the experimental time) in nucleation caused by rapid expansion of a vapor (Andres and Boudart, 1965; Kwauk and Debenedetti, 1993), or in the case of glasses and deeply supercooled liquids (Angell et al., 1984; Shi and Seinfeld, 1991). The discussion in Sections 3.1.1–3.1.5 assumes that steady state exists. Here we address, briefly, the approach to steady state. To this end, we combine the continuum version of Equations (3.3) and (3.4), to yield a differential equation for the time-dependent embryo distribution, $f(n, t)$,

$$\frac{\partial f}{\partial t} = \frac{\partial}{\partial n} \left[\beta F N \frac{\partial (f/N)}{\partial n} \right]$$

$$= \frac{\partial}{\partial n} \left(\beta F \frac{\partial f}{\partial n} \right) + \frac{1}{kT} \frac{\partial}{\partial n} \left(\beta F f \frac{d W_{\min}}{dn} \right), \qquad (3.126)$$

where β is the thermal flux, and use has been made of the Boltzmann form for the equilibrium distribution given by (3.16). Subject to the usual conditions $f = 0$ for $n \gg n^*$ and $(f/N) \to 1$ for $n \to 1$, it describes the evolution of f towards steady state.

In the approximate treatment of Wakeshima (1954), integration of the continuum analogue of Equation (3.3) subject to the boundary conditions mentioned above yields

$$\int_{-\infty}^{\infty} \frac{J}{\beta F N} d(n - n^*) = - \int_{1}^{0} d(f/N) = 1. \qquad (3.127)$$

Expanding J and N about n^* and replacing $F(n)$ and $\beta(n)$ by $F(n^*)$ and $\beta(n^*)$ as in (3.24)–(3.26) leads to

$$J_* - J_{ss} = -\frac{1}{4\pi Z^2} \cdot \left(\frac{\partial^2 J}{\partial n^2} \right)_*, \qquad (3.128)$$

where J_* is the non-steady-state nucleation rate corresponding to the critical-sized nucleus, J_{ss} is the steady state nucleation rate, and Z is the Zeldovich factor, $[-W''(n^*)/2\pi kT]^{1/2}$. Differentiation of the continuum analogues of (3.3) and (3.4) gives

$$J_* = -\beta F(n^*) \left(\frac{\partial f}{\partial n} \right)_*, \qquad (3.129)$$

$$\left(\frac{\partial^2 J}{\partial n^2} \right)_* = -\frac{d}{dt} \left(\frac{\partial f}{\partial n} \right)_*. \qquad (3.130)$$

Combining Equations (3.128)–(3.130) yields

$$\frac{d J_*}{dt} + \tau^{-1} J_* = \tau^{-1} J_{ss} \qquad (3.131)$$

whose solution, subject to the initial condition $J_*(0) = 0$, is

$$J_*(t) = J_{ss} \left[1 - \exp(-t/\tau) \right]. \qquad (3.132)$$

The characteristic time for the attainment of steady state is τ, commonly referred to as the time lag. In Wakeshima's treatment, this quantity is given by

$$\tau = \frac{1}{4\pi} \cdot \frac{1}{\beta F(n^*) Z^2} = \frac{1}{4} \cdot \frac{\delta^2}{\beta F(n^*)}, \qquad (3.133)$$

Table 3.5. Time Lag Prefactors

Prefactor[a]	Reference
$1/4\pi$	Wakeshima (1954)
$1/4$	Collins (1955)
$1/2$	Feder et al. (1966)
$[E_1(x) + 0.5772 + \ln x]/4\pi$	Shi et al. (1990)[b,c]

[a]$\tau = $ Prefactor$/[\beta F(n^*)Z^2]$
[b]Time lag for the establishment of a steady state concentration of critical nuclei.
[c]$E_1(x) = \int_x^\infty t^{-1}e^{-t}\,dt$; $x = \left\{3n^*\left[1 - (n^*)^{-1/3}\right]\delta^{-1}\right\}^2 \exp\left\{2\left[(n^*)^{-1/3} - 1\right]\right\}.$

where $\delta = \left(Z\sqrt{\pi}\right)^{-1/2}$ is the width of the region about n^* where the difference between $W(n)$ and $W(n^*)$ is less than kT,

$$|W(n) - W(n^*)| \leq kT \tag{3.134}$$

(Binder, 1987). For incompressible embryos in an ideal-gas continuum, (3.133) can be rewritten as

$$\tau = \frac{\sigma}{\beta kT(\ln S)^2} \tag{3.135}$$

where S is the supersaturation. This equation indicates that fast attainment of steady-state conditions is favored by large supersaturations and low interfacial tensions. As an example, for an interfacial tension of 0.03 N m^{-1}, a supersaturation of 10, a thermal flux of 10^{21} cm^{-2} sec^{-1}, and $T = 300$ K, the relaxation time is 0.14 μsec.

Other time lag estimates differ in the prefactor multiplying the term $1/(\beta FZ^2)$ in (3.133); this quantity by itself provides an order-of-magnitude estimate of the time lag. Some prefactors are listed in Table 3.5.

Figure 3.16 shows a numerical calculation of the evolution of the nucleation rate towards its steady-state value (Shi et al., 1990). The closed-form results listed in Table 3.5 involve different numerical approximations to the transient problem (e.g., singular perturbation: Shi et al., 1990). On the other hand, formally exact treatments of both the discrete [i.e., (3.3) and (3.4)] and continuum [i.e., (3.126)] versions of the transient problem that require numerical evaluation of integrals or summations for the calculation of the time lag include the work of Andres and Boudart (1965) (discrete), Frisch and Carlier (1971) (discrete and continuum), Shizgal and Barrett (1989) (discrete and continuum), Wu (1992) (discrete and continuum), and Shneidman and Weinberg (1992a,b) (discrete and continuum).

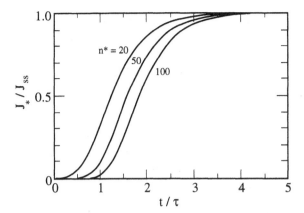

Figure 3.16. Evolution of the transient nucleation rate towards its steady-state value. The symbol τ here denotes $\left[2\beta F Z^2\right]^{-1}$, where β is the thermal flux, F is the surface area of the critical nucleus, and Z is the Zeldovich factor. n^* is the number of molecules in the critical nucleus. (Adapted from Shi et al., 1990)

3.2 SPINODAL DECOMPOSITION

Spinodal decomposition is the initial mechanism by which an unstable liquid phase relaxes towards equilibrium. It is a spontaneous process. In contrast, nucleation is an activated process: a free energy barrier must be overcome in order to form a critical nucleus. The height of this barrier decreases with penetration into the coexistence region, but, according to classical nucleation theory, it never vanishes. Thus, in classical nucleation theory, there is no such thing as an unstable state. The type of perturbation considered by nucleation theory, embryo formation, is small in extent but large in intensity. In contrast, the type of perturbation considered in thermodynamic stability theory, such as a small density fluctuation occurring over a macroscopic volume, is large in extent but small in intensity. In the thermodynamic viewpoint, loss of stability with respect to such perturbations occurs at a sharply defined condition: the spinodal. The classic work of Cahn and Hilliard (Cahn and Hilliard, 1958, 1959, 1971; Cahn, 1959, 1961, 1962, 1965) reconciles the kinetic and thermodynamic viewpoints. The basis of this approach is a theory of interfaces (Cahn and Hilliard, 1958) first formulated by van der Waals (1893; Rowlinson, 1979), and now commonly referred to as the van der Waals–Cahn–Hilliard theory of inhomogeneous fluids. This treatment of interfaces is closely related to the density functional method discussed in Section 3.1.2. For excellent discussions of the equilibrium theory of inhomogeneous fluids, interested readers should consult Abraham (1974, 1979), Davis and Scriven (1982), McCoy and Davis (1979), Bongiorno et al. (1976), and Rowlinson and Widom (1982).

The theory of Cahn and Hilliard explains the second fundamental mecha-nism of phase separation: spinodal decomposition. It addresses both thermody-namic and kinetic aspects of the problem. It is discussed in this chapter because it is intimately linked to nucleation theory, and important aspects of it deal with time-dependent phenomena. In the van der Waals–Cahn–Hilliard approach, the interface between two homogeneous phases is treated as a finite region within which densities[18] exhibit a spatial variation between the limits corresponding to the two bulk phases. Consider for simplicity a plane vapor-liquid interface (Figure 3.17), which can be characterized by a single space-varying density. Then, with z denoting the direction normal to the interface, the Helmholtz energy per unit volume (Helmholtz energy density) a at a point z within the interface is written as

$$a(z) = a[\rho(z)] + K[\rho'(z)]^2, \qquad (3.136)$$

where $a[\rho(z)]$ is the Helmholtz energy density of a homogeneous fluid of number density ρ, constrained to remain homogeneous inside the coexistence region; ρ' denotes $d\rho/dz$, and K is a positive constant. In the original van der Waals theory, K is given by

$$K = -\frac{\pi}{3} \int r^4 \phi(r)\, dr \qquad (3.137)$$

where $\phi(r)$ is the intermolecular pair potential (Rowlinson and Widom, 1982). The above expression for K is based on the assumption that there are no con-tributions to the local entropy density due to density gradients. Additional expressions for K according to other assumptions are discussed by Rowlinson and Widom (1982).

Equation (3.136) is a truncated expansion of the Helmholtz energy density in powers of the density gradient. That the lowest-order term is quadratic in the gradient follows from the requirement that, in the absence of an external field, the Helmholtz energy should not depend on the direction of the interface. There is no a priori reason for truncating the infinite series in powers of the density gradient. Abraham (1975, 1979) used perturbation theory to generalize the van der Waals–Cahn–Hilliard treatment and showed that the truncated expansion (3.136) is recovered in the limit of small interfacial density gradients (i.e., close to the critical point). Within the van der Waals formalism, the Helmholtz energy of a volume V is then given by

$$A = \int_V \left\{ a[\rho(z)] + K[\rho'(z)]^2 \right\} dV. \qquad (3.138)$$

---

[18]We use the term density in the general sense of Griffiths and Wheeler (1970), to denote an intensive property that is not equal in two coexisting phases.

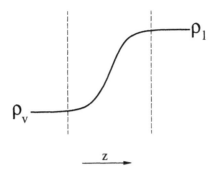

Figure 3.17. Schematic representation of the density profile across a planar vapor-liquid interface.

The interfacial tension is the difference between the actual Helmholtz energy per unit interfacial area, measured on a plane normal to z, and that which the system would have if the properties of the bulk phases on both sides of the interface were constant up to a plane located so as to satisfy a condition of zero surface excess of matter:

$$\int_{-\infty}^{+\infty} [\rho(z) - \rho_b] \, dz = 0, \qquad (3.139)$$

where ρ_b denotes the bulk density, which has different values on either side of the interface. Thus we write

$$\sigma = \int_{-\infty}^{+\infty} \{a[\rho(z)] + K[\rho'(z)]^2 - a_b\} \, dz, \qquad (3.140)$$

where a_b denotes the Helmholtz energy density in the bulk phase of density ρ_b. The theory assumes that the local Helmholtz energy density across the interface can be represented by a mean-field, van der Waals-like continuous function $a(\rho)$.[19] Because of Equation (3.139), we can write for a flat interface between equilibrium phases,

$$\sigma = \int_{-\infty}^{+\infty} \left[\Delta a(\rho) + K\rho'^2\right] \, dz, \qquad (3.141)$$

where Δa is the difference between the mean-field Helmholtz energy density of a uniform system and the Helmholtz energy density of an equilibrium two-phase system of the same overall density, as shown in Figure 3.18.

[19] Fisk and Widom (1969) extended the theory by studying the behavior of a fluid interface near the critical point. This was done by using an $a(\rho)$ functionality consistent with the experimentally observed behavior of fluids close to the critical point.

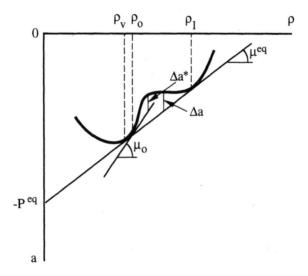

Figure 3.18. Density dependence of the Helmholtz energy per unit volume, $a = \mu\rho - P$. ρ_v and ρ_l are the vapor and liquid densities in equilibrium at a pressure P^{eq} and having a chemical potential μ^{eq}. ρ_0 and μ_0 are the density and chemical potential corresponding to an initially uniform metastable condition.

The equilibrium density profile is that which minimizes the Helmholtz energy, (3.138), subject to the material balance constraint, (3.139). Equivalently, the density distribution is obtained from the minimization of the integral in Equation (3.141), a formulation which already incorporates the material balance constraint. Using the standard techniques of variational calculus (Sokolnikoff and Redheffer, 1966), the following differential equation is obtained for the equilibrium density distribution for the case of a density-independent K:

$$2K\frac{d^2\rho}{dz^2} = \frac{d\Delta a}{d\rho}. \qquad (3.142)$$

Integrating once and using the boundary conditions $d\rho/dz = 0$, $\Delta a = 0$ at infinity, we have

$$K\left(\frac{d\rho}{dz}\right)^2 = \Delta a. \qquad (3.143)$$

This differential equation allows the calculation of the density profile across the interface, and in light of (3.141) it implies the following equivalent expressions for the interfacial energy:

$$\sigma = \int_{-\infty}^{+\infty} 2K\rho'^2\, dz = \int_{-\infty}^{+\infty} 2\Delta a\, dz = \int_{\rho_b^\alpha}^{\rho_b^\beta} 2\,[K\,\Delta a]^{1/2}\, d\rho, \qquad (3.144)$$

where ρ_b^α and ρ_b^β are the bulk densities, vapor and liquid in this case (Rowlinson and Widom, 1982). Note that (3.144) has been written for the more general case where K depends on ρ.

The above treatment can be applied to the calculation of the energetics of nucleation (Cahn and Hilliard, 1959). To this end, one considers the formation of an embryo within a bulk phase of initial density ρ_0. At constant temperature and volume, the reversible work of embryo formation is simply the change in Helmholtz energy associated with the above process,

$$W = \int_V \left[a - a_0 + K(\nabla\rho)^2 \right] dV \qquad (3.145)$$

or, equivalently (see Figure 3.18),

$$W = \int_V \left[\Delta a^* + K(\nabla\rho)^2 \right] dV + \mu_0 \int_V (\rho - \rho_0) \, dV, \qquad (3.146)$$

where Δa^* is the distance from the $a(\rho)$ curve to the tangent at ρ_0, and μ_0 is the chemical potential of the initial, metastable uniform phase. The second integral vanishes since the system's mass is constant. Therefore, for a spherical embryo,

$$W = 4\pi \int_0^\infty r^2 \left[\Delta a^* + K \left(\frac{d\rho}{dr} \right)^2 \right] dr. \qquad (3.147)$$

The unstable critical nucleus will be characterized by a density profile that maximizes (3.147). Taking variations, and for a density-independent K, one obtains the following differential equation for the density profile:

$$2K\rho'' + (4K/r)\rho' = \frac{d\Delta a^*}{d\rho}, \qquad (3.148)$$

where primes and double primes denote differentiation with respect to r.

The critical nucleus that results from the above treatment resembles that of classical nucleation theory when the degree of penetration into the coexistence region is small. In particular, as ρ_0 tends to ρ_b, the saturated density of the bulk phase at the given temperature, the density at the center of the nucleus tends to the saturated density of the new phase in equilibrium with the original bulk phase at the given temperature, and the interfacial tension tends to that of a flat interface (Cahn and Hilliard, 1959). However, as ρ_0 approaches ρ_s, the density at which the bulk phase becomes unstable at the given temperature, or spinodal density, the work of nucleus formation and the size of the critical nucleus scale as

$$W_{min} \propto (\rho_s - \rho_0)^{3/2}, \qquad (3.149)$$

$$r_{1/2} \propto (\rho_s - \rho_0)^{-1/2} \qquad (3.150)$$

(Cahn and Hilliard, 1959), where $r_{1/2}$ is the distance from the center of the critical nucleus to a point where the density is $(\rho_{ctr} - \rho_0)/2$, with ρ_{ctr} the density at the center, satisfying

$$\rho_{ctr}/\rho_0 = 8.1 \, (\rho_s/\rho_0) - 7.1. \tag{3.151}$$

At the spinodal, then, the work of nucleus formation vanishes, and the critical nucleus becomes arbitrarily large.

The significance of these results follows from considering the stability of an initially uniform state to infinitesimal isothermal density fluctuations. The Helmholtz energy change per unit volume due to such fluctuations is given by

$$\frac{\Delta A}{V} = \frac{1}{V} \int_V \left[a - a_0 + K(\nabla\rho)^2 \right] dV, \tag{3.152}$$

where the subscript 0 denotes the initial homogeneous condition. Without loss of generality, we assume a sinusoidal density perturbation,

$$\rho - \rho_0 = \alpha \cos Bz, \tag{3.153}$$

and expand the local free energy density a about the initial homogeneous value,

$$a = a_0 + (\rho - \rho_0) \left(\frac{\partial a}{\partial \rho} \right)_0 + \frac{1}{2} (\rho - \rho_0)^2 \left(\frac{\partial^2 a}{\partial \rho^2} \right)_0 + \cdots, \tag{3.154}$$

where the partial derivatives remind us of the functionality $a(\rho, T)$. Upon substituting into (3.152), and integrating between 0 and π/B (conservation of mass), we obtain for the one-dimensional case

$$\frac{\Delta A}{V} = \frac{\alpha^2}{4} \left[\left(\frac{\partial^2 a}{\partial \rho^2} \right)_0 + 2KB^2 \right]. \tag{3.155}$$

The homogeneous system is stable ($\Delta A > 0$) with respect to density fluctuations of all wavelengths as long as the curvature of $a(\rho)$, $(a'')_0$, is positive. When this quantity is negative, the system is unstable to fluctuations of wavelength greater then λ_c,

$$\lambda_c = \frac{2\pi}{B_c} = \sqrt{\frac{8\pi^2 K}{-(a'')_0}}. \tag{3.156}$$

Thus the critical wavelength diverges at the spinodal, where $a'' = v \, (\partial P/\partial \rho)_T$ vanishes. Equation (3.156) is a classic result first derived by Cahn (1961). In this theory, then, the spinodal is a sharp boundary dividing metastable states (in which phase separation occurs by the activated process of nucleation and growth) from unstable states, where phase separation occurs spontaneously by spinodal decomposition. At the spinodal, the free energy cost of nucleus formation vanishes; however, the size of the critical nucleus, and the critical wavelength for spontaneous phase split, diverge.

The evolution in time of phase separation by spinodal decomposition is described by a diffusion equation for the density (pure fluid) or species concentration (mixture) (Cahn, 1961, 1965; Cahn and Hilliard, 1971). To illustrate the basic aspects of the theory, we consider the one-dimensional, pure fluid case. The diffusion equation reads

$$\frac{\partial \rho(z)}{\partial t} = \frac{\partial}{\partial z} \left\{ M \frac{\partial}{\partial z} \mu \left[\rho(z) \right] \right\}, \tag{3.157}$$

where M (> 0) is a molecular mobility, and μ, the generalized chemical potential, is defined by

$$\delta A = \int \mu[\rho(z)] \delta \rho(z) \, dV. \tag{3.158}$$

The key assumption is that the chemical potentials defined thermodynamically in (3.158), and phenomenologically in (3.157), are the same quantity; a reasonable approximation so long as equilibrium considerations can be applied locally. The chemical potential follows from the variational derivative in (3.158),

$$\mu[\rho(z)] = \frac{\partial a[\rho(z)]}{\partial \rho} - 2K \frac{\partial^2 \rho}{\partial z^2}. \tag{3.159}$$

Substituting in (3.157) and neglecting terms not linear in ρ, the diffusion equation reads

$$\frac{\partial \rho}{\partial t} = M \frac{\partial^2 a[\rho(z)]}{\partial \rho^2} \frac{\partial^2 \rho}{\partial z^2} - 2KM \frac{\partial^4 \rho}{\partial z^4}, \tag{3.160}$$

where, for small density variations, the mobility is independent of density. The solution to the linearized diffusion equation is

$$\rho(z, t) - \rho_0 = \sum_B C(B, t) \cos(Bz), \tag{3.161}$$

$$C(B, t) = C(B, 0) \exp[\Lambda(B)t], \tag{3.162}$$

where B ($=2\pi/\lambda$) is a wave number, and Λ, the amplification factor, is given by

$$\Lambda(B) = -MB^2 \left[\frac{\partial^2 a}{\partial \rho^2} + 2KB^2 \right]. \tag{3.163}$$

When $a'' > 0$, the amplification factor is negative, and density fluctuations are damped. When $a'' < 0$, density fluctuations exceeding a critical wavelength grow, and spinodal decomposition occurs. This critical wavelength is given by

(3.156). The amplification factor exhibits a sharp maximum at a wave number B_{max} that can be obtained from (3.163) by differentiation,

$$B_{max} = B_c/\sqrt{2}. \tag{3.164}$$

The sharpness of the maximum at B_{max}, and the fact that the amplification factor enters the problem as an exponential, imply that one can ignore all density fluctuations with wave numbers different from B_{max}. Under these conditions, the morphology of the phase-separating system consists of a superposition of waves with fixed wavelength but random phase and amplitude (Cahn, 1965). Cahn's theory predicts an exponential growth of density fluctuations of sufficiently long wavelengths. The amplification factor and the wavelength of the most rapidly growing density fluctuation are independent of time. Since this linearized theory is valid for small deviations from the initial unstable uniform state, it describes only the initial stages of spinodal decomposition.

Figure 3.19 shows the time dependence of the wavelength of the fastest growing fluctuation during spinodal decomposition in the three-dimensional Lennard-Jones fluid (Mruzik et al., 1978). Note that B_{max} ($= 2\pi/\lambda_{max}$) is at first independent of time, in agreement with Cahn's theory, but deviates from this behavior after the initial stages of spinodal decomposition. Figure 3.20 shows spinodal decomposition in a two-dimensional Lennard-Jones fluid (Abraham et al., 1982; Koch et al., 1983). Note the highly interconnected domains of high and low densities, a morphological signature of spinodal decomposition. In this particular example, spinodal decomposition persists up to roughly 40×10^{-12} sec. Thereafter, wave breakup and domain growth take over. High- and low-density interconnected domains are also evident in Figure 3.21, a computer simulation of spinodal decomposition in the three-dimensional Lennard-Jones fluid (Abraham et al., 1976). In this figure, high-density regions ($\rho\sigma^3 > 0.35$) are shown solid, and low-density regions ($\rho\sigma^3 < 0.35$) are shown open. Figure 3.22 shows unmixing by spinodal decomposition in a B_2O_3-PbO-Al_2O_3 glass (Zarzycki and Naudin, 1969). Note the interconnected-domain morphology.

The experimental investigation of phase separation in unstable liquid mixtures was pioneered by Goldburg and co-workers (Huang et al., 1974; Chou and Goldburg, 1979, 1981), and by Knobler and co-workers (Wong and Knobler, 1978, 1981; Knobler and Wong, 1981). More recent studies include those of Guenoun et al. (1990), Chan et al. (1991), Cumming et al. (1992), Katzen and Reich (1993), and Beysens et al. (1994). Huang et al. (1974) quenched a critical binary mixture of methanol and cyclohexane 2 mK below the critical temperature and studied the resulting phase separation by light scattering. They found exponential growth of the scattered intensity over a period of 1 minute following the quench, in agreement with the Cahn-Hilliard prediction (3.161)–(3.163). Chou and Goldburg (1979) studied the growth and coalescence of domains in critically quenched binary liquid mixtures by light scattering. These

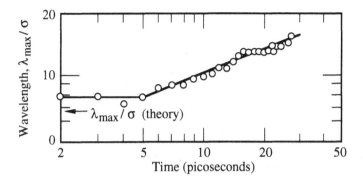

Figure 3.19. Time dependence of the wavelength of the fastest-growing density fluctuation (λ_{max}) during a molecular dynamics simulation of isothermal liquid-vapor spinodal decomposition in the three-dimensional Lennard-Jones fluid $(kT/\varepsilon = 0.8;\ \rho\sigma^3 = 0.35)$. λ_{max} was determined from the wave number corresponding to which the structure factor exhibited the fastest growth. The theoretical value was calculated using Abraham's generalized theory (Abraham, 1976) of spinodal decomposition. [Reprinted, with permission, from Mruzik et al., *J. Chem. Phys.* 69: 3462 (1978)]

processes follow the initial growth of delocalized concentration fluctuations described by the Cahn-Hilliard theory. Chou and Goldburg observed the power law $k_{max} \propto t^{-1/3}$, and, at later stages, $k_{max} \propto t^{-1}$; in other words, the characteristic size of domains, k_{max}^{-1}, grows as $t^{1/3}$ (Lifshitz and Slyozov, 1961; Binder and Stauffer, 1974; Binder, 1977) and then as t (Siggia, 1979). Here, k is the scattering wave vector, and k_{max} is the wave vector corresponding to maximum scattered intensity. It is assumed that, at short times, $k_{max} \approx B_{max}$ [see (3.164)], and at all times, k_{max}^{-1} is representative of the observable characteristic domain size. Wong and Knobler (1978) used a pressure jump technique to quench the isobutyric acid–water system into the unstable region, and studied the resulting phase separation by light scattering. They found a power-law growth of the scattered intensity even at the smallest times, in contradiction with the exponential growth predicted by the Cahn-Hilliard theory. Figure 3.23 shows the dependence of the scattered intensity on time and wave vector. Note that the wave vector corresponding to maximum intensity is not constant.

Cahn and Hilliard's theory has been refined by incorporating Brownian motion (Cook, 1970); by recasting it in a more rigorous statistical mechanical framework (Langer, 1971, 1973; Langer et al., 1975); by reformulating it in terms of generalized cluster dynamics (Binder et al., 1978); and by incorporating stress tensor and heat flux vector fluctuations (Koch et al., 1982). These treatments have extended the applicability of the original theory beyond

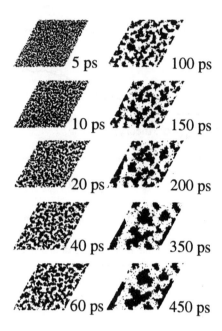

Figure 3.20. Spinodal decomposition in the two-dimensional Lennard-Jones fluid $(kT/\varepsilon = 0.45;\ \rho\sigma^3 = 0.325)$. Numbers indicate the time elapsed after the initiation of the isothermal simulation, in which atoms were placed in a triangular lattice. [Reprinted, with permission, from Koch et al., *Phys. Rev. A*. 27: 2152 (1983)]

Figure 3.21. Perspectives of isodensity surfaces of the Lennard-Jones fluid undergoing isothermal spinodal decomposition $(kT/\varepsilon = 0.8;\ \rho\sigma^3 = 0.35)$. High-density regions $(\rho\sigma^3 > 0.35)$ are shown solid; low-density regions $(\rho\sigma^3 < 0.35)$ open. (Adapted from Abraham et al., 1976)

the short-time regime (Langer, 1971), allowing the quantitative description of domain growth and coalescence. The theory of Langer (1971) predicts non-exponential kinetics, even at short times, in contrast to (3.162) and (3.163). Both nonexponential (Bortz et al., 1974; Marro et al., 1975; Rao et al., 1976;

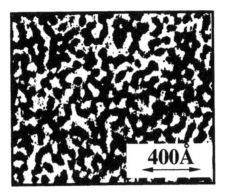

Figure 3.22. Transmission electron micrograph of a B_2O_3-PbO-Al_2O_3 glass after a quench from the melt into the unstable region. (Adapted from Zarzycki and Naudin, 1969)

Sur et al., 1977; Heerman, 1984a) and early-time-exponential kinetics (Mruzik et al., 1978) have been observed in computer simulation studies of spinodal decomposition.

The Cahn-Hilliard theory is of fundamental importance, because it explains successfully the initial mechanism of phase separation in an unstable system. It also provides a single conceptual framework within which the stability of matter to localized, high-intensity fluctuations (which are important in metastable systems) and to small-amplitude, long-ranged fluctuations (which are important in unstable systems) can be understood. Note, however, that the Cahn-Hilliard theory assumes the existence of a continuous free-energy function with negative curvature inside the coexistence region. As discussed in Chapter 2, such functions can only result from the imposition of constraints, which the Cahn-Hilliard theory does not incorporate. Accordingly, this treatment predicts a sharp spinodal singularity where a sudden change from nucleation to spinodal decomposition occurs. As discussed in Chapter 2, this picture is strictly valid only in the limit of infinitely long-ranged interactions.

3.3 THE TRANSITION FROM NUCLEATION TO SPINODAL DECOMPOSITION

A rigorous treatment of the transition between nucleation and spinodal decomposition does not yet exist. However, progress has been made, mainly by Binder and Klein, and their respective co-workers. Binder (1984, 1986) considered the

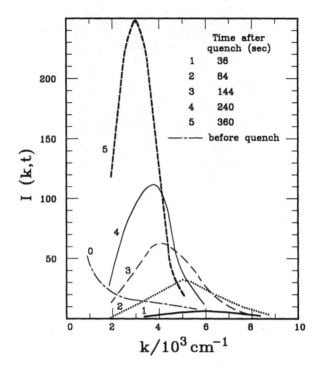

Figure 3.23. Scattered light intensity I (arbitrary units) as a function of wave vector k and time t for a critical mixture of isobutyric acid and water quenched into the unstable region. $k = (4\pi n/\lambda) \sin(\theta/2)$, where n is the refractive index, λ is the wavelength of the incident beam, and θ is the scattering angle. (Adapted from Wong and Knobler, 1978)

formation of a spherical liquid droplet in a supercooled vapor.[20] The minimum work required to form such a droplet can be written, in the usual fashion, as

$$W_{min} = 4\pi r^2 \sigma + n\left[\mu_l(P) - \mu_v(P)\right], \qquad (3.165)$$

where r is the radius of a droplet containing n molecules, and P is the bulk pressure. At small supersaturations, the chemical potentials can be expanded about their value at the equilibrium pressure. This leads to

$$W_{min} = 4\pi r^2 \sigma + \frac{(4\pi/3)r^3}{(-K_T^v)} \cdot (v_l - v_v)\,\delta v, \qquad (3.166)$$

---

[20]In Binder's treatment, dimensionality is a parameter. The present discussion is written for three-dimensional systems.

where the isothermal compressibility is that of the saturated vapor, $\delta v\,(< 0)$ denotes the difference in specific volume between the supercooled and saturated vapors, and v_l and v_v denote the specific volumes of the saturated liquid and vapor phases. Therefore the reversible work (free energy) of formation of the critical nucleus is

$$W_{\min}(r^*) = \frac{16\pi}{3} \cdot \frac{\sigma^3 \left(K_T^v\right)^2}{\left[\frac{\Delta v_{cx}^2}{v_l v_v}\right]^2} \cdot \left[\frac{\delta\,|v|}{\Delta v_{cx}}\right]^{-2}, \tag{3.167}$$

where Δv_{cx} denotes the difference in specific volume between the saturated vapor and liquid phases. Binder investigated the behavior of Equation (3.167) both close to, and far from, the critical point. To this end, he first proposed a Ginzburg criterion[21] for nucleation: mean-field theory is accurate when the mean squared density fluctuation within the nonhomogeneous interfacial region between the droplet and the supercooled vapor is much smaller than the square of the density difference between the liquid and vapor densities. This criterion takes the form

$$\lambda^3 \left(1 - T/T_c\right)^{1/2} \gg 1, \tag{3.168}$$

where λ is the range of intermolecular interactions in units of a characteristic molecular size (Binder, 1984, 1986). When the above inequality is satisfied, mean-field theory applies, and we can write

$$K_T \propto (1 - T/T_c)^{-1}\,,\, \Delta v_{cx} \propto (1 - T/T_c)^{1/2}\,,\, \sigma \propto \lambda\,(1 - T/T_c)^{3/2}\,, \tag{3.169}$$

whereupon (3.167) now reads

$$W_{\min}/kT_c \approx \lambda^3 (1 - T/T_c)^{1/2} \left[\frac{\delta\,|v|}{\Delta v_{cx}}\right]^{-2}. \tag{3.170}$$

The validity of the Ginzburg criterion was assumed in the derivation of (3.170); furthermore, $(-\delta v) \ll \Delta v_{cx}$ for small supersaturations. Hence for nucleation away from the critical point and close to coexistence, the free energy barrier associated with the formation of a critical nucleus is always large, and classical nucleation theory applies. Close to criticality, Equation (3.169) is replaced by

$$K_T \propto (1 - T/Tc)^{-\gamma}\,, \quad \Delta v_{cx} \propto (1 - T/Tc)^{\beta}\,,$$

$$\sigma \propto (1 - T/T_c)^{2v}\,, \quad 3v = \gamma + 2\beta, \tag{3.171}$$

---

[21] A criterion for determining when a system's behavior is dominated by fluctuations (density fluctuations for a pure substance, concentration fluctuations for a mixture) is called a Ginzburg criterion (Ginzburg, 1961; Anisimov et al., 1992). It takes the form of an inequality; if the inequality is satisfied, fluctuations are unimportant.

where $\gamma \neq 1$, $\beta \neq 1/2$, $\nu \neq 1/2$. Therefore, up to finite prefactors,

$$W_{\min}/kT_c \propto \left[\frac{\delta |v|}{\Delta v_{cx}} \right]^{-2}. \tag{3.172}$$

In this case, nucleation barriers are small only at small supersaturations. At large supersaturations, however, there exists a broad region where the free energy cost of forming a critical nucleus is comparable to the thermal energy. Under these conditions, it is meaningless to speak of free energy barriers, since these are crossed spontaneously. Thus, close to the critical point, there exists a broad region where a gradual transition from nucleation to spinodal decomposition occurs.

At large supersaturations, the linearizations used to arrive at Equation (3.167) are no longer valid. In this case one has

$$W_{\min}/kT_c \propto \lambda^3 (1 - T/T_c)^{1/2} \left[\frac{v_m - v_{sp}}{\Delta v_{cx}} \right]^{3/2}, \tag{3.173}$$

where v_m and v_{sp} denote the specific volumes of the vapor at bulk (metastable) conditions and at the spinodal (Binder, 1984). This equation can be used to calculate the width ($v_m - v_{sp}$) of the transition region between nucleation and spinodal decomposition. In the transition region, nucleation barriers are of the same order as the thermal energy; hence

$$\left[\frac{\delta v_{sp}}{\Delta v_{cx}} \right] \approx \left[\lambda^3 (1 - T/T_c)^{1/2} \right]^{-2/3} \tag{3.174}$$

with δv_{sp} ($= v_m - v_{sp}$) the width of the transition region. The quantity in brackets in the right-hand side defines the Ginzburg criterion; hence the transition between nucleation and spinodal decomposition becomes sharper upon moving away from the critical point. In other words, the spinodal becomes an increasingly plausible approximation the deeper one moves into the region of the phase diagram where mean-field approximations are valid (Corti and Debenedetti, 1994). Conversely, the transition between nucleation and spinodal decomposition becomes increasingly smeared as the critical point is approached. The picture that follows from Binder's analysis is illustrated in Figure 3.24 (qualitatively identical conclusions apply to a superheated liquid). It shows that classical nucleation applies at small supersaturations, and away from criticality. Also shown is the transition region between nucleation and spinodal decomposition, and its gradual broadening as the critical point is approached. Classical nucleation theory predicts that the size of the critical nucleus decreases monotonically with supersaturation, and it does not distinguish metastability from instability. This is incorrect. Accordingly, the region labeled "spinodal" nucleation denotes the high-supersaturation regime where phase changes must be initiated by a different mechanism.

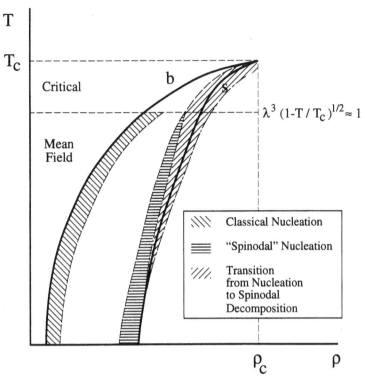

Figure 3.24. Regions where classical and "spinodal" nucleation occurs in a super-cooled vapor; transition between nucleation and spinodal decomposition; and transition between mean-field and critical nucleation regimes according to Binder's analysis. *b* and *s* are the binodal and spinodal loci. (Adapted from Binder, 1984, 1986).

Although a predictive treatment of nucleation at very high supersaturations does not yet exist, Klein and Unger (1983) and Unger and Klein (1984) have proposed a phenomenological approach to this problem. Their treatment involves writing a free energy density expression that incorporates the cost of creating nonhomogeneities through the usual square-gradient term. Evaluation of the partition function in the vicinity of the homogeneous mean-field free energy's spinodal leads to the three important results:

$$r_c \propto \lambda \, |\Delta\mu|^{-1/4} \, , \tag{3.175}$$

$$W_{min}/kT_c \propto \lambda^3 \, |\Delta\mu|^{3/4} \, , \tag{3.176}$$

$$\tau \propto |\Delta\mu|^{-23/8} \exp\left\{ C\lambda^3 \, |\Delta\mu|^{3/4} \right\} \, , \tag{3.177}$$

where r_c is the radius of the critical nucleus, $\Delta\mu$ is the difference in bulk chemical potential between the metastable and spinodal conditions at the same temperature, τ is the lifetime of the metastable state, and C is a temperature-dependent but finite constant. Equations (3.175) and (3.176) imply that the critical nucleus diverges at the spinodal, as in the Cahn-Hilliard theory, and that the free energy barrier to nucleation vanishes at the spinodal. A vanishing barrier to nucleation means that the spinodal cannot be reached because of spontaneous thermal fluctuations. Free energy barriers to nucleation increase as λ^3; hence the longer the range of the interactions, the deeper into the coexistence region the system can be quenched before thermal fluctuations nucleate the new phase.

The lifetime of the metastable state is the product of an exponential term due to the free energy barrier to nucleation times a prefactor that accounts, among other things, for the diffusive motion of molecules towards the nucleus. The free energy contribution decreases towards a constant value as the spinodal is approached, but the prefactor, and hence the lifetime, diverge at the spinodal. This is the result of cooperativity and sluggish dynamics in a system with diverging correlation length. Although spontaneous fluctuations destabilize the metastable system, the dynamics of such increasingly cooperative fluctuations becomes slow close to the spinodal. For a given quench, the lifetime of the metastable state grows with the interaction range.

By mapping their treatment onto a percolation problem[22] Unger and Klein argued that the critical nucleus is not compact but ramified (i.e., $m \propto R^{d_f}$, $d_f < 3$, where m is the mass of the nucleus, R its radius of gyration, and d_f the fractal dimension).[23] The radius of gyration is given by

$$R^2 = n^{-1} \sum_{i=1}^{n} r_i^2 \qquad (3.178)$$

where n is the number of basic units in a cluster (e.g., sites, atoms, molecules), and r_i is the distance from the center of mass to site i. Unger and Klein showed that the free energy for the percolation problem is identical to the mean-field, uniform free energy of the field-theoretic problem, and thus were able to map the latter's spinodal onto the former's percolation line. They therefore argued that a nucleating droplet of diameter ξ must have a structure similar to that of percolation clusters of connected length ξ. In particular, the fractal dimension of the two objects must be the same. As the depth of the quench increases, ramified nuclei become important because free energy barriers to their formation vanish rapidly (Klein, 1981).

Computer simulations of nucleation and growth in lattice models (Heerman and Klein, 1983a,b) are not inconsistent with the idea that ramified droplets

---

[22]The basic problem in percolation theory (see, e.g., Ziman, 1979) is to calculate the probability that a unit element of a given system (for example, sites distributed at random on a lattice) belongs to an infinite, system-spanning cluster.

[23]Unger and Klein mapped their field-theoretic model onto the correlated-site, random-bond Ising model. In this model, occupied nearest-neighbor sites on a lattice form a bond with probability p_b. As p_b is decreased, clusters become smaller and more ramified.

are important in deep quenches. Heerman and Klein studied the equivalent-neighbor model, in which an occupied site on a lattice interacts with any other occupied site within a given interaction range. Two occupied sites belong to a cluster if they are within the interaction range of each other and if there is an active bond between them, bonds being active with a probability p_b ($p_b \to 1$ for $T \to 0$; $p_b \to 0$ for $T \to \infty$ or $\rho \to 1$). Their Monte Carlo results following deep quenches inside the low-density ("vapor" side) metastable region in a simple cubic lattice are illustrated in Figure 3.25. It shows the relationship between the number of sites S in a cluster and its radius of gyration R. Points labeled 1 denote a "critical" nucleus, and successive numbers denote its evolution. In order for a nucleus to qualify as critical, Heerman and Klein required that at least one nucleus grow monotonically, and that the density increase monotonically until the stable phase is reached. The spread in the size of the nucleating cluster implies that there is no well-defined critical nucleus. There appear to be two growth mechanisms. Initially, the critical nucleus grows in mass though not in extension, the radius of gyration remaining relatively constant until an effective dimension d_+ is reached,

$$d_+ = \frac{\ln S}{\ln R} = 3 + \frac{B}{\ln R}, \qquad (3.179)$$

where B is the y-axis intercept of the straight line drawn through the asymptotic, late-time data. Note that the effective dimension equals the fractal dimension in the large-R limit. Thereafter, droplets grow as compact, three-dimensional objects. They are now, however, past the nucleating stage, and into the growth regime. Compactification of ramified droplets was also observed in Monte Carlo simulations of a deeply quenched equivalent-neighbor binary lattice system[24] by Heerman (1984b). This is illustrated in Figure 3.26, which shows that the radius of gyration of the largest droplet formed after a deep quench decreases even as its mass increases.

The formulation of a rigorous theory of nucleation for deeply quenched systems remains a major challenge. Binder's treatment is insightful but qualitative. The work of Klein and Unger suggests that ramified, noncompact nuclei become important as the depth of penetration into the coexistence region increases. Computer simulations on highly idealized models are not inconsistent with this concept. However, Unger and Klein's treatment is phenomenological: it does not explain how ramified nuclei are formed, nor does it arrive at an expression for the nucleation rate that is based on a mechanism for their formation. Incorporating existing insights from computer simulations, order-of-magnitude analysis, and scaling arguments into a predictive and microscopically based the-

---

[24]This is similar to the single-component, lattice-gas case, except that the total number of sites belonging to each species is conserved. During a simulation, the system evolves by random interchanges of unlike first neighbors.

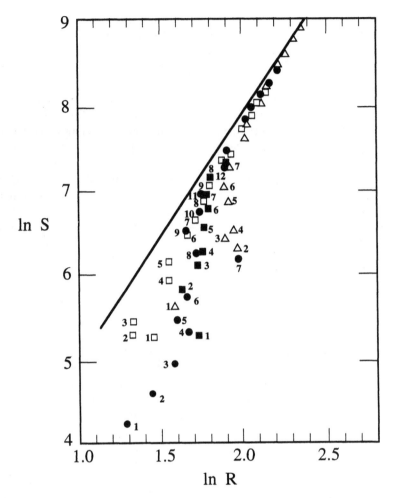

Figure 3.25. Evolution of the size S and radius of gyration R of critical clusters in Monte Carlo simulations of the equivalent-neighbor model. Numbers indicate a time sequence, and symbols denote separate quenches into the coexistence region. (Adapted from Heerman and Klein, 1983b)

ory of phase change in nearly unstable systems is an important and still open problem.

3.4 HETEROGENEOUS NUCLEATION

In most practical circumstances, suspended impurities or imperfectly wetted surfaces provide the interface on which the growth of a new phase is initiated.

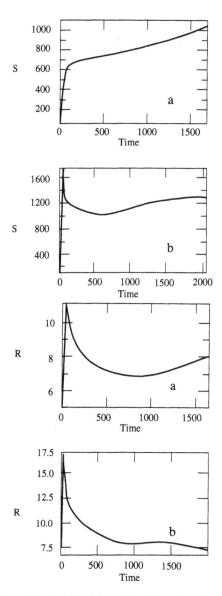

Figure 3.26. Evolution of the size S and the radius of gyration R of the largest droplet in a Monte Carlo simulation of the equivalent-neighbor binary system, following quenches into the coexistence region. Time is measured in Monte Carlo steps. a denotes a quench into the metastable region; b denotes a quench up to the mean-field spinodal. (Adapted from Heerman, 1984b)

This process, heterogeneous nucleation, limits the extent of penetration into the metastable region. The avoidance of heterogeneous nucleation is the most important criterion in the design of techniques for the study of metastable liquids.

Consider (Figure 3.27) the formation of a vapor embryo at a solid-liquid interface. The minimum work associated with this process is

$$W_{\min} = \sigma_{gl} F_{gl} + (\sigma_{gs} - \sigma_{ls}) F_{gs} + (P - P')V' + [\mu'(T, P') - \mu(T, P)]n, \quad (3.180)$$

where subscripts denote the appropriate interface, n denotes the number of molecules in the embryo, and primed and unprimed quantities denote embryo and bulk quantities, as in Section 3.1.1. At equilibrium, the following condition must be satisfied:

$$\sigma_{gl} \cos(\pi - \theta) + \sigma_{gs} = \sigma_{ls}, \quad (3.181)$$

which, together with geometry, allows the surface contribution to the minimum work to be written as

$$\sigma_{gl} F_{gl} + (\sigma_{gs} - \sigma_{ls}) F_{gs} = \pi R^2 \sigma \left[2(1 + \cos\theta) + \sin^2\theta \cos\theta \right], \quad (3.182)$$

where the unsubscripted surface tension refers hereafter to liquid-gas contact. The minimum work required to form a critical nucleus consequently reads

$$W_{\min} = \frac{16\pi\sigma^3}{3(P' - P)^2} \cdot \frac{(1 + \cos\theta)^2(2 - \cos\theta)}{4}. \quad (3.183)$$

This expression is the product of the minimum work required to form a critical nucleus homogeneously times a geometric correction factor. The latter varies monotonically from 1 to 0 as θ increases from 0 to 180°. Therefore a well-wetted surface[25] (small contact angle) has little effect on the energetics of nucleus formation [Figure 3.27(b)]. The geometric correction factor in Equation (3.183) is the ratio of the volume of the truncated nucleus to that of a sphere with the same radius. Thus the poorer the wetting, the smaller the volume of the truncated bubble, and the smaller the free energy cost associated with its formation.

From the energetics of embryo formation, and following arguments entirely analogous to those used for the homogeneous case, the following expression for the rate of heterogeneous nucleation results (Blander and Katz, 1975):

$$J = N_{\text{tot}}^{2/3} \, a \sqrt{\frac{2\sigma}{\pi m B(2 - \cos\theta)}} \cdot \exp\left[-\frac{16\pi\sigma^3}{3kT(P^e - P)^2} \cdot \psi(\theta) \right], \quad (3.184)$$

[25] It is understood that wetting refers to the liquid-solid contact. Thus in Figure 3.27(b) the wetting of the solid by the liquid is better than in Figure 3.27(a).

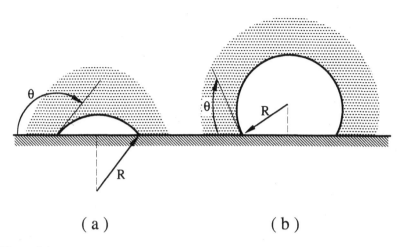

Figure 3.27. Formation of a vapor bubble on a solid surface under poor (a) and good (b) wetting conditions. Wetting refers to solid-liquid contact.

where a is the surface available for heterogeneous nucleation per unit bulk volume of liquid phase, B is defined in Equation (3.94), P^e is the equilibrium vapor pressure, and $\psi\,(\theta)$ is the geometric factor in Equation (3.183). Equation (3.184) assumes ideal-gas behavior for the vapor inside the bubble, and $\delta \approx 1$ [see Equation (3.93)]. Neglecting the trigonometric factor in the preexponential term, comparison with Equation (3.92), its homogeneous counterpart, yields

$$\frac{J_{\text{het}}}{J_{\text{hom}}} \approx \frac{a}{N_{\text{tot}}^{1/3}} \exp\left\{\frac{W_{\min}}{kT}\,[1 - \psi\,(\theta)]\right\}. \tag{3.185}$$

Although the presence of solid surfaces always lowers the free energy barrier to embryo formation, a minimum value of the contact angle is required in order for heterogeneous nucleation to become the predominant mechanism. As an example, the ratio of heterogeneous to homogeneous nucleation grows from 3.2×10^{-8} to 3.2×10^3 when θ increases from $10°$ to $70°$, for N_{tot}, W/kT, and a values of 3.3×10^{22} cm^{-3}, 100, and 1 cm^{-1}, respectively. In practice, Equation (3.185) is of limited use because of the highly irregular nature of the surfaces on which heterogeneous nucleation commonly takes place.

Apfel (1970), Winterton (1977), and Trevena (1987) have studied the role of heterogeneous nucleation in limiting the degree of penetration into the metastable region in studies of liquids under tension. The following discussion is based on their work. Consider an idealized crevice [Figure 3.28(a)], part of the irregular solid-liquid interface due to the presence of a suspended impurity or the container's walls. If the crevice is imperfectly wetted, a vapor cavity can be stabilized when the liquid is subject to positive pressure. Subsequent application of tension may result in the release of the vapor, and hence in an artificially low value of a liquid sample's cavitation threshold. In Figure 3.28(a), a is the radius of the crevice at the point of contact between the solid, liquid, and vapor phases. θ_a is the so-called advancing contact angle: once reached, the liquid-vapor interface will move towards the apex, maintaining this angle, as the applied pressure increases (Trevena, 1987). The corresponding situation when the liquid is under tension is shown in Figure 3.28(b), where θ_r is now the so-called receding contact angle. At equilibrium, with r denoting the radius of curvature of the vapor-liquid interface, σ the vapor-liquid interfacial tension, P_i the pressure inside the bubble, and P the imposed (bulk) pressure, we must have

$$P = P_i + \frac{2\sigma}{r}. \tag{3.186}$$

If the contact angle between the interface and the crevice is exactly equal to θ_a when the external pressure P is imposed, the crevice is referred to as a critical crevice (Apfel, 1970). For such a crevice [Figure 3.29(a)],

$$a_c = r \left|\cos\left(\theta_a - \beta\right)\right| = \frac{2\sigma}{(P - P_i)} \left|\cos\left(\theta_a - \beta\right)\right|, \tag{3.187}$$

while for crevices larger and smaller than the critical size [Figures 3.29(b) and 3.29(c)] we have, respectively,

$$a = a_c = \frac{2\sigma \left|\cos\left(\theta_a - \beta\right)\right|}{P - P_i},$$

$$a_0 = r \left|\cos\left(\theta - \beta\right)\right|, \theta < \theta_a. \tag{3.188}$$

A large positive pressure P is often applied prior to subjecting the sample to tension. This forces residual amounts of dissolved gas into crevices, as depicted in Figures 3.28 and 3.29. In order for the gas to be stabilized in the crevice, we must have

$$\theta_a > \beta + \pi/2 > \theta_r. \tag{3.189}$$

Consider now the situation when the pressure is reduced. If the reduction is large enough to cause θ_r to be reached or the interface to become hemispherical, the interface will recede from the crevice's apex, causing the eventual release of a bubble from the crevice. The sample's cavitation threshold will thus be determined by the nucleation of bubbles at the solid-liquid interface. For crevices larger than the critical size, this condition occurs when

$$\pi = P_i - (P - P_i) \frac{\left|\cos\left(\theta_r - \beta\right)\right|}{\left|\cos\left(\theta_a - \beta\right)\right|}; \quad (\theta_r > \beta),$$

$$\pi = P_i - (P - P_i) \left|\cos\left(\theta_a - \beta\right)\right|^{-1}; \quad (\theta_r < \beta), \tag{3.190}$$

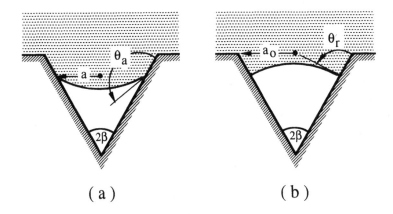

Figure 3.28. Schematic representation of a vapor bubble in a crevice under positive (a) and negative (b) bulk pressure. (Adapted from Apfel, 1970)

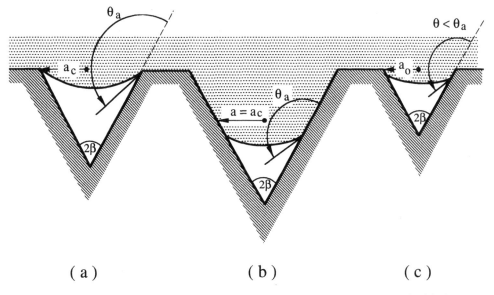

Figure 3.29. A vapor bubble trapped in a critical (a), supercritical (b), and subcritical (c) crevice. (Adapted from Apfel, 1970)

while, for subcritical crevices, bubble release occurs when

$$\pi = P_i - \frac{2\sigma \, |\cos(\theta_r - \beta)|}{a_0}; \quad (\theta_r > \beta),$$

$$\pi = P_i - \frac{2\sigma}{a_0} \qquad\qquad (\theta_r < \beta), \qquad (3.191)$$

where π is the bulk (generally negative) liquid pressure. Equations (3.190) and (3.191) imply very different behavior. For large crevices, π is unaffected by σ, whereas for subcritical crevices, the tension at which the sample will cavitate increases linearly with σ. Taking the size of the crevice to be indicative of the dimensions of suspended impurities, the above results indicate that π is insensitive to the dimensions of such impurities, provided they exceed a minimum size. For small particles, on the other hand, Equation (3.191) indicates that the maximum attainable tension will be larger the smaller the dimensions of suspended impurities remaining after the sample is pretreated. For large crevices, the presence of dissolved gas impurities, which cause P_i to increase, can be very significant if the crevice is poorly wetted (θ_a close to $\pi/2 + \beta$; the interface close to horizontal). Finally, prepressurization is only effective in allowing higher tensions to be attained if crevices are sufficiently large.

References

Abraham, F.F. 1974. *Homogeneous Nucleation Theory. The Pretransition Theory of Vapor Condensation*. Chap. 5. Academic Press: New York.

Abraham, F.F. 1975. "A Theory for the Thermodynamics and Structure of Nonuniform Systems, with Application to the Liquid-Vapor Interface and Spinodal Decomposition." *J. Chem. Phys.* 63: 157.

Abraham, F.F. 1976. "A Generalized Diffusion Equation for Nonuniform Fluid Systems, with Application to Spinodal Decomposition." *J. Chem. Phys.* 64: 2660.

Abraham, F.F. 1979. "On the Thermodynamics, Structure, and Phase Stability of the Nonuniform Fluid State." *Phys. Rep.* 53: 95.

Abraham, F.F., and G.M. Pound. 1968. "Re-examination of Homogeneous Nucleation Theory: Statistical Thermodynamics Aspects." *J. Chem. Phys.* 48: 732.

Abraham, F.F., J.K. Lee, and J.A. Baker. 1974. "Physical Cluster Free Energy from Liquid-State Perturbation Theory." *J. Chem. Phys.*. 60: 246.

Abraham, F.F., D.E. Schreiber, M.R. Mruzik, and G.M. Pound. 1976. "Phase Separation in Fluid Systems by Spinodal Decomposition: A Molecular-Dynamics Simulation." *Phys. Rev. Lett.* 36: 261.

Abraham, F.F., S.W. Koch, and R. C. Desai. 1982. "Computer-Simulation of an Unstable Two-Dimensional Fluid: Time-Dependent Morphology and Scaling." *Phys. Rev. Lett.* 49: 923.

Adams, G.W., J.L. Schmitt, and R.A. Zalabsky. 1984. "The Homogeneous Nucleation of Nonane." *J. Chem. Phys.* 81: 5074.

Akhtar, D.,V.D. Vankar, T.C. Goel, and K.C. Chopra. 1979. "Stabilization and Transformation Kinetics of the Metastable Phases of Liquid-Quenched Antimony." *J. Mater. Sci.* 14: 2422.

Andres, R.P. 1965. "Homogeneous Nucleation from the Vapor Phase." *Ind. Eng. Chem.* 57: 24.

Andres, R.P., and M. Boudart. 1965. "Time Lag in Multistate Kinetics: Nucleation." *J. Chem. Phys.* 42: 2057.

Angell, C.A. 1982. "Supercooled Water." In *Water: A Comprehensive Treatise*, F. Franks, ed., vol. 7, chap. 1. Plenum: New York.

Angell, C.A., D.R. MacFarlane, and M. Oguni. 1984. "The Kauzmann Paradox, Metastable Liquids, and Glasses." In *Dynamic Aspects of Structural Change in Liquids and Glasses*, C.A. Angell and M. Goldstein, eds. *Ann. N.Y. Acad. Sci.* 484: 241.

Anisimov, M.A., S.B. Kiselev, J.V. Sengers, and S. Tang. 1992. "Crossover Approach to Global Critical Phenomena in Fluids." *Physica A*. 188: 487.

Apfel, R.E. 1970. "The Role of Impurities in Cavitation-Threshold Determination." *J. Acoust. Soc. Amer.* 48: 1179.

Apfel, R.E. 1972. "Water Superheated to 279.5°C at Atmospheric Pressure." *Nature*. 238: 63.

Avedisian, C.T. 1985. "The Homogeneous Nucleation Limits of Liquids." *J. Phys. Chem. Ref. Data* 14: 695.

Avedisian, C.T., and I. Glassman. 1981. "Superheating and Boiling of Water in Hydrocarbons at High Pressures." *Int. J. Heat Mass Transf.* 24: 695.

Avedisian, C.T., and J.R. Sullivan. 1984. "A Generalized Corresponding States Method for Predicting the Limits of Superheat of Mixtures. Application to the Normal Alcohols." *Chem. Eng. Sci.* 39: 1033.

Becker, R., and W. Döring. 1935. "Kinetische Behandlung der Keimbildung in Übersättigten Dampfen." *Ann. Phys. (Leipzig)*. 24: 719.

Beysens, D., P. Guenoun, P. Sibille, and A. Kumar. 1994. "Dimple and Nose Coalescences in Phase-Separation Processes." *Phys. Rev. E*. 50: 1299.

Binder, K. 1977. "Theory of the Dynamics of 'Clusters' II. Critical Diffusion in Binary Systems and the Kinetics of Phase Separation." *Phys. Rev. B*. 15: 4425.

Binder, K. 1984. "Nucleation Barriers, Spinodals, and the Ginzburg Criterion." *Phys. Rev. A*. 29: 341.

Binder, K. 1986. "Decay of Metastable and Unstable States: Mechanisms, Concepts and Open Problems." *Physica A*. 140: 35.

Binder, K. 1987. "Theory of First-Order Phase Transitions." *Rep. Prog. Phys*. 50: 783.

Binder, K., and D. Stauffer. 1974. "Theory for the Slowing Down of the Relaxation and Spinodal Decomposition of Binary Mixtures." *Phys. Rev. Lett*. 33: 1006.

Binder, K., C. Billotet, and P. Mirold. 1978. "On the Theory of Spinodal Decomposition in Solid and Liquid Binary Mixtures." *Z. Phys. B*. 30: 183.

Blander, M., and J.L. Katz. 1972. "The Thermodynamics of Cluster Formation in Nucleation Theory." *J. Stat. Phys*. 4: 55.

Blander, M., and J.L. Katz. 1975. "Bubble Nucleation in Liquids." *Amer. Inst. Chem. Eng. J*. 21: 833.

Blander, M., D. Hengstenberg, and J.L. Katz. 1971. "Bubble Nucleation in n-Pentane, n-Hexane, n-Pentane + Hexadecane Mixtures, and Water." *J. Phys. Chem*. 75: 3613.

Bongiorno, V., L.E. Scriven, and H.T. Davis. 1976. "Molecular Theory of Fluid Interfaces." *J. Colloid Interf. Sci*. 57: 462.

Bortz, A.B., M.H. Kalos, J.L. Lebowitz, and M.A. Zandejas. 1974. "Time Evolution of a Quenched Binary Alloy: Computer Simulation of a Two-Dimensional Model System." *Phys. Rev. B*. 10: 535.

Bosio, L., A. Defrain, and M. Dupont. 1971. "Liste des Distances Réticulaires des Phases Cristallines γ et δ du Gallium." *J. Chim. Phys*. 68: 542.

Buckle, E.R. 1961. "Studies on the Freezing of Pure Liquids. II. The Kinetics of Homogeneous Nucleation in Supercooled Liquids." *Proc. Roy. Soc. A* 261: 189.

Buckle, E.R., and A.R. Ubbelohde. 1960. "Studies on the Freezing of Pure Liquids. I. Critical Supercooling in Molten Alkali Halides." *Proc. Roy. Soc. A*. 259: 325.

Buivid, M.G., and M.V. Sussman. 1978. "Superheated Liquids Containing Suspended Particles." *Nature*. 275: 203.

Cahn, J. W. 1959. "Free Energy of a Nonuniform System. II. Thermodynamic Basis." *J. Chem. Phys*. 30: 1121.

Cahn, J.W. 1961. "On Spinodal Decomposition." *Acta Metall*. 9: 795.

Cahn, J.W. 1962. "On Spinodal Decomposition in Cubic Crystals." *Acta Metall.* 10: 179.

Cahn, J.W. 1965. "Phase Separation by Spinodal Decomposition in Isotropic Systems." *J. Chem. Phys.* 42: 93.

Cahn, J.W., and J.E. Hilliard. 1958. "Free Energy of a Nonuniform System. I. Interfacial Free Energy." *J. Chem. Phys.* 28: 258.

Cahn, J.W., and J.E. Hilliard. 1959. "Free Energy of a Nonuniform System. III. Nucleation in a Two-Component Incompressible Fluid." *J. Chem. Phys.* 31: 688.

Cahn, J.W., and J.E. Hilliard. 1971. "Spinodal Decomposition: a Reprise." *Acta Metall.* 19: 151.

Chan, C.K., F. Perrot, and D. Beysens. 1991. "Experimental Study and Model Simulation of Spinodal Decomposition in a Binary Mixture Under Shear." *Phys. Rev. A.* 43: 1826.

Chandrasekhar, S. 1943. "Stochastic Problems in Physics and Astronomy." *Rev. Mod. Phys.* 15: 1.

Chiang, P., M.D. Donohue, and J.L. Katz. 1988. "A Kinetic Approach to Crystallization from Ionic Solution." *J. Colloid Interf. Sci.* 122: 251.

Chou, Y.C., and W.I. Goldburg. 1979. "Phase Separation and Coalescence in Critically Quenched Isobutyric Acid–Water and 2,6-Lutidine–Water Mixtures." *Phys. Rev. A.* 20: 2105.

Chou, Y.C., and W.I. Goldburg. 1981. "Angular Distribution of Light Scattered from Critically Quenched Mixtures." *Phys. Rev. A.* 23: 858.

Cohen, E.R. 1970. "The Accuracy of the Approximations in Classical Nucleation Theory." *J. Stat. Phys.* 2: 147.

Collins, F.C. 1955. "Time Lag in Spontaneous Nucleation Due to Non-Steady State Effects." *Z. Elektrochem.* 59: 404.

Cook, H.E. 1970. "Brownian Motion in Spinodal Decomposition." *Acta Metall.* 18: 297.

Coriell, S.R., S.C. Hardy, and R.F. Sekerka. 1971. "A Non-Linear Analysis of Experiments on the Morphological Stability of Ice Crystals Freezing from Aqueous Solutions." *J. Cryst. Growth.* 11: 53.

Corti, D.S., and P.G. Debenedetti. 1994. "A Computational Study of Metastability in Vapor-Liquid Equilibrium." *Chem. Eng. Sci.* 49: 2717.

Courtney, W.G. 1961. "Remarks on Homogeneous Nucleation." *J. Chem. Phys.* 35: 2249

Cumming, A., P. Wiltzius, F.S. Bates, and J.H. Rosedale. 1992. "Light Scattering Experiments on Phase-Separation Dynamics in Binary Fluid Mixtures." *Phys. Rev. A.* 45: 885.

Cwilong, B.M. 1945. "Sublimation in a Wilson Chamber." *Nature.* 155: 361.

Cwilong, B.M. 1947. "Sublimation in a Wilson Chamber." *Proc. Roy. Soc. A.* 190: 137.

Danilov, N.N., and Ye.N. Sinitsyn. 1980. "Radiation-Triggered Flashing of Superheated Binary Solutions." *Heat Transf. Sov. Res.* 12: 71.

Davis, H.T., and L.E. Scriven. 1982. "Stress and Structure in Fluid Interfaces." *Adv. Chem. Phys.* 59: 357.

DeNordwall, H.J., and L.A.K. Staveley. 1954. "Further Studies of the Supercooling of Drops of Some Molecular Liquids." *J. Chem. Soc.* p. 224.

Doyle, G.J. 1961. "Self-Nucleation in the Sulfuric Acid–Water System." *J. Chem. Phys.* 35: 795.

Dufour, L., and R. Defay. 1963. *Thermodynamics of Clouds.* pp. 179–181. Academic Press: New York.

Eberhart, J.G., W. Kremsner, and M. Blander. 1975. "Metastability Limits of Super-heated Liquids: Bubble Nucleation Temperatures of Hydrocarbons and their Mixtures." *J. Colloid Interf. Sci.* 50: 369.

Ellerby, H.M., and H. Reiss. 1992. "Toward a Molecular Theory of Vapor-Phase Nucleation. II. Fundamental Treatment of the Cluster Distribution." *J. Chem. Phys.* 97: 5766.

Ellerby, H.M., C.L. Weakliem, and H. Reiss. 1991. "Toward a Molecular Theory of Vapor-Phase Nucleation. I. Identification of the Average Embryo." *J. Chem. Phys.* 95: 9209.

Ermakov, G.V., and V.P. Skripov. 1967. "Saturation Line, Critical Parameters, and the Maximum Degree of Superheating of Perfluoro-Paraffins." *Russ. J. Phys. Chem.* 41: 39.

Farkas, L. 1927. "Keimbildungsgeschwindigkeit in Übersättigten Dämpfen" *Z. Phys. Chem. (Leipzig).* A125: 236.

Feder, J., K.C. Russell, J. Lothe, and G.M. Pound. 1966. "Homogeneous Nucleation and Growth of Droplets in Vapours." *Adv. Phys.* 15: 111.

Fisk, S., and B. Widom. 1969. "Structure and Free Energy of the Interface Between Fluid Phases in Equilibrium near the Critical Point." *J. Chem. Phys.* 50: 3219.

Flageollet-Daniel, C., J.P. Garnier, and P. Mirabel. 1983. "Microscopic Surface Tension and Binary Nucleation." *J. Chem. Phys.* 78: 2600.

Follstaedt, D.M., P.S. Peercy, and J.H. Perepezko. 1986. "Phase Selection During Pulsed Laser Annealing of Manganese." *Appl. Phys. Lett.* 48: 338.

Forest, T.W., and C.A. Ward. 1977. "Effect of a Dissolved Gas on the Homogeneous Nucleation Pressure of a Liquid." *J. Chem. Phys.* 66: 2322.

Forest, T.W., and C.A. Ward. 1978. "Homogeneous Nucleation of Bubbles in Solutions at Pressures Above the Vapor Pressure of the Pure Liquid." *J. Chem. Phys.* 69: 2221.

Franks, F. 1982. "The Properties of Aqueous Solutions at Subzero Temperature." In *Water: A Comprehensive Treatise*, F. Franks, ed., vol. 7, chap. 3. Plenum: New York.

Frenkel, J. 1955. *Kinetic Theory of Liquids*. Chap 7. Dover: New York.

Friedlander, S.K. 1977. *Smoke, Dust, and Haze: Fundamentals of Aerosol Behavior*. Chaps. 8 and 9. Wiley: New York.

Frisch, H.L., and C.C. Carlier. 1971. "Time Lag in Nucleation." *J. Chem. Phys.* 54: 4326.

Gibbs, J.W. 1875–76 and 1877–78. "On the Equilibrium of Heterogeneous Substances." *Trans. Conn. Acad.* III: 108 (Oct. 1875–May 1876) and III: 343 (May 1877–July, 1878).

Gibbs, J.W. 1961. *The Scientific Papers of J. Willard Gibbs, Ph.D., LL.D. I. Thermodynamics*. pp. 219–331. Dover: New York.

Ginzburg, V.L. 1961. "Some Remarks on Phase Transitions of the Second Kind and the Microscopic Theory of Ferroelectric Materials." *Sov. Phys. Sol. State.* 2: 1824.

Girshick, S.L. 1991. "Comment of 'Self-Consistency Correction to Homogeneous Nucleation Theory.'" *J. Chem. Phys.* 94: 826.

Girshick, S.L., and C.-P. Chiu. 1990. "Kinetic Nucleation Theory: A New Expression of the Rate of Homogeneous Nucleation From an Ideal Supersaturated Vapor." *J. Chem. Phys.* 93: 1273.

Griffiths, R.B., and J.C. Wheeler. 1970. "Critical Points in Multicomponent Systems." *Phys. Rev. A.* 2: 1047.

Guenoun, P., D. Beysens, and M. Robert. 1990. "Dynamics of Wetting and Phase Separation." *Phys. Rev. Lett.* 65: 2406.

Harrowell, P., and D.W. Oxtoby. 1984. "A Molecular Theory of Crystal Nucleation from the Melt." *J. Chem. Phys.* 80: 1639.

Haymet, A.D.J., and D. W. Oxtoby. 1981. "A Molecular Theory of the Solid-Liquid Interface." *J. Chem. Phys.* 74: 2559.

Heerman, D.W. 1984a. "Test of the Validity of the Classical Theory of Spinodal Decomposition." *Phys. Rev. Lett.* 52: 1126.

Heerman, D.W. 1984b. "Dynamical Spinodal: the Transition Between Nucleation and Spinodal Decomposition." *Z. Phys. B. Cond. Matt.* 55: 309.

Heerman, D.W., and W. Klein. 1983a. "Percolation and Droplets in a Medium-Range Three-Dimensional Ising Model." *Phys. Rev. B.* 27: 1732.

Heerman, D.W., and W. Klein. 1983b. "Nucleation and Growth of Nonclassical Droplets." *Phys. Rev. Lett.* 50: 1062.

Heist, R.H., and H. He. 1994. "Review of Vapor to Liquid Homogeneous Nucleation Experiments from 1968 to 1992." *J. Phys. Chem. Ref. Data.* 23: 781.

Hill, T.L. 1956. *Statistical Mechanics. Principles and Selected Applications.* Ch. 3. McGraw-Hill: New York.

Holden, B.S., and J.L. Katz. 1978. "The Homogeneous Nucleation of Bubbles in Superheated Binary Liquid Mixtures." *Amer. Inst. Chem. Eng. J.* 24: 260.

Huang, J.S., W.I. Goldburg, and A.W. Bjerkaas. 1974. "Study of Phase Separation in a Critical Binary Liquid Mixture: Spinodal Decomposition." *Phys. Rev. Lett.* 32: 921.

Hung, C.-H., M.J. Krasnopoler, and J.L. Katz. 1989. "Condensation of a Supersaturated Vapor. VIII. The Homogeneous Nucleation of n-Nonane." *J. Chem. Phys.* 90: 1856.

Jackson, K.A. 1965. "Nucleation from the Melt." *Ind. Eng. Chem.* 57: 28.

Kacker, A., and R. Heist. 1985. "Homogeneous Nucleation Rate Measurements. I. Ethanol, n-Propanol, and i-Propanol." *J. Chem. Phys.* 82: 2734.

Kagan, Yu. 1960. "The Kinetics of Boiling of a Pure Liquid." *Russ. J. Phys. Chem.* 34: 42.

Katz, J.L., and M. Blander. 1973. "Condensation and Boiling: Corrections to Homogeneous Nucleation Theory for Nonideal Gases." *J. Colloid Interf. Sci.* 42: 496.

Katz, J.L., and M.D. Donohue. 1979. "A Kinetic Approach to Homogeneous Nucleation Theory." *Adv. Chem. Phys.* 40: 137.

Katz, J.L., and H. Wiederisch. 1977. "Nucleation Theory without Maxwell Demons." *J. Colloid Interf. Sci.* 61: 351.

Katzen, D., and S. Reich. 1993. "Image Analysis of Phase Separation in Polymer Blends." *Europhys. Lett.* 21: 55.

Klein, W. 1981. "Percolation, Droplet Models, and Spinodal Points." *Phys. Rev. Lett.* 47: 1569.

Klein, W., and C. Unger. 1983. "Pseudospinodals, Spinodals, and Nucleation." *Phys. Rev. B.* 28: 445.

Knobler, C.M., and N.-C. Wong. 1981. "Light Scattering Studies of Phase Separation in Isobutyric Acid and Water Mixtures. 2. Test of Scaling." *J. Phys. Chem.* 85: 1972.

Koch, S.W., R.C. Desai, and F.F. Abraham. 1982. "Spinodal Decomposition of a One-Component Fluid: A Hydrodynamic Fluctuation Theory and Comparison with Computer Simulation." *Phys. Rev. A.* 26: 1015.

Koch, S.W., R.C. Desai, and F.F. Abraham. 1983. "Dynamics of Phase Separation in

Two-Dimensional Fluids: Spinodal Decomposition." *Phys. Rev. A.* 27: 2152.

Kreidenweis, S.M., and J.H. Seinfeld. 1988. "Nucleation of Sulfuric Acid–Water and Methanesulfonic Acid–Water Solution Particles: Implication for the Atmospheric Chemistry of Organosulfur Species." *Atmosph. Environ.* 22: 283.

Kwauk, X., and P.G. Debenedetti. 1993. "Mathematical Modeling of Aerosol Formation by Rapid Expansion of Supercritical Solutions in a Converging Nozzle." *J. Aerosol Sci.* 24: 445.

Laaksonen, A., and D.W. Oxtoby. 1995. "Gas-Liquid Nucleation of Nonideal Binary Mixtures. I. A Density Functional Study." *J. Chem. Phys.* 102: 5803.

Laaksonen, A., V. Talanquer, and D. Oxtoby. 1995. "Nucleation: Measurements, Theory, and Atmospheric Applications." *Annu. Rev. Phys. Chem.* 46: 489.

Langer, J.S. 1971. "Theory of Spinodal Decomposition in Alloys." *Ann. Phys.* 65: 53.

Langer, J.S. 1973. "Statistical Methods in the Theory of Spinodal Decomposition." *Acta Metall.* 21: 1649.

Langer, J.S., M. Bar-on, and H.D. Miller. 1975. "New Computational Method in the Theory of Spinodal Decomposition." *Phys. Rev. A.* 11: 1417.

Lee, J.K., J.A. Barker, and F.F. Abraham. 1973. "Theory and Monte Carlo Simulation of Physical Clusters in the Imperfect Vapor." *J. Chem. Phys.* 58: 3166.

Lifshitz, E.M., and L.P. Pitaevskii. 1981. *Physical Kinetics.* Vol. 10 of Course of Theoretical Physics, by L.D. Landau and E.M. Lifshitz. Chap. 12. Pergamon Press: Oxford.

Lifshitz, E.M., and V.V. Slyozov. 1961. "The Kinetics of Precipitation from Supersaturated Solid Solutions." *J. Phys. Chem. Sol.* 19: 35.

Lothe, J., and G.M. Pound. 1962. "Reconsiderations of Nucleation Theory." *J. Chem. Phys.* 36: 2080.

Lothe, J., and G.M. Pound. 1966. "On the Statistical Mechanics of Nucleation Theory." *J. Chem. Phys.* 45: 630.

Lothe, J., and G. M. Pound. 1968. "Concentration of Clusters and the Classical Phase Integral." *J. Chem. Phys.* 48: 1849.

Marro, J., A.B. Bortz, M.H. Kalos, and J.L. Lebowitz. 1975. "Time Evolution of a Quenched Binary Alloy. II. Computer Simulation of a Three-Dimensional Model System." *Phys. Rev. B.* 12: 2000.

McCoy, B.F., and H.T. Davis. 1979. "Free-Energy of Inhomogeneous Fluids." *Phys. Rev. A.* 20: 1201.

McDonald, J.E. 1962."Homogeneous Nucleation of Vapor Condensation. I. Thermodynamic Aspects." *Amer. J. Phys.* 30: 870.

McDonald, J.E. 1963. "Homogeneous Nucleation of Vapor Condensation. II. Kinetic Aspects." *Amer. J. Phys.* 31: 31.

McGraw, R. 1981. "A Corresponding States Correlation of the Homogeneous Nucleation Thresholds of Supercooled Vapors." *J. Chem. Phys.* 75: 5514.

Miller, R.C., R.J. Anderson, J.L. Kassner, and D.E. Hagen. 1983. "Homogeneous Nucleation Rate Measurements for Water Over a Wide Range of Temperature and Nucleation Rate." *J. Chem. Phys.* 78: 3204.

Mirabel, P. and J.L. Clavelin. 1978. "On the Limiting Behavior of Binary Homogeneous Nucleation Theory." *J. Aerosol Sci.* 9: 219.

Mirabel, P., and J.L. Katz. 1974. "Binary Homogeneous Nucleation as a Mechanism for the Formation of Aerosols." *J. Chem. Phys.* 60: 1139.

Mori, Y., K. Hijikata, and T. Nagatani. 1976. "Effect of Dissolved Gas on Bubble

Nucleation." *Int. J. Heat Mass Transf.* 19: 1153.

Mruzik, M.R., F.F. Abraham, and G.M. Pound. 1978. "Phase Separation in Fluid Systems by Spinodal Decomposition. II. A Molecular Dynamics Computer Simulation." *J. Chem. Phys.* 69: 3462.

Myerson, A.S., and A.F. Izmailov. 1993. "The Structure of Supersaturated Solutions." In *Handbook of Crystal Growth*, D.T.J. Hurle, ed., vol. 1, chap. 5. Elsevier: Amsterdam.

Narsimhan, G., and E. Ruckenstein. 1989. "A New Approach for the Prediction of the Rate of Nucleation in Liquids." *J. Colloid Interf. Sci.* 128: 549.

Nowakowski, B., and E. Ruckenstein. 1990. "Rate of Nucleation in Liquids for FCC and Icosahedral Clusters." *J. Colloid Interf. Sci.* 139: 500.

Nowakowski, B., and E. Ruckenstein. 1991a. "A Kinetic Approach to the Theory of Nucleation in Gases." *J. Chem. Phys.* 94: 1397.

Nowakowski, B., and E. Ruckenstein. 1991b."Homogeneous Nucleation in Gases: A Three-Dimensional Fokker-Planck Equation for Evaporation from Clusters." *J. Chem. Phys.* 94: 8487.

Oxtoby, D.W. 1984. "Nucleation of Crystals from the Melt." In *Dynamic Aspects of Structural Change in Liquids and Glasses*, C.A. Angell and M. Goldstein, eds. *Ann. N.Y. Acad. Sci.* 484: 26.

Oxtoby, D.W. 1988. "Nucleation of Crystals from the Melt." *Adv. Chem. Phys.* 70: 263.

Oxtoby, D.W. 1991. "Crystallization of Liquids: A Density-Functional Approach." In *Liquids, Freezing, and Glass Transition*, J.P. Hansen, D. Levesque, and J. Zinn-Justin, eds., Chap. 3. NATO Advanced Study Institute Session LI. North-Holland: Amsterdam.

Oxtoby, D.W. 1992a. "Nucleation." In *Fundamentals of Inhomogeneous Fluids*, D. Henderson, ed., chap. 10. Marcel Dekker: New York.

Oxtoby, D.W. 1992b. "Homogeneous Nucleation: Theory and Experiment." *J. Phys. Cond. Matt.* 4: 7627.

Oxtoby, D.W., and R. Evans. 1988."Nonclassical Nucleation Theory for the Gas-Liquid Transition." *J. Chem. Phys.* 89: 7521.

Oxtoby, D.W., and A.D.J. Haymet. 1982. "A Molecular Theory of the Solid-Liquid Interface. II. Study of bcc Crystal-Melt Interfaces." *J. Chem. Phys.* 76: 6262.

Patrick-Yeboah, J.R., and R.C. Reid. 1981. "Superheat-Limit Temperatures of Polar Liquids." *Ind. Eng. Chem. Fundam.* 20: 315.

Pavlov, P.A., and V.P. Skripov. 1970. "Kinetics of Spontaneous Nucleation in Strongly Heated Liquids." *High Temp.* 8: 540.

Peng, D.-Y., and D.B. Robinson. 1976. "A New Two-Constant Equation of State." *Ind. Eng. Chem. Fundam.* 15: 59.

Perepezko, J.H. 1984. "Nucleation in Undercooled Liquids." *Mater. Sci. Eng.* 65: 125.

Perepezko, J.H., and I.E. Anderson. 1980. "Metastable Phase Formation in Undercooled Liquids." In *Synthesis and Properties of Metastable Phases*, T.J. Rowland and E.S. Machlin, eds., p. 13. Metallurgical Society of American Institute of Mechanical Engineers: Warrendale, Pa.

Peters, F., and B. Paikert. 1989. "Experimental Results on the Rate of Nucleation in Supersaturated n-Propanol, Ethanol, and Methanol Vapors." *J. Chem. Phys.* 91: 5672.

Porteous, W., and M. Blander. 1975. "Limits of Superheat and Explosive Boiling of Light Hydrocarbons, Halocarbons, and Hydrocarbon Mixtures." *Amer. Inst. Chem. Eng. J.* 21: 560.

Rao, M., M.H. Kalos, J.L. Lebowitz, and J. Marro. 1976. "Time Evolution of a Quenched Binary Alloy. III. Computer Simulation of a Two-Dimensional Model System." *Phys. Rev. B*. 13: 4328.

Ramakrishnan, T.V., and M. Yussouf. 1979. "First-Principles Order-Parameter Theory of Freezing." *Phys. Rev. B*. 19: 2775.

Rasmussen, D.H. 1986a. "Clustering and Nucleation in Supersaturated Regular Solutions." *J. Chem. Phys*. 85: 2277.

Rasmussen, D.H. 1986b. "Dynamic Surface Tension and Classical Nucleation Theory." *J. Chem. Phys*. 85: 2272.

Rasmussen, D.H., and A.P. MacKenzie. 1972. In *Water Structure and the Water Polymer Interface*, H.H.G. Jellinek, ed., p. 126. Plenum: New York.

Rasmussen, D.H., A.P. MacKenzie, C.A. Angell, and J.C. Tucker. 1973. "Anomalous Heat Capacities of Supercooled Water and Heavy Water." *Science*. 181: 342.

Redlich, O., and J.N.S. Kwong. 1949. "On the Thermodynamics of Solutions. V: An Equation of State. Fugacities of Gaseous Solutions." *Chem. Rev*. 44: 234.

Reid, R.C. 1976. "Superheated Liquids." *Amer. Sci*. 64: 146.

Reid, R.C., J.M. Prausnitz, and B.E. Poling. 1987. *The Properties of Gases and Liquids*. 4th ed. Appendix A. McGraw-Hill: New York.

Reiss, H. 1950. "The Kinetics of Phase Transitions in Binary Systems." *J. Chem. Phys*. 18: 840.

Reiss, H. 1970. "The Treatment of Droplike Clusters by Means of the Classical Phase Integral in Nucleation Theory." *J. Stat. Phys*. 2: 83.

Reiss, H., and J.L. Katz. 1967. "Resolution of the Translation-Rotation Paradox in the Theory of Irreversible Condensation." *J. Chem. Phys*. 46: 2496.

Reiss, H., and M. Shugard. 1976. "On the Composition of Nuclei in Binary Systems." *J. Chem. Phys*. 65: 5280.

Reiss, H., J.L. Katz, and E.R. Cohen. 1968. "Translation-Rotation Paradox in the Theory of Nucleation." *J. Chem. Phys*. 48: 5553.

Reiss, H., A. Tabazadeh, and J. Talbot. 1990. "Molecular Theory of Vapor Phase Nucleation: The Physically Consistent Cluster." *J. Chem. Phys*. 92: 1266.

Renner, T.A., G.H. Kucera, and M. Blander. 1975. "Explosive Boiling in Light Hydrocarbons and Their Mixtures." *J. Colloid Interf. Sci*. 52: 391.

Rowlinson, J.S. 1979. "Translation of J.D. van der Waals' 'The Thermodynamic Theory of Capillarity Under the Hypothesis of a Continuous Variation in Density.'" *J. Stat. Phys*. 20: 197.

Rowlinson, J.S., and B. Widom. 1982. *Molecular Theory of Capillarity*. Chaps. 1–5. Oxford University Press: Oxford.

Ruckenstein, E., and B. Nowakowski. 1990. "A Kinetic Theory of Nucleation in Liquids." *J. Colloid Interf. Sci*. 137: 583.

Ruckenstein, E., and B. Nowakowski. 1991. "A Unidimensional Fokker-Planck Approximation in the Treatment of Nucleation in Gases." *Langmuir*. 7: 1537.

Schmitt, J.L. 1981. "Precision Expansion Chamber for Homogeneous Nucleation Studies." *Rev. Sci. Instrum*. 52: 1749.

Schmitt, J.L., G.W. Adams, and R.A. Zalabsky. 1982. "Homogeneous Nucleation of Ethanol." *J. Chem. Phys*. 77: 2089.

Seinfeld, J.H. 1986. *Atmospheric Chemistry and Physics of Air Pollution*. Chap. 9. Wiley: New York.

Shi, G., and J.H. Seinfeld. 1990. "Kinetics of Binary Nucleation: Multiple Paths and the Approach to Stationarity." *J. Chem. Phys.* 93: 9033.

Shi, G., and J.H. Seinfeld. 1991. "Transient Kinetics of Nucleation and Crystallization: Part I. Nucleation." *J. Mater. Res.* 6: 2091.

Shi, G., J.H. Seinfeld, and K. Okuyama. 1990. "Transient Kinetics of Nucleation." *Phys. Rev. A.* 41: 2101.

Shizgal, B., and J.C. Barrett. 1989. "Time Dependent Nucleation." *J. Chem. Phys.* 91: 6505.

Shneidman, V., and M.C. Weinberg. 1992a. "Induction Time in Transient Nucleation Theory." *J. Chem. Phys.* 97: 3621.

Shneidman, V., and M.C. Weinberg. 1992b. "Transient Induction Time from the Birth-Death Equations." *J. Chem. Phys.* 97: 3629.

Shugard, W.J., R.H. Heist, and H. Reiss. 1974. "Theory of Vapor Phase Nucleation in Binary Mixtures of Water and Sulfuric Acid." *J. Chem. Phys.* 61: 5298.

Siggia, E.D. 1979. "Late Stages of Spinodal Decomposition in Binary Mixtures." *Phys. Rev. A.* 20: 595.

Sinitsyn, E.N., N.N. Danilov, and V.P. Skripov. 1971. "Nucleation in Superheated Liquid Cyclohexane-Benzene Mixtures." *Teplofizika.* 425: 22.

Skripov, V.P. 1974. *Metastable Liquids.* Israel Program for Scientific Translations: Jerusalem; Wiley: New York.

Skripov, V.P., and G.V. Ermakov. 1964. "Pressure Dependence of the Limiting Super-heating of a Liquid." *Russ. J. Phys. Chem.* 38: 208.

Skripov, V.P., and V.I. Kukushkin. 1961. "Apparatus for Observing the Superheating Limits of Liquids." *Russ. J. Phys. Chem.* 35: 1393.

Skripov, V.P., V.G. Baidakov, S.P. Protsenko, and V.V. Maltsev. 1973. "Metastable States of Liquid Nitrogen and Limit of Thermodynamic Stability." Translated from *Teplofiz. Vys. Temp.* 11: 682.

Skripov, V.P., E.N. Sinitsyn, P.A. Pavlov, G.V. Ermakov, G.N. Muratov, N.V. Bulanov, and V.G. Baidakov. 1988. *Thermophysical Properties of Liquids in the Metastable (Superheated) State.* Gordon and Breach: New York.

Sokolnikoff, I.S., and R.M. Redheffer. 1966. *Mathematics of Physics and Modern Engineering.* 2nd ed. Chap. 5. McGraw-Hill: New York.

Springer, G.S. 1978. "Homogeneous Nucleation." *Adv. Heat Transf.* 14: 281.

Stauffer, D. 1976. "Kinetic Theory of Two-Component ('Heteromolecular') Nucleation and Condensation." *J. Aerosol. Sci.* 7: 319.

Stillinger, F.H. 1968. "Comment on the Translation-Rotation Paradox in the Theory of Irreversible Condensation." *J. Chem. Phys.* 48: 1430.

Strey, R., P.E. Wagner, and T. Schmeling. 1986. "Homogeneous Nucleation Rates for n-Alcohol Vapors Measured in a Two-Piston Expansion Chamber." *J. Chem. Phys.* 84: 2325.

Sur, A., J.L. Lebowitz, J. Marro, and M.H. Kalos. 1977. "Time Evolution of a Quenched Binary Alloy. IV. Computer Simulation of a Three-Dimensional Model System." *Phys. Rev. B.* 15: 3014.

Talanquer, V., and D. W. Oxtoby. 1993. "Nucleation in Dipolar Fluids: Stockmayer Fluids." *J. Chem. Phys.* 99: 4670.

Talanquer, V., and D.W. Oxtoby. 1994. "Dynamical Density Functional Theory of Gas-Liquid Nucleation." *J. Chem. Phys.* 100: 5190.

Talanquer, V., and D.W. Oxtoby. 1995. "Nucleation of Bubbles in Binary Fluids." *J. Chem. Phys.* 102: 2156.

Thomas, D.G., and L.A.K. Staveley. 1952. "A Study of the Supercooling of Drops of Some Molecular Liquids." *J. Chem. Soc.* p. 4569.

Tiller, W.A. 1991. *The Science of Crystallization. Microscopic Interfacial Phenomena.* Chap. 8. Cambridge University Press: Cambridge.

Trevena, D.H. 1987. *Cavitation and Tension in Liquids.* Chap. 2. Adam Hilger: Bristol.

Turnbull, D. 1969. "Under What Conditions Can a Glass be Formed?" *Contemp. Phys.* 10: 473.

Turnbull, D., and R.E. Cech. 1950. "Microscopic Observation of Solidification of Small Metal Droplets." *J. Appl. Phys.* 21: 804.

Turnbull, D., and J.C. Fisher. 1949. "Rate of Nucleation in Condensed Systems." *J. Chem. Phys.* 17: 71.

Unger, C., and W. Klein. 1984. "Nucleation Theory near the Classical Spinodal." *Phys. Rev. B.* 29: 2698.

van der Waals, J.D. 1893. "Thermodynamische Theorie de Capillariteit in de Onderstelling van Continue Dichtheidsverandering." *Verhand. Konink. Akad. Wetensch. Amsterdam. Sect. 1.* Deel I: 2.

Viisanen, Y., R. Strey, and H. Reiss. 1993. "Homogeneous Nucleation Rates for Water." *J. Chem. Phys.* 99: 4680.

Volmer, M., and A. Weber. 1926. "Keimbildung in Übersättigten Gebilden." *Z. Phys. Chem.* 119: 277.

Wagner, P.E., and R. Strey. 1984. "Measurements of Homogeneous Nucleation Rates for n-Nonane Vapor Using a Two-Piston Expansion Chamber." *J. Chem. Phys.* 80: 5266.

Wakeshima, H. 1954. "Time Lag in Self Nucleation." *J. Chem. Phys.* 22: 1614.

Weakliem, C.L., and H. Reiss. 1993. "Toward a Molecular Theory of Vapor-Phase Nucleation. III. Thermodynamic Properties of Argon Clusters from Monte Carlo Simulations and a Modified Liquid Drop Theory." *J. Chem. Phys.* 99: 5374.

Weakliem, C.L., and H. Reiss. 1994. "Toward a Molecular Theory of Vapor-Phase Nucleation. IV. Rate Theory Using the Modified Liquid Drop Model." *J. Chem. Phys.* 101: 2398.

Wiederisch, H., and J.L. Katz. 1979. "The Nucleation of Voids and Other Irradiation-Produced Defect Aggregates." *Adv. Colloid Interf. Sci.* 10: 33.

Wilemski, G. 1975. "Binary Nucleation. I. Theory Applied to Water-Ethanol Vapors." *J. Chem. Phys.* 62: 3763.

Wilemski, G. 1984. "Composition of the Critical Nucleus in Multicomponent Vapor Nucleation." *J. Chem. Phys.* 80: 1370.

Wilemski, G. 1987. "Revised Classical Binary Nucleation Theory for Aqueous Alcohol and Acetone Vapors." *J. Phys. Chem.* 91: 2492.

Wilemski, G. 1995. "The Kelvin Equation and Self-Consistent Nucleation Theory." *J. Chem. Phys.* 103: 1119.

Winterton, R.H.S. 1977. "Nucleation of Boiling and Cavitation." *J. Phys. D. Appl. Phys.* 10: 2041.

Wong, N.-C., and C.M. Knobler. 1978. "Light Scattering Study of Phase Separation in Isbutyric Acid + Water Mixtures." *J. Chem. Phys.* 69: 725.

Wong, N.-C., and C.M. Knobler. 1981. "Light-Scattering Studies of Phase Separation in

Isobutyric Acid + Water Mixtures: Hydrodynamic Effects." *Phys. Rev. A*. 24: 3205.

Woodruff, D.P. 1973. *The Solid-Liquid Interface*. Chap. 2. Cambridge University Press: Cambridge.

Wu, D.T. 1992. "The Time Lag in Nucleation Theory." *J. Chem. Phys.* 97: 2644.

Wyslouzil, B.E., J.H. Seinfeld, R.C. Flagan, and K. Okuyama. 1991a. "Binary Nucleation in Acid–Water Systems. I. Methanesulfonic Acid–Water." *J. Chem. Phys.* 94: 6827.

Wyslouzil, B.E., J.H. Seinfeld, R.C. Flagan, and K. Okuyama. 1991b. "Binary Nucleation in Acid–Water Systems. II. Sulfuric Acid–Water and a Comparison with Methanesulfonic Acid–Water." *J. Chem. Phys.* 94: 6827.

Yoon, W., J.S. Paik, D. LaCourt, and J.H. Perepezko. 1986. "The Effect of Pressure of Phase Selection during Nucleation of Undercooled Bismuth." *J. Chem. Phys.* 60: 3489.

Zarzycki, J., and F. Naudin. 1969. "Spinodal Decomposition in the B_2O_3-PbO-Al_2O_3 System." *J. Non-Cryst. Sol.* 1: 215.

Zeldovich, Ya.B. 1942. "On the Theory of New Phase Formation: Cavitation." *Zhur. Eksper. Teor. Fiz.* 12: 525. Translated in *Selected Works of Yakov Borisovich Zeldovich. Volume I. Chemical Physics and Hydrodynamics*, J.P. Ostriker, G.I. Barenblatt, and R.A. Sunayev, eds. p. 120. Princeton University Press: Princeton (1992).

Zeng, X.C., and D.W. Oxtoby. 1991a. "Binary Homogeneous Nucleation Theory for the Gas-Liquid Transition: A Nonclassical Approach." *J. Chem. Phys.* 95: 5940.

Zeng, X.C., and D.W. Oxtoby. 1991b. "Gas-Liquid Nucleation in Lennard-Jones Fluids." *J. Chem. Phys.* 94: 4472.

Ziman, J.M. 1979. *Models of Disorder. The Theoretical Physics of Homogeneously Disordered Systems*. Chap. 9. Cambridge University Press: Cambridge.

Note Added in Proof

The nucleation theorem is an exact relation between the derivative of the work of formation of the critical nucleus with respect to the chemical potential of a component in the bulk metastable phase, and the size and composition of the critical nucleus. This derivative can be related to experimentally measurable quantities. Hence, it allows the determination of the size and composition of the critical nucleus from experiments. Thorough discussions are presented by Oxtoby and Kashchiev, *J. Chem. Phys.* 100, 7665, 1994, and Oxtoby and Laaksonen, *J. Chem. Phys.* 102, 6846, 1995.

Recent applications of density functional theory can be found in Shen and Oxtoby, *J. Chem. Phys.* 104, 4233, 1996, and Talanquer and Oxtoby, *J. Chem. Phys.* 104, 1993, 1996.

A powerful biased Monte Carlo simulation method has been applied by Frenkel and coworkers to study free energy barriers, critical nucleus size, and nucleation rates in moderately supercooled liquids. See ten Wolde, Ruiz-Montero, and Frenkel, *Phys. Rev. Lett.* 75, 2714, 1995, and *J. Chem. Phys.* 104, 9932, 1996.

4

Supercooled Liquids

Supercooled liquids are metastable with respect to the crystalline state. However, crystallization is not the only possible outcome of supercooling. At sufficiently low temperatures, the characteristic time for structural relaxation becomes comparable to the duration of a macroscopic experiment (say, 10^2 sec). On this and shorter time scales, the supercooled liquid is structurally arrested. It has a modulus, but lacks long-range order: it becomes a glass. This transformation can only be observed if nucleation is suppressed during cooling, for example, by performing a sufficiently rapid quench. Therefore, different cooling strategies can lead to the formation of an ordered or a disordered solid phase.[1] The former process is discontinuous, because it is accompanied by the sudden evolution of heat and by a volume discontinuity. In contrast, glass formation (vitrification) is continuous, because it is not accompanied by heat effects or volume discontinuities. This cooling-strategy-dependent possibility for a supercooled liquid either to crystallize or to fall out of equilibrium is the distinguishing feature of low-temperature liquid metastability. Of the two phenomena, crystallization and glass formation, only the latter involves new concepts that are not discussed elsewhere in this book. The emphasis in this chapter, accordingly, is on the interplay of kinetics and thermodynamics that underlies the glass transition. No attempt is made here to cover solid-state properties systematically, be they those of crystals or glasses.

The competition between glass formation and crystallization is discussed in Section 4.1. Section 4.2 describes the elementary phenomenology of the glass transition. Theories of the glass transition fall into two categories: thermodynamic and kinetic. They are discussed in Sections 4.3 and 4.4, respectively. According to the thermodynamic viewpoint, the experimentally observable glass transition is but the kinetically controlled manifestation of an underlying singularity. This singularity can be described in purely static terms. In the nonthermodynamic approach, on the other hand, structural arrest is viewed as a dynamic singularity in the relaxation of the supercooled liquid, which is not accompanied by thermodynamic singularities. The classification of liquids into strong and fragile, discussed in Section 4.5, provides a useful framework

---

[1]The term solid is used here in a mechanical sense, to denote a body for which stress and deformation are proportional to each other over some range of deformations.

for relating complex relaxation behavior to molecular architecture. Section 4.6 addresses the interesting properties of supercooled water, and their relationship to this substance's different glassy phases. Finally, Section 4.7 discusses those features of computer simulations that are relevant to the study of supercooled liquids.

4.1 CRYSTALLIZATION AND VITRIFICATION

The phenomena that occur during supercooling are the result of the interplay between two characteristic times τ_1 and τ_2 (Angell, 1988a). τ_1 is the time required for a given volume fraction of a sample to crystallize. τ_2 is a structural relaxation time. The competition between crystallization and vitrification follows from the fact that τ_2 increases upon supercooling, whereas τ_1 has a non-monotonic temperature dependence. To explain the latter, it suffices to consider the ideal case of noninteracting spherical crystallites formed by homogeneous nucleation, and which grow with a constant linear velocity u. The volume fraction ϕ that crystallizes after a time τ_1 is then given by

$$\phi(\tau_1) = \int_0^{\tau_1} J \frac{4\pi}{3} \left[\int_0^{t'} u \, dt'' \right]^3 dt', \tag{4.1}$$

where J is the homogeneous nucleation rate. For isothermal conditions, this becomes

$$\phi = \frac{\pi}{3} J u^3 \tau_1^4 \tag{4.2}$$

(see, e.g., Avrami, 1939, 1940, 1941). Since (4.2) assumes noninteracting crystallites, it applies to the early stages of the growth process. The growth of crystals in a pure liquid is limited by the rate at which molecules cross the liquid-crystal interface, in which case the linear growth rate u can be written as

$$u = \frac{f \, kT}{3\pi \, a^2 \, \eta} \left\{ 1 - \exp\left[-\frac{\Delta h_m}{kT} \left(1 - \frac{T}{T_m} \right) \right] \right\} \tag{4.3}$$

where f is the fraction of sites on the interface available for the incorporation of molecules, a is a length of the order of a molecular diameter, η is the viscosity, Δh_m (> 0) is the latent heat of fusion, and T_m is the equilibrium melting temperature (Hillig and Turnbull, 1956). The linear growth rate u exhibits a nonmonotonic dependence on T, at first increasing as the temperature falls below T_m, only to fall abruptly upon further cooling due to the corresponding viscosity increase. Since J also depends nonmonotonically upon the extent of supercooling (see Section 3.1.5), it follows that τ_1, the time required to crystallize a given volume fraction of liquid following an instantaneous quench to the desired temperature, must first decrease, reach a minimum, and then increase

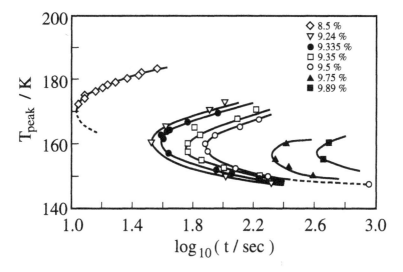

Figure 4.1. Time-temperature-transformation (TTT) diagram for aqueous LiCl solutions of different compositions (wt. % LiCl). The curves give the temperature dependence of the time required to reach the maximum crystallization rate. (Adapted from MacFarlane et al., 1983)

with the extent of supercooling (Uhlmann, 1972). The nonmonotonic relation between crystallization rate and extent of supercooling is therefore the result of the competition between the thermodynamic driving force for nucleation and the kinetics of growth. Figure 4.1 shows the experimentally determined relationship between T and τ_1 for aqueous LiCl solutions (MacFarlane et al., 1983). Such curves are known as TTT (time, temperature, transformation) diagrams. Theoretical TTT diagrams can be constructed by solving Equation (4.2) for τ_1 as a function of T, for given physical properties and crystallized fraction, ϕ (Uhlmann, 1972).

The second characteristic time of relevance to the fate of a supercooled liquid, τ_2, is the internal (molecular) relaxation time.[2] As the temperature is lowered below the freezing point, τ_2 increases very sharply from a high-temperature limiting value of ca. 10^{-13} sec corresponding to a quasilattice vibration time for liquids above their melting point (Angell, 1988a), to a point where it becomes comparable to the duration of a macroscopic experiment, ca. 10^2 sec. Thus, if a liquid can be prevented from crystallizing down to a temperature at which the characteristic time for molecular rearrangement becomes comparable to the experimental time scale, it appears structurally arrested on that time scale. The

[2]Strictly speaking, there is a spectrum of internal relaxation times. For the present purposes, τ_2 can be taken as any suitable mean of the distribution, such as the geometric mean (Zallen, 1983).

Table 4.1. Calculated Critical Cooling Rates for Several Substances[a]

Substance	$(dT/dt)_c$ (K sec^{-1})
SiO_2	7×10^4
GeO_2	7×10^2
Phenyl salicylate	50
Water	10^7
Ag	10^{10}

[a] After Uhlmann (1972).

liquid is then said to have vitrified: it is a glass.[3] The relationship between τ_1 and τ_2 is illustrated in Figure 4.2. Of course, structural arrest is a meaningless concept without reference to the time scale being investigated. In general, the shorter the time scale being probed, the higher the temperature at which falling out of equilibrium is observed due to the intersection of the internal relaxation and observation time scales (Angell and Torell, 1983).

Returning now to TTT curves, if ϕ is small enough to represent a limit of detection (e.g., 10^{-6}; Zarzycki, 1990), the extremum, or nose, of the TTT diagram, with coordinates (τ_{1n}, T_n), defines an effective critical cooling rate,

$$\left| \frac{dT}{dt} \right|_c = \frac{T_m - T_n}{\tau_{1n}} . \tag{4.4}$$

When the latter is exceeded, the sample cannot crystallize (since $\tau_1 \propto \phi^{-1/4}$, the exact value of ϕ has little effect on the calculated critical cooling rate). Table 4.1 lists calculated critical cooling rates for different substances. The fastest cooling rate that can be currently attained in the laboratory is roughly 10^7 K sec^{-1}.[4]

A glass can be formed provided the liquid is cooled fast enough to prevent crystallization. The question then becomes not whether a given substance can vitrify, but under what conditions it can do so. Confirmation of this perspective, according to which the vitreous state can always be attained for fast enough cooling rates follows from the great diversity of substances that are known to vitrify. Some examples are listed in Table 4.2.

---

[3] The rapid quenching of liquids is not the only route to the formation of glasses. Other methods include reactive precipitation, electrolytic deposition (starting from a liquid); quenching a vapor, cathodic sputtering, ion implantation, chemical vapor deposition (starting form a vapor); and radiation or cold compression of a crystal (Zarzycki, 1990; Angell, 1995).

[4] This applies to bulk samples and μm-sized droplets. Very thin ($\approx$ 100 nm) films can be cooled much faster, e.g., 10^{14} K sec^{-1} (Greer, 1995).

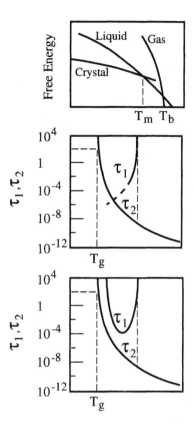

Figure 4.2. Schematic relationship between the characteristic time for crystallization of a given volume fraction, τ_1, and the internal relaxation time, τ_2, for a supercooled liquid (times are in seconds). The upper diagram is a schematic of the isobaric relationship between the Gibbs energy and temperature for the gaseous, liquid, and crystalline phases of a pure substance. T_b is the boiling temperature, and T_m the melting temperature. The hypothetical intersection of the τ_1 and τ_2 curves (middle diagram) does not occur because τ_1 increases sharply, as crystallization is controlled by viscous effects in the highly supercooled liquid. T_g is the glass transition temperature, where $\tau_2 \approx 10^2$ sec. (Adapted from Angell, 1988a)

Conditions that favor vitrification are those that hinder crystallization:[5] high surface tension and high entropy of fusion. In particular, substances with $\alpha^3 \Gamma' > 0.73$ [see Equation (3.123) and Figure 3.14] are virtually impossi-

[5]Many polymers, such as statistical copolymers or atactic vinyl polymers, are incapable of crystallizing because of their irregular structure.

Table 4.2. Examples of Substances that Form Glasses[a]

Type	Example
Elements	P, S, Se
Oxides	SiO_2, GeO_2, B_2O_3, P_2O_5, As_2O_3, Sb_2O_3
Chalcogenides	As_2S_3
Halides	BeF_2, $ZnCl_2$
Molten salts	$KNO_3 + Ca(NO_3)_2$, $K_2CO_3 + MgCO_3$
Aqueous solutions of salts, acids, and bases	H_2SO_4(aq.), KOH(aq.), LiCl(aq.)
Organic compounds	Methanol, ethanol, glycerol, glucose, toluene, o-terphenyl, m-xylene, fructose, sucrose
Polymers	Poly(ethylene oxide), polystyrene, poly(vinyl choride)
Metal-metalloid alloys	Pd + Si, Fe + B, Fe + Ni + P + B
Metal-metal alloys	Ni + Nb, Cu + Zn

[a] Adapted from Zarzycki (1990).

ble to crystallize upon supercooling, unless seeded (Turnbull, 1969), and they therefore vitrify easily. Although Γ' $(= \Delta h_m / kT_m)$, the dimensionless entropy of fusion, is in general known, α $[= \sigma (v^s)^{2/3} / \Delta h_m$, with v^s the volume per molecule of the solid, and Δh_m the heat of fusion per molecule] is not, depending as it does on σ, which is not directly measurable (Section 3.1.5). Typically, $1 \leq \Gamma' \leq 10$, while α, when calculated from nucleation experiments, is less than 1 (e.g., $\alpha \approx 0.4$ for Bi; Turnbull, 1969).

The increased sluggishness in the exploration of configurations that characterizes vitrification is manifested macroscopically by a sharp viscosity increase. Therefore, for a given viscosity-temperature relationship, vitrification is favored by low values of T_m. Since $T_m = \Delta h_m / \Delta s_m$, it follows that increased molecular asymmetry among a group of structurally similar substances tends to lower T_m and increase glass-forming tendency. This is well illustrated by the xylenes (MacFarlane and Angell, 1982; Zarzycki, 1990). The three isomers have very similar boiling points (similar cohesive energies) but very different melting points. The highly symmetric p-xylene melts at 13.3°C and does not vitrify at experimentally attainable cooling rates. The m-isomer, on the other hand, melts at −47.9°C, and vitrifies, as does o-xylene, which melts at −25°C.[6] Since the boiling temperature obeys corresponding states but the melting point does

[6] o- and m-xylene can be vitrified if they are emulsified, but not, apparently, in bulk form (MacFarlane and Angell, 1982). See Table 4.3.

Table 4.3. Relationship between T_b/T_m and Glass-Forming Tendency

Liquid	T_b (°C)	T_m (°C)	T_b/T_m	Glass-Forming Tendency
Water	100	0	1.37	Marginal[a]
p-Xylene	138.3	13.3	1.44	No[b]
o-Xylene	144.4	−25.2	1.68	Marginal[c]
Bromobenzene	156	−30.8	1.77	Marginal[c]
Chlorobenzene	132	−45.2	1.78	Marginal[c]
m-Xylene	139.1	−48	1.83	Marginal[c]
Methanol	65	−93.9	1.89	Yes
Toluene	110.6	−95	2.15	Yes
Ethanol	78.5	−117.3	2.26	Yes

[a] Cooling at $> 10^5$ K sec^{-1} (Mayer, 1985a).

[b] Does not vitrify at experimentally attainable cooling rates.

[c] Requires emulsification (MacFarlane and Angell, 1982).

not, the ratio T_b/T_m varies greatly from one type of substance to another, and is a useful empirical indicator of glass-forming tendency (Turnbull and Cohen, 1958; MacFarlane and Angell, 1982). It is found that $T_b/T_m > 2$ for many organic liquids that are easy to vitrify (Turnbull and Cohen, 1958). Examples of the relationship between T_b/T_m and glass-forming tendency are given in Table 4.3.

4.2 ELEMENTARY PHENOMENOLOGY OF VITRIFICATION UPON SUPERCOOLING

Figure 4.3 illustrates the relationship between the specific volume and the temperature of a liquid as it is rapidly cooled isobarically. As the temperature is lowered below the freezing point T_m, the liquid contracts provided its thermal expansion coefficient α_p is positive. Upon further supercooling, a temperature is reached at which α_p starts to change rapidly. Beyond a narrow temperature range, the slope of the volume vs. temperature curve attains a second constant value, which is generally smaller than that corresponding to the liquid branch, and comparable to that of a crystalline solid. The material that results from the isobaric quench is a glass. The solidification process, however, is continuous. In contrast, if the liquid is cooled slowly, or if the sample is seeded with crystals, its volume will change discontinuously as it crystallizes at (or near) T_m.

The temperature defined by the intersection of the liquid and vitreous portions of the volume vs. temperature curve is the glass transition temperature, T_g.

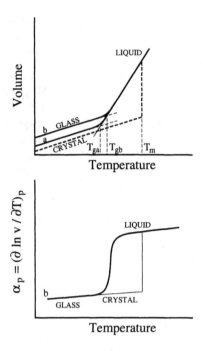

Figure 4.3. Isobaric relationship between the volume and temperature in the liquid, glassy, and crystalline states. T_m is the melting temperature, and T_{ga} and T_{gb} are the glass transition temperatures corresponding to low (a) and high (b) cooling rates. The lower diagram shows the behavior of the thermal expansion coefficient for curve b.

Contrary to T_m, it is not a true transition temperature, the vitrification process occurring instead over a narrow temperature interval. Furthermore, T_g depends on the cooling rate, so that the slower a sample is cooled, the lower its glass transition temperature will be, provided that cooling is still fast enough to avert crystallization (Scherer, 1986). Data are often correlated in Arrhenius form, with the logarithm of the cooling rate related linearly to T_g^{-1} (e.g., Moynihan et al., 1976), but deviations occur at low cooling rates (Brüning and Samwer, 1992). The formation of a glass by rapid quenching of a liquid is thus not an equilibrium process, and leads to a solid phase whose properties depend on its cooling history. This is shown in Figure 4.3, where the faster-quenched glass (branch b) has a lower density than glass a at the same temperature and pressure. Whereas the state of a pure liquid or crystal is specified completely by fixing T and P (or T and ρ), this is not true for a glass. Given the nonequilibrium nature of the glass transition as it is commonly observed in the laboratory, it would appear that kinetics alone governs the phenomenon, and that thermodynamic arguments are here either irrelevant or incorrect. As we shall see, however,

Table 4.4. Thermal Expansion Coefficients for Some Supercooled Liquids and Glasses

Substance	$10^4 \alpha_{p,\text{liq}}$ ($^\circ\text{C}^{-1}$)	$10^4 \alpha_{p,\text{glass}}$ ($^\circ\text{C}^{-1}$)	Reference
Propylene glycol	6.2	2	Kauzmann (1948)
Glycerol	4.83	2.4	Kauzmann (1948)
$Na_2S_2O_3 \cdot 5H_2O$	3.62	2.1	Kauzmann (1948)
Sucrose	5.02	2.54	Kauzmann (1948)
B_2O_3	6.1	0.5	Kauzmann (1948)
Se	4.2	1.7	Kauzmann (1948)
o-Terphenyl	7.49	2	Laughlin and Uhlmann (1972)
α-Phenyl–o-cresol	7.53	2.45	Laughlin and Uhlmann (1972)
Polystyrene	5.5	1.8	Sperling (1986)
Poly(methyl methacrylate)	4.6	2.15	Sperling (1986)

thermodynamic questions arise quite naturally once care is taken not to confuse the kinetically controlled phenomenon that is observed in experiments (the laboratory glass transition) with the idealized condition that would presumably result in the limit of infinitely slow cooling in the absence of crystallization.

Table 4.4 lists typical values of α_p for supercooled liquids and their corresponding glasses.

Although in general $\alpha_{p,\text{liq}} > \alpha_{p,\text{glass}}$, the opposite situation is possible. $CH_3COOLi \cdot 10H_2O$ is an example of a compound whose thermal expansion coefficient increases upon cooling across the glass transition, although the reported difference in α_p for this substance is so small [$\alpha_{p,\text{liq}} - \alpha_{p,\text{glass}} = -(0.4 \times 10^{-4})^\circ\text{C}^{-1}$ at 1 bar] as to be on the threshold of measurability (Williams and Angell, 1977; see also Section 4.3.3).

Figure 4.4 illustrates the thermal behavior that characterizes the cooling of a liquid through the glass transition (Scherer, 1986). As is the case for the volumetric behavior, the enthalpy changes continuously across T_g, but the corresponding isobaric temperature derivative, $c_p \ [= (\partial h/\partial T)_p = T(\partial s/\partial T)_p]$, changes steeply. Contrary to the volumetric derivative, however, since a system's c_p increases in relation to the number of its degrees of freedom, $\Delta c_p = c_{p,\text{liq}} - c_{p,\text{glass}} > 0$, always.[7] The vertical line at T_m corresponds to the ef-

---

[7]Goldstein and co-workers (Goldstein, 1976a,b; Gujrati and Goldstein, 1980) have shown that in many cases the heat capacities of the glassy and crystalline states of a substance can differ appreciably because of vibrational differences between the amorphous and crystalline phases (anharmonicity), and because of the temperature-dependent change in the number of molecular units participating in relaxation processes in the glass. Figure 4.4 is therefore a simplification.

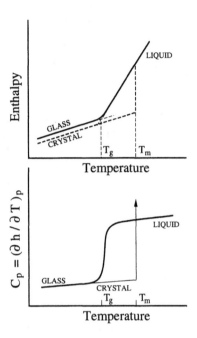

Figure 4.4. Isobaric relationship between the enthalpy and temperature in the liquid, glassy, and crystalline states. T_m is the melting temperature, and T_g the glass transition temperature. The lower diagram shows the corresponding behavior of the isobaric heat capacity when a liquid is cooled through T_g; an overshoot occurs when a glass is heated across T_g. The arrow indicates the δ-function singularity due to latent heat at a first-order phase transition.

fectively infinite heat capacity of any material undergoing a first order phase transition, in which a finite entropy change occurs at a given temperature. The continuity of enthalpy means that vitrification is not accompanied by heat effects. Thus a glass differs from a liquid because of the absence in the former of certain degrees of freedom that are "frozen in" at T_g (Kauzmann, 1948). The absence of these degrees of freedom manifests itself not in the volume, enthalpy, or entropy, but in their temperature derivatives.

Liquids such as SiO_2 and GeO_2 form tetrahedrally coordinated networks that resist thermal degradation across T_g (Stixrude and Bukowinski, 1990; Zallen, 1983). Such substances, therefore, have a small Δc_p upon vitrification, since fewer degrees of freedom are "frozen in" upon cooling below T_g (Angell and Sichina, 1976). At the other extreme, simple molecular liquids exhibit relatively large Δc_p across T_g. Table 4.5 gives Δc_p values for several liquids. Given the variety of bonds that are susceptible to thermal rupture near T_g in

Table 4.5. Heat Capacity Change at the Glass Transition for Some Supercooled Liquids[a]

Liquid	T_g (°C)	Δc_p (J/mol K)
Methanol	−170	26
Ethanol	−178	36.6
1-Propanol	−177	40.2
Glycerol	−88	23
Cyclohexanol	−123	24
Cyclohexene	−192	11.4
Toluene	−160	53.1
Ethylbenzene	−162	61.9

[a] Data from Privalko (1980).

different substances (e.g., every bond in As_2S_3; none in typical organic liquids), comparisons of Δc_p are of limited value. A more appropriate quantity is the ratio $c_{p,\text{liq}}/c_{p,\text{glass}}$ (Angell and Sichina, 1976). It equals roughly 1.1 for GeO_2 (stable network), and 1.9 for 2-methylbutane (no stable liquid structure).

The narrow transformation interval commonly referred to as the glass transition is the temperature range where the internal relaxation time τ_2 becomes of the order of 10^2 sec. A property that is intimately related to the internal relaxation time is the viscosity η. It is a measure of a material's resistance to flow when sheared. Shear-induced deformation takes place via the sliding of molecular layers past each other under the action of an applied force. A material's viscosity is a direct measure of the characteristic time τ for molecular rearrangement in response to an applied shear stress. Specifically, $\eta = G_\infty \tau$, where G_∞ is the instantaneous (elastic) shear modulus. For many substances η is approximately 10^{12} Pa sec $= 10^{13}$ poise (P) at the glass transition (see Angell, 1991a for exceptions). This follows from the concept of T_g as indicative of structural arrest due to the crossing of the molecular relaxation and experimental time scales, and consequently from fixing G_∞ at a solidlike value, 10^{10} N m^{-2}, and τ at 10^2 sec. Thus, over time scales shorter than τ, a glass behaves mechanically like a solid. The calorimetric T_g (Figure 4.4) is usually higher than the temperature at which $\eta = 10^{13}$ P. Equivalently, η at the calorimetric T_g is lower than 10^{13} P, typically 10^9 P (Angell, 1990a). Figure 4.5 shows the viscosity-temperature relationship of several glass-forming materials at atmospheric pressure. Note the wide spectrum of T_g.

In general, the faster the relaxation mechanism, the higher the temperature at which the crossing between the experimental and relaxation time scales occurs.

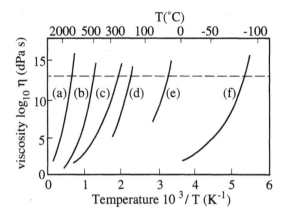

Figure 4.5. Relationship between viscosity and temperature of several glass-forming materials at atmospheric pressure. The horizontal line corresponds to a viscosity of 10^{13} dPa sec ($= 10^{13}$ P), which is commonly associated with the glass transition. (a) SiO_2 ($T_g = 1430$ K; Zallen, 1983); (b) soda lime–SiO_2 ($T_g = 810$ K; Zarzycki, 1990); (c) B_2O_3 ($T_g = 500$ K; Zarzycki, 1990); (d) As_2S_3 ($T_g = 470$ K; Zallen, 1983); (e) Se ($T_g = 310$ K; Zallen, 1983); (f) glycerol ($T_g = 185$ K; Zarzycki, 1990). (Adapted from Zarzycki, 1990)

This can be seen clearly in the novel specific heat spectroscopy technique developed by Birge and Nagel (Birge and Nagel, 1985; Birge, 1986; Grest and Nagel, 1987). By measuring the temperature response to an oscillating heat flux, these authors showed that the higher the frequency, the higher the temperature at which the real part of the isobaric heat capacity drops from liquidlike to glasslike values.

4.3 THERMODYNAMIC VIEWPOINT OF THE GLASS TRANSITION

Theories of the glass transition are of two classes. Thermodynamic ones view the experimentally observable glass transition as a kinetically controlled manifestation of an underlying singularity. This singularity can be described in purely static (thermodynamic) terms. This viewpoint underlies both the entropy and free volume theories, which are discussed in Sections 4.3.2 and 4.3.3, respectively. In contrast, Section 4.4 discusses the nonthermodynamic viewpoint, according to which vitrification occurs as a result of a purely dynamic singularity upon deep supercooling.

4.3.1 Kauzmann's Paradox

A consequence of the difference between liquid and crystalline heat capacities, first pointed out by Kauzmann (1948), underlies most thermodynamic thinking about the glass transition. The isobaric heat capacity of a supercooled liquid exceeds that of the corresponding crystal. Hence, the entropy of the liquid decreases faster than that of the crystal upon isobaric cooling. Thus one can calculate a pressure-dependent temperature T_K, the Kauzmann temperature, at which the entropy difference between the liquid and crystalline phases vanishes. To this end, we write ($T < T_m$)

$$s_l(T_m) = s_l(T) + \int_T^{T_m} \frac{c_{p,l}}{T} dT,$$

$$s_{cr}(T_m) = s_{cr}(T) + \int_T^{T_m} \frac{c_{p,cr}}{T} dT, \tag{4.5}$$

where subscripts l and cr denote the liquid and crystalline phases. T_K is then given by

$$\Delta s_m = \frac{\Delta h_m}{T_m} = \int_{T_K}^{T_m} \frac{\Delta c_p}{T} dT, \tag{4.6}$$

where Δc_p denotes the positive difference between the supercooled liquid and crystalline heat capacities. Below T_K, the hypothetical supercooled liquid would have a lower entropy than the crystal at the same temperature and pressure. It follows from (4.6) that, for given T_m and Δs_m, T_K will be higher (closer to T_g) the larger the difference between liquid and crystalline heat capacities. Thus, for substances that show appreciable heat capacity changes across T_g, the difference between T_g and the calculated T_K can be quite small [e.g., $T_g = 213$ K, $T_K = 198$ K for Cd(NO$_3$)$_2 \cdot$ 4H$_2$O; Angell and Tucker, 1974].

The difference in chemical potential between the supercooled liquid and crystalline phases at T_K is given by

$$\Delta\mu(T_K) = \mu_l(T_K) - \mu_{cr}(T_K) = \int_{T_K}^{T_m} \Delta c_p \left(\frac{T_m}{T} - 1\right) dT \tag{4.7}$$

which is necessarily positive. The isobaric temperature dependence of the supercooled liquid and crystalline entropies and of the chemical potential difference between these two phases is shown schematically in Figure 4.6.

Figure 4.7 illustrates the calorimetric route to calculating the Kauzmann temperature. T_K is determined by matching the area corresponding to the entropy of fusion with that spanned by the integral of the heat capacity difference between the supercooled liquid and the crystal. This calorimetric calculation of T_K is based on the use of a heat capacity for the supercooled liquid obtained

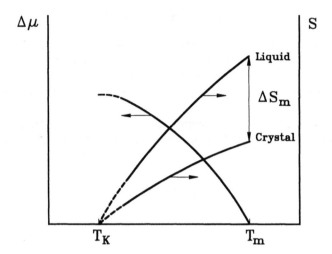

Figure 4.6. Schematic representation of the isobaric temperature dependence of the entropy of a supercooled liquid and a crystal, and of the chemical potential difference between these two phases, between T_m, the melting point, and T_K, the Kauzmann temperature.

by extrapolating stable and metastable behavior below T_g. If one accepts the validity of this extrapolation, the vanishing of the entropy difference between the liquid and the crystal becomes the unavoidable condition that would be attained if the liquid were cooled sufficiently slowly (so that it would not fall out of equilibrium), yet without crystallizing.

In practice, T_K is not attainable experimentally because vitrification intervenes ($T_g > T_K$). However, the behavior that would apparently result should structural equilibration occur down to T_K merits careful consideration. Upon further isobaric cooling below T_K, the entropy of the liquid would be lower than that of the corresponding crystal at the same temperature and pressure. This is certainly unusual;[8] thermodynamics, however, places no restriction on the sign of the entropy difference between stable and metastable phases at fixed temperature and pressure. Consider, on the other hand, the extrapolation of the crystalline and amorphous entropies below T_K. The entropy of the crystal, which we assume perfect, approaches zero as $T \to 0$. That of the disordered phase, eventually a glass, would approach negative values unless its heat capacity dropped below that of the crystal. Negative entropies are inconsistent with Boltzmann's formula $S = k \ln \Omega$, where Ω denotes the number of quan-

---

[8]At fixed volume, a system of hard spheres will freeze to a solid that has a higher entropy than the liquid. This follows from the fact that the Helmholtz energy A and the entropy S of a hard sphere system are related by $A = -TS$; hence minimization of A at fixed T implies maximization of S. However, at constant pressure, the entropy of fusion of hard spheres is positive ($\Delta S/Nk = 1.16$).

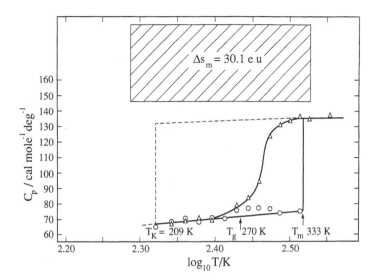

Figure 4.7. Isobaric heat capacities of the crystalline, glassy, and liquid states of $(CH_3CO_2)_2Mg\cdot4H_2O$. The shaded area is proportional to the entropy of fusion. T_m is the melting temperature, and T_g the glass transition temperature. T_K is the Kauzmann temperature, at which the extrapolated entropy of the supercooled liquid becomes equal to the entropy of the crystal. [Reprinted, with permission, from Angell and Tucker, *J. Phys. Chem.* 78: 278 (1974). Copyright 1974, American Chemical Society]

tum states corresponding to a given energy, volume, and mass. Of course, it is impossible to assign an absolute value to the entropy. In saying that the entropy of the crystal approaches zero as $T \to 0$ we neglect intranuclear and isotopic degrees of freedom (Denbigh, 1981; Guggenheim, 1986). Because these do not depend on temperature and pressure at ordinary terrestrial conditions, they contribute an additive constant to the entropy whose exact value is irrelevant (Schrödinger, 1989). Thus the above argument still holds, that is to say, we must have $k \ln \Omega > 0$, always.

It is therefore not the crossing of the liquid and crystalline entropy curves at T_K (unusual as this undoubtedly is), but their extrapolation to $T \to 0$ after crossing, that leads to unphysical results. This scenario has come to be known as the Kauzmann paradox. In addition to the above-mentioned implausible decrease of the amorphous phase's heat capacity below that of the crystal, there are two resolutions of the Kauzmann paradox. One was put forward by Kauzmann (1948). He argued that when a liquid is supercooled, molecular motion becomes progressively constrained, and free energy barriers to molecular rearrangements increase. At the same time, free energy barriers to crystallization decrease, because supercooling causes a decrease in the size of the critical nucleus. Eventually, the two types of free energy barriers become comparable. This is the situation depicted in Figure 4.2, should the curves labeled τ_1 and τ_2 intersect (middle graph). Thus, if measurements are to be made on the liquid before it crystallizes, so too must they be made before the structure can equilibrate. But this means that the liquid behaves like a glass. If enough time is allowed for the liquid to sample other liquidlike configurations, it will crystallize. Spontaneous crystallization, then, prevents the low-temperature extrapolation of liquid properties on which the existence of the Kauzmann paradox is based.

Alternatively, the Kauzmann temperature can be viewed as an absolute limit, an underlying singularity below which the liquid cannot exist. Having thus far avoided crystallization, the supercooled liquid can only escape the impending entropy crisis by undergoing a sharp glass transition at T_K. This transformation is called the ideal glass transition (Angell and Tucker, 1974). Contrary to the experimentally observed phenomenon occurring at T_g, the formation of an ideal glass is a true thermodynamic transition that occurs at a precise temperature. By definition, the ideal glass has the same entropy as the crystal. Therefore it must correspond to a state in which the system has settled into the deepest of all amorphous potential energy minima. Other comparably deep minima, if they exist, must be few in number, and they must all be mutually inaccessible. It is this mutual inaccessibility that makes the entropy of the disordered ideal glass equal to that of the ordered crystal (it is assumed that differences in vibrational entropy between the crystal and glassy phases are negligible at low temperatures). In this alternative resolution of the Kauzmann paradox, an ideal glass transition intervenes to thwart the entropy crisis.

Kauzmann's resolution of the impending entropy crisis via homogeneous nucleation cannot apply universally. As was discussed in Section 4.1, the intersection of the crystallization and structural relaxation time scales generally does not occur, because of the viscosity-driven increase in τ_1 at low temperatures. Nevertheless, some computer simulations are not inconsistent with Kauzmann's proposed resolution of the paradox, in the specific case of the hard sphere fluid. Extensive molecular dynamics calculations (Woodcock, 1981) showed that the exploration of the metastable supercooled liquid becomes impossible at volume fractions[9] exceeding $\phi = 0.565$, due to spontaneous crystallization.[10] Equation-of-state calculations (Woodcock, 1981) predict that the Kauzmann point for the hard sphere fluid occurs at $\phi = 0.630$. Speedy (1994) avoided crystallization

---

[9]The volume fraction is the ratio of occupied to total volume. It is also called the packing fraction. For hard spheres, $\phi = \pi \rho \sigma^3 / 6$, where σ is the diameter and ρ is the number density.

[10]The volume fractions of the coexisting liquid and solid phases at the hard sphere transition (Alder and Wainwright, 1957, 1962; Wood and Jacobson, 1957; Hoover and Ree, 1968), of the close-packed face-centered cubic crystal, and of a randomly close-packed assembly of hard spheres (Jodrey and Tory, 1981, 1985) are 0.494 ± 0.002, 0.545 ± 0.002, $\pi\sqrt{2}/6 (= 0.74)$, and 0.648, respectively.

by rapid compression across the range $0.548 \leq \phi \leq 0.577$. He found that the resulting hard sphere glasses had reproducible properties, and calculated the glass transition to occur at $\phi = 0.562$, where the fluid and glassy branches of the equation of state intersect. He found that the Kauzmann paradox was avoided by spontaneous crystallization in the density range $0.548 \leq \phi \leq 0.577$, and by spontaneous vitrification if $\phi > 0.577$. To date, it has not been possible to supercool the hard sphere fluid beyond $\phi > 0.562$ in simulations; it is found that Kauzmann's paradox is always resolved, by either crystallization or vitrification. For colloidal hard spheres, the glass transition occurs around $\phi \approx 0.56$ (e.g., Pusey and van Megen, 1987; van Blaaderen and Wiltzius, 1995).

It has already been mentioned that, in general, the viscosity-driven increase in the crystallization time prevents the crossing of the crystallization and structural relaxation time scales on which Kauzmann's proposed resolution of the entropy paradox is predicated. Another argument against the general validity of such a resolution was put forward by Angell et al. (1984). These authors presented calculations for supercooled lithium disilicate ($Li_2O \cdot 2SiO_2$), and showed that the induction time required for nucleation to occur can be greater than the structural relaxation time by several orders of magnitude. This would imply that the supercooled liquid can equilibrate long before homogeneous nucleation takes place, making it impossible to avoid the entropy crisis by homogeneous nucleation should the liquid not vitrify. Angell and Donella (1977) noted that in several aqueous ionic solutions, the highest homogeneous nucleation temperature, which corresponds to the most likely crystallization process [precipitation of pure ice in their $Ca(NO_3)_2$ example] decreases markedly with increased solute content, and is on a collision course with the locus of Kauzmann temperatures determined as illustrated in Figure 4.7. When the highest homogeneous nucleation temperature becomes lower than T_K, resolution of the paradox by homogeneous nucleation is not possible. In special cases like that of $Ca(NO_3)_2$ in water, where $(T_g - T_K)$ is small, it should then be possible to find evidence of the high-temperature remnant of the ideal glass transition by calorimetric determination of enthalpy loss in emulsified samples annealed below T_g (Angell and Donnella, 1977). Such experiments have not been done to date.

A penetrating critique of the validity of liquid-property extrapolations below T_g, which lie at the heart of the Kauzmann paradox and the concept of an ideal glass transition, has been formulated by Stillinger (1988a). He considered a supercooled liquid's $(3N + 1)$-dimensional potential energy hypersurface, which is generated by assigning coordinates to every one of the N atoms in a macroscopic system (Goldstein, 1969). To each such assignment there corresponds a value of the potential energy U, and hence a point $(U, \mathbf{r}_1, \ldots, \mathbf{r}_N)$ on the hypersurface. Atomic motion in a deeply supercooled liquid consists of thermally driven vibrations about deep potential energy minima, plus infrequent visitations of different minima. This means that vibrational and configurational contributions to the metastable liquid's properties are separable, a condition that is only meaningful when the region of the potential energy hypersurface being sampled consists of deep and well separated local minima ($|U/N| \gg kT$). The key quantity in Stillinger's argument is a basin enumeration function, $\gamma\,(U/N)$, defined in such a way that the number of potential energy basins whose depth per atom lies in the range $U/N \pm d\,(U/N)/2$ is $\exp(N\gamma)d\,(U/N)$. Assuming plausible forms for $\gamma\,(U/N)$ and for the vibrational contribution to the Gibbs energy, Stillinger concluded that in nonpolymeric supercooled liquids the rate of entropy loss predicted by extrapolating liquid properties from above T_g cannot persist indefinitely. The entropy of the supercooled liquid would instead decay smoothly to zero as T approaches zero, requiring a sharp change in curvature of the entropy-temperature relationship. Such a curvature change cannot be detected using extrapolations from above T_g, and Stillinger's argument in fact calls into question the validity of such extrapolations.

Neither arguments in favor (e.g., Angell and Donnella, 1977), nor those against (e.g., Stillinger, 1988a) the possible existence of an ideal glass transition can be considered definitive. The former merely suggest possible exceptions to Kauzmann's resolution of the paradox. Stillinger's analysis, though entirely plausible, is not a rigorous proof. The basic question of whether there exists a thermodynamic transition due to the vanishing of the entropy difference between a supercooled liquid and a crystal is still open.

We now derive further useful equations related to T_K. We first write the specific entropy of a crystal and of a liquid at T_g as

$$s_l\,(T_g) = s_{gl}\,(0) + \int_0^{T_g} \frac{c_{p,gl}}{T}\,dT,$$

$$s_{cr}\,(T_g) = \int_0^{T_g} \frac{c_{p,cr}}{T}\,dT, \tag{4.8}$$

where subscript gl denotes the glass and $s_{gl}(0)$ is the specific entropy at absolute zero of the glass formed at T_g (i.e., its zero-point entropy). Therefore, the entropy difference between the liquid and crystalline phases at T_g is given by

$$\Delta s\,(T_g) = s_{gl}\,(0) + \int_0^{T_g} \frac{\left(c_{p,gl} - c_{p,cr}\right)}{T}\,dT. \tag{4.9}$$

Thus, when the glassy and crystalline heat capacities are similar, the zero-point entropy of the glass is approximately equal to the entropy excess which is "frozen in" at T_g. We can also write

$$\left(\frac{\partial\,\Delta s}{\partial T}\right)_P = \frac{c_{p,l} - c_{p,cr}}{T} = \frac{\Delta c_p}{T} \tag{4.10}$$

and, therefore,

$$\Delta s\,(T_g) = \int_{T_K}^{T_g} \frac{\Delta c_p}{T}\,dT. \tag{4.11}$$

The zero-point entropy can be readily obtained by writing equations for the specific liquid entropy referred to the crystal and glass at absolute zero, respectively,

$$s_l(T) = \int_0^{T_m} \frac{c_{p,cr}}{T} dT + \frac{\Delta h_m}{T_m} + \int_{T_m}^T \frac{c_{p,l}}{T} dT,$$

$$s_l(T) = s_{gl}(0) + \int_0^{T_g} \frac{c_{p,gl}}{T} dT + \int_{T_g}^{T_m} \frac{c_{p,l}}{T} dT + \int_{T_m}^T \frac{c_{p,l}}{T} dT, \quad (4.12)$$

whereupon we obtain

$$s_{gl}(0) = \frac{\Delta h_m}{T_m} + \int_0^{T_g} \frac{(c_{p,cr} - c_{p,gl})}{T} dT - \int_{T_g}^{T_m} \frac{\Delta c_p}{T} dT$$

$$\approx \frac{\Delta h_m}{T_m} - \int_{T_g}^{T_m} \frac{\Delta c_p}{T} dT, \quad (4.13)$$

where the approximate equality applies when the crystalline and glassy heat capacities are similar.

4.3.2 Cooperative Relaxations and the Entropy Viewpoint

When a liquid is supercooled isobarically, its entropy decreases sharply (see Figure 4.6). Underlying this progressive loss of entropy is a paucity of accessible configurations that affects both the equilibrium and the transport properties of the supercooled liquid. The experimentally observable glass transition can then be viewed as the kinetically controlled manifestation of an underlying entropy crisis at the Kauzmann temperature. The entropy-based viewpoint of the glass transition aims at quantifying the above picture.

An entropy theory that addresses equilibrium thermodynamic properties exists only for polymers (Gibbs, 1956; Gibbs and DiMarzio, 1958). Briefly, in the Gibbs-DiMarzio theory, the formation of an ideal glass is a true second-order phase transition (Ehrenfest, 1933), characterized by continuity of the first derivatives of the Gibbs energy (entropy and volume), and discontinuity of the second derivatives (isobaric heat capacity and thermal expansion coefficient). Laboratory glass transitions are viewed as observable manifestations of this underlying thermodynamic singularity, to which they tend in the limit of infinitely slow cooling. Gibbs and DiMarzio used lattice statistics to calculate the partition function of an assembly of linear chains whose stiffness they quantified in terms of a fraction of strained bonds and a corresponding energy cost per strained bond. Their treatment yields a second-order transition at the point where the configurational entropy of the chain assembly vanishes. Gibbs and DiMarzio assumed that the entropy of a deeply supercooled polymer liquid can be separated into configurational and vibrational contributions; and, furthermore, that the latter is equal in the perfectly ordered solid and in the highly supercooled liquid. Underlying this assumption is the idea that molecules in a deeply supercooled liquid execute vibrations about their local equilibrium positions most of the time, and the slight anharmonicity in the liquid-phase vibrations is all that distinguishes fluid- and solid-phase vibrational entropies. The configurational entropy is then simply the difference between the entropies of the supercooled liquid and the ordered crystal. In the Gibbs-DiMarzio theory, this quantity vanishes at a nonzero temperature.

For nonpolymeric liquids, the focus of this book, there exists no entropy-based treatment of the glass transition that addresses equilibrium thermodynamic properties. However, the entropy-based theory of cooperative relaxations due to Adam and Gibbs (1965) has proved extremely useful for the interpretation of transport and relaxation in supercooled liquids. It is a thermodynamic theory because it interprets structural arrest in terms of the supercooled liquid's configurational entropy (Mohanty, 1988, 1990a,b, 1991, 1994, 1995; Mohanty et al., 1994).

Central to the theory of Adam and Gibbs is the notion of a cooperatively rearranging region, defined as a group of molecules that can rearrange itself into a different configuration independently of its environment as a result of an energy fluctuation. The basic physics is appealingly simple: as the liquid is supercooled, cooperatively rearranging regions grow, and relaxations require the concerted participation of progressively larger groupings of molecules. This increased cooperativity is reflected in a loss of configurational entropy, which the Adam-Gibbs theory relates to an increase in the molecular relaxation time τ_2 (see Figure 4.2). In this section we derive the Adam-Gibbs theory, and we discuss the important insights into the properties of supercooled liquids that follow from the concept of a cooperatively rearranging region. This fruitful idea is common to all of the work reviewed below.

To quantify the relationship between entropy loss and sluggish relaxation, we first write, following Adam and Gibbs, the isothermal-isobaric partition function Δ of a cooperatively rearranging region composed of z molecules:

$$\Delta\,(z, P, T) = \sum_{E,V} \omega\,(z, E, V)\,\exp\,(-\beta H)\,, \qquad (4.14)$$

where the sum is over energies (E) and volumes (V) sampled by the cooperatively rearranging region, ω is the cooperatively rearranging region's number of distinguishable configurations of given energy and volume, and H is the region's instantaneous enthalpy. If the sum is restricted to energies and volumes that allow rearrangement, there results a constrained partition function for rearrangeable subsystems. Calling such a restricted partition function Δ', the fraction of subsystems that are rearrangeable, f, is given by

$$f = \Delta'/\Delta = \exp\left[-\beta\left(G' - G\right)\right], \qquad (4.15)$$

where $G\,(= -kT \ln \Delta)$ is the cooperatively rearranging region's Gibbs energy. The transition probability of a z-sized region, p, is proportional to f, and can be written as

$$p(T) = A \exp[-\beta z \delta \mu], \tag{4.16}$$

where $z\mu = G$ (μ is the chemical potential per molecule; hence $\delta\mu = \mu' - \mu$ and $\delta\mu > 0$) and we have neglected the pressure dependence of $\delta\mu$. The average transition probability $\overline{p}$ is then simply

$$\overline{p(T)} = \sum_{z=z*}^{\infty} A \exp[-\beta z \delta \mu] = \frac{A \exp[-\beta z^* \delta \mu]}{1 - \exp(-\beta \delta \mu)}$$

$$\approx \overline{A} \exp(-\beta z^* \delta \mu), \tag{4.17}$$

where z^* is a lower limit to the size of cooperatively rearranging regions that have nonvanishing transition probabilities, and $\overline{A}$ is only weakly temperature dependent ($\beta\delta\mu \gg 1$) compared to $\exp(-\beta z^* \delta \mu)$. Equation (4.17) implies that the overwhelming number of transitions is undergone by regions having the smallest possible size, z^*, that can give rise to a cooperative rearrangement.

We assume that the partition function of the deeply supercooled liquid can be written as a product of two independent partition functions: one is vibrational; the other, configurational (Goldstein, 1969, 1976a,b). The former is associated with the thermal vibrations that molecules execute about a local potential energy minimum. The latter is associated with the exploration of the various local minima. Let $\Omega(E_{\mathrm{pot}}, N, V)$ denote the number of distinguishable potential energy minima of depth E_{pot} for an N-molecule system occupying a volume V. The configurational entropy of the whole system is given by

$$S_{\mathrm{conf}} = k \ln \Omega(E_{\mathrm{pot}}, V, N) \approx n \, s_{\mathrm{conf}}, \tag{4.18}$$

where n is the number of cooperatively rearranging regions, assumed of equal size, each making an additive contribution s_{conf} to the total configurational entropy. This additivity follows from the fact that a cooperatively rearranging region is by definition independent of its surroundings. It is further assumed that the supercooled liquid, having the ability to explore different potential energy minima, possesses configurational entropy. In contrast, the crystal, being confined to a single minimum, possesses none. As in the Gibbs-DiMarzio theory, the entropy difference between the supercooled liquid and the crystal is purely configurational (i.e., we neglect differences in vibrational entropy), and vanishes at T_K. In one mole of molecules there are $n = N_A/z$ cooperatively rearranging regions (N_A = Avogadro's number). Therefore, with S_{conf} denoting the molar configurational entropy,

$$z^* = \frac{s_{\mathrm{conf}}^* N_A}{S_{\mathrm{conf}}}. \tag{4.19}$$

Equation (4.19) relates the minimum size of a cooperatively rearranging region to the system's configurational entropy. The configurational entropy of the critically sized cooperatively rearranging region, s_{conf}^*, is a number of order $k \ln 2$, since two distinct configurations are the minimum that is required for a rearrangement to occur. Substituting in Equation (4.17), we have

$$\overline{p(T)} = \overline{A} \exp\left(-\frac{\beta \, s_{conf}^* \, \delta\mu}{S_{conf}}\right) = \overline{A} \exp\left(-\frac{C}{T \, S_{conf}}\right), \qquad (4.20)$$

where the chemical potential difference is now on a per mole basis. This chemical potential difference plays the role of a free energy barrier to cooperative rearrangement. Since the characteristic molecular relaxation time τ is inversely proportional to $\overline{p}$, the Adam-Gibbs theory leads to the important result

$$\tau \propto \exp\left(\frac{C}{T \, S_{conf}}\right) \approx \exp\left(\frac{C}{T \, \Delta s}\right), \qquad (4.21)$$

where Δs is the molar entropy difference between the supercooled liquid and the crystal. As discussed above, a central assumption of the theory is the validity of the equality $S_{conf} = \Delta s$. Equation (4.21) relates the growth in molecular relaxation time to the dearth of configurational entropy that accompanies deep supercooling on account of the growth of cooperatively rearranging regions.[11] The importance of (4.21) follows from the fact that the coefficient of self-diffusion, D, and the fluidity, η^{-1}, are inversely proportional to a molecular relaxation time $\left(\eta \propto D^{-1} \propto \tau\right)$. Accordingly, the Adam-Gibbs theory provides a relationship between transport coefficients and configurational entropy.

The separation of vibrational and configurational contributions to the partition function of a deeply supercooled liquid underlies not only the work of Adam and Gibbs, but also several other theories discussed in this section. Goldstein (1969) was the first to articulate the viewpoint according to which molecular motions in supercooled liquids consist of anharmonic vibrations about deep potential energy minima, and of infrequent visitations of different such minima. Building on Goldstein's ideas, Stillinger and co-workers developed the concept of an inherent structure, which leads, among other things, to a useful descriptive viewpoint of supercooled liquids (Stillinger and Weber, 1982, 1983a,b, 1984a,b; Stillinger and LaViolette, 1985; LaViolette and Stillinger, 1985, 1986; Stillinger, 1985, 1988a,b, 1995). Stillinger's novel idea was to map the actual trajectory executed by a system onto the successive potential energy minima (inherent structures) about which the system vibrates. Thus, each configuration is assigned uniquely to one inherent structure. Computationally, the local minimum is found by steepest-descent quenching of configurations generated by molecular dynamics simulation. Using this technique, Stillinger and co-workers showed that, at constant density, the pair correlation function of model atomic systems calculated from the inherent structures is very nearly

---

[11] Note that the theory does not provide a prescription for calculating the size of a cooperatively rearranging region. This is an important limitation.

independent of temperature (Stillinger and Weber, 1984a,b). They also investigated melting (Stillinger and Weber, 1982), freezing (LaViolette and Stillinger, 1985), and the dynamics of transitions between inherent structures in supercooled atomic systems (Stillinger and Weber, 1983a).

We now discuss the implications of (4.21). A functional relationship that describes the temperature dependence of the difference between liquid and crystalline heat capacities in many simple liquids is $\Delta c_p = K/T$ (Alba et al., 1990). Using this in Equation (4.11), and substituting in (4.21), we obtain a Vogel-Tamman-Fulcher (VTF) form for the temperature dependence of the viscosity, $\eta = B \exp[A/(T - T_0)]$,

$$\tau = \text{const.} \times \exp\left(\frac{CT_K/K}{(T - T_K)}\right) = \text{const.} \times \exp\left(\frac{A}{T - T_K}\right) \propto \eta \quad (4.22)$$

(Vogel, 1921; Tamman and Hesse, 1926; Fulcher, 1925). The VTF equation is commonly used to correlate viscosities of supercooled liquids (e.g., Angell and Smith, 1982; Angell et al., 1982a; Angell 1990a). It is often observed that VTF behavior breaks down close to T_g, with the viscosity reverting to Arrhenius behavior. This is illustrated in Figure 4.8, taken from a study by Laughlin and Uhlmann (1972), who also found low-temperature VTF breakdown in salol, α-phenyl-o-cresol, and tri-α-naphthylbenzene. These authors obtained good VTF fits only for viscosities smaller than 10^4 P. In contrast, good agreement with the VTF equation over eleven orders of magnitude in the relaxation time, with no detectable low-temperature reversion to Arrhenius behavior, was found by Angell and co-workers (Angell and Smith, 1982; Angell et al., 1982a) for the polyhydric alcohols sorbitol, glycerol, ethylene glycol, and propylene glycol.[12]

A stringent test of the soundness of the Adam-Gibbs theory is provided by comparing relaxation-based determinations of T_0 with calorimetric calculations in which T_K is determined as shown in Figure 4.7. Such a comparison is made in Table 4.6.

Table 4.6 shows cases of striking agreement, as well as some of striking disagreement, between T_0 and T_K. Taken together with Figure 4.8 and the preceding discussion, these results point simultaneously to the usefulness of the Adam-Gibbs theory as a basis for transport property correlations, and to its limitations. That the theory's predictions are neither universally applicable nor in general quantitatively accurate, is hardly surprising in light of the approximations involved in the derivation of Equation (4.21).

The derivation of a relationship between configurational entropy and relaxation kinetics being the main result of the Adam-Gibbs theory, it is natural to

---

[12]In that work, VTF behavior was tested by direct viscosity measurements, using Equation (4.22); by dielectric relaxation measurements, using $\tau \propto \exp[A/(T - T_0)]$, where τ is the dielectric relaxation time; and by using literature values for the viscosity and the high-frequency shear modulus G_∞ to obtain $\tau_s = \eta/G_\infty$, the shear relaxation time, which was then fitted to VTF form.

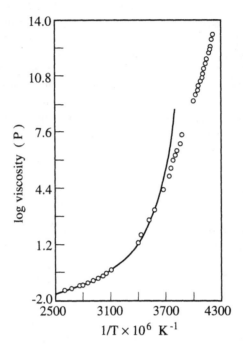

Figure 4.8. Viscosity-temperature relationship for supercooled o-terphenyl. The line is a Vogel-Tamman-Fulcher fit. [Reprinted, with permission, from Laughlin and Uhlmann, *J. Phys. Chem.* 76: 2317 (1972). Copyright 1972, American Chemical Society]

interpret Equation (4.21) so as to associate the locus of glass transitions, $T_g(P)$, with a constant relaxation time, and therefore a constant value of $T S_{\mathrm{conf}}$. Strictly speaking, there is no such thing as a $T_g(P)$ line, but rather a narrow region in the (T, P) plane where sharp but not discontinuous changes in derivative properties occur (e.g., Figures 4.3 and 4.4). For the purposes of the present discussion, we consider the locus of glass transitions associated with a given cooling rate, and assume that it can be described by a $T_g(P)$ line (Goldstein, 1975). Expanding $T S_{\mathrm{conf}}$, we have

$$d\left(T S_{\mathrm{conf}}\right) = 0 = \left[S_{\mathrm{conf}} + T \left(\frac{\partial S_{\mathrm{conf}}}{\partial T} \right)_P \right] dT + T \left(\frac{\partial S_{\mathrm{conf}}}{\partial P} \right)_T dP \quad (4.23)$$

and, therefore,

$$\frac{dT_g}{dP} = \frac{-T_g \left(\partial S_{\mathrm{conf}} / \partial P \right)_T}{S_{\mathrm{conf}} + T_g (\partial S_{\mathrm{conf}} / \partial T)_P} \approx \frac{T_g v_g \Delta \alpha_p}{S_{\mathrm{conf}} + \Delta c_p}, \quad (4.24)$$

where $\Delta \alpha_p$ and Δc_p refer to differences between liquid and crystalline deriva-

Table 4.6. Comparisons between the VTF Singular Temperature T_0 and the Kauzmann Temperature T_K

Substance	$T_K{}^a$ (K)	$T_0{}^b$ (K)	$T_0{}^c$ (K)	$T_0{}^d$ (K)	$T_g/T_K{}^a$	$T_g/T_0{}^b$
2-Methyl pentane[e]	58	59				
Ethanol[e]	63	58				
Glycerol[e]	134	132				
Glycerol[f]					1.26	1.45
Glycerol[g]	135	124		137		
$H_2SO_4 \cdot 3H_2O^e$	135		128			
$Ca(NO_3)_2 \cdot 4H_2O^e$	202	205	201			
$Ca(NO_3)_2 \cdot 4H_2O^h$					1.08	1.07
$KNO_3{}^e$	224		219			
$ZnCl_2{}^e$	250	260				
$B_2O_3{}^e$	335	402				
n-Propanol[f]					1.37	3.93
$Cd(NO_3)_2 \cdot 4H_2O^h$					1.08	3.93
Propylene glycol[g]	115		125			
Methanol[g]	64	60				

[a] T_K from calorimetric measurements (see, e.g., Figure 4.7).
[b] T_0 from viscosity measurements fitted to $\eta \propto \exp[A/(T - T_0)]$.
[c] T_0 from conductance measurements fitted to $\Lambda \propto \exp[-A/(T - T_0)]$.
[d] T_0 from dielectric relaxation measurements fitted to $\tau_d \propto \exp[A/(T - T_0)]$.
[e] Angell and Rao (1972).
[f] Adam and Gibbs (1965).
[g] Angell et al. (1982).
[h] Angell and Tucker (1974).

tive properties (or between liquid and glassy derivative properties if anharmonic contributions are small; Goldstein, 1976a,b). The second form results from neglecting the specific volume difference between the glass and the crystal at T_g and P.

If we assume, somewhat arbitrarily, that observable glass transitions occur not at a fixed value of the relaxation time, but rather when the entropy difference between the supercooled liquid and the ideal glass[13] reaches a critical value, we can derive another equation for the locus $T_g(P)$. To do this, we write, for

[13] The ideal glass (Section 4.3.1) would result in the limit of infinitely slow cooling.

entropy changes in the supercooled liquid and in the ideal glass,

$$ds_l = \frac{c_{p,l}}{T} dT - v\alpha_{p,l} dP,$$

$$ds_{gl} = \frac{c_{p,gl}}{T} dT - v\alpha_{p,gl} dP. \tag{4.25}$$

Therefore, if we neglect the difference in volume between the actual and ideal glasses, the locus along which $(s_l - s_{gl})$ is constant satisfies

$$\frac{dT_g}{dP} = \frac{T_g v_g \Delta\alpha_p}{\Delta c_p}. \tag{4.26}$$

Invoking Equations (4.24)–(4.26) we obtain

$$\left(\frac{dT_g}{dP}\right)_\tau - \left(\frac{dT_g}{dP}\right)_{\Delta s} \approx T_g v_g \Delta\alpha_p \left[\frac{1}{S_{conf} + \Delta c_p} - \frac{1}{\Delta c_p}\right] < 0, \tag{4.27}$$

where Δs now denotes the difference between liquid and ideal-glass entropies, and the subscripts denote the quantity whose assumed invariance defines the locus of glass transitions (Angell et al., 1977). Equation (4.26) is generally accurate to within ca. 30% for a wide variety of substances (see, however, Angell and Sichina, 1976, and Angell et al., 1977, for two exceptions). It is not possible to distinguish experimentally Equation (4.26) from the approximate form of Equation (4.24), since typically $S_{conf}/\Delta c_p < 0.3$, and the resulting differences are within the experimental uncertainty of $\Delta\alpha_p$ and Δc_p (Angell et al., 1977). Figure 4.9 shows experimental and calculated $T_g(P)$ lines, the latter using Equation (4.26) and the known T_g and physical properties at atmospheric pressure.

Equation (4.26) is identical to the Ehrenfest expression that would result from the assumption that $T_g(P)$ is a locus of second-order phase transitions along which the entropy difference between the liquid and the glass vanishes. However, experimentally observable glass transitions are not equilibrium phase transitions. T_g is not a sharply defined temperature, and the changes in derivative properties, though steep (see Figures 4.3 and 4.4), are not discontinuous. Furthermore, at a true second-order transition, both entropy and volume are continuous. Expanding the liquid and glassy volumes as was done in Equation (4.25) for the entropy, and setting $d\Delta v = 0$, we obtain for the hypothetical equilibrium $T_g(P)$ locus defined by volume continuity between liquid and glass

$$\frac{dT_g}{dP} = \frac{\Delta K_T}{\Delta\alpha_p} \tag{4.28}$$

and, therefore, comparing (4.26) and (4.28),

$$\frac{\Delta K_T}{\Delta\alpha_p} = \frac{T_g v_g \Delta\alpha_p}{\Delta c_p}. \tag{4.29}$$

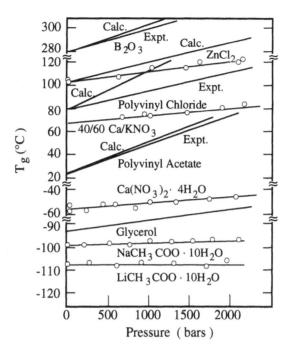

Figure 4.9. Experimental and theoretical pressure dependence of the glass transition for various liquids. Calculations are based on Equation (4.26). (Adapted from Angell and Sichina, 1976)

Experimentally, Equations (4.28) and (4.29) never hold for glasses. Instead, it is found that

$$\frac{\Delta K_T \, \Delta c_p}{T_g v_g \, (\Delta \alpha_p)^2} > 1 \qquad (4.30)$$

(Davies and Jones, 1953a,b) where, typically, this quantity, called the Prigogine-Defay ratio, is close to 2 (Goldstein, 1965, 1975; O'Reilly, 1962).

Although the Adam-Gibbs theory is based on the idea of cooperative relaxations, it provides no means for calculating the size of a cooperatively rearranging region. Several investigators have proposed definitions for this characteristic length scale and addressed the possible molecular mechanisms that underlie the growth of cooperatively rearranging regions. Frank (1952) noted that the thirteen-atom icosahedron formed by a central atom surrounded by twelve radially equidistant nearest neighbors gives rise to a very deep potential-energy minimum. This polyhedron composed of twenty identical tetrahedra is shown in Figure 4.10. Hoare (1976, 1978) noted that for atoms interacting through central potentials, non-space-filling arrangements, which he called amorphons,

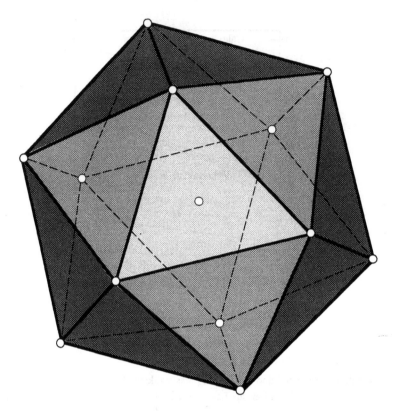

Figure 4.10. A thirteen-atom icosahedron. (Adapted from Steinhardt et al., 1983)

are energetically favored over space-filling clusters[14] for cluster sizes of up to several hundred atoms. Among the small-sized clusters, the thirteen-atom icosahedron is especially stable, and is separated from alternative arrangements by high potential energy barriers. Hoare proposed that such energetically favored amorphons should arise naturally and grow as the temperature of the supercooled liquid is lowered. The amorphons are local regions where noncrystallographic symmetry is energetically favored. The gradual growth of locally coherent regions with predominantly icosahedral symmetry is limited by two processes. First, the amorphons are not space filling; hence their growth must of necessity be self-limiting (Sachdev, 1992). Furthermore, such regions, embedded as they are in a continuum provided by the supercooled liquid, are initially noninteracting. As they grow, however, they gradually impinge on each other; the ensuing jamming is reflected macroscopically by a sudden increase in the viscosity and, eventually, by structural arrest. Since these interactions

---

[14]For example, the face-centered or body-centered cubic packings.

occur across domain walls, Stillinger (1988b) proposed that the cooperatively rearranging region should be the domain wall (surface), rather than its contents (volume). This leads to a modified Adam-Gibbs expression with $\log \tau \propto S_{\text{conf}}^{-2/3}$ [compare with (4.21)].

Hoare's amorphons endow the cooperatively rearranging regions of Adam and Gibbs' theory with plausible shapes, sizes, and mechanisms for growth. This descriptive viewpoint of the glass transition was investigated by Nelson and co-workers (Steinhardt et al., 1981, 1983; Nelson, 1983), using molecular dynamics. These authors considered two atoms to be nearest neighbors if their centers are separated by a distance smaller than a suitably defined cutoff. To each such nearest-neighbor pair there corresponds a "bond," the midpoint of which defines a vector $\mathbf{r}$ in an arbitrary coordinate reference system. Thus, with each nearest-neighbor bond one can associate a set of spherical harmonics Y_{lm} (Courant and Hilbert, 1989; Nelson, 1983), of which the thirteen harmonic polynomials corresponding to $l = 6$ are appropriate for the quantification of icosahedral symmetry ($m = -6, -5, \ldots, 6$),

$$Q_{6m}(\mathbf{r}) = Y_{6m}\left[\theta(\mathbf{r}), \phi(\mathbf{r})\right], \tag{4.31}$$

where θ and ϕ are the bond's polar angles defined with respect to the above-mentioned coordinate system. The bond-orientational order parameter Q_6 is a rotationally invariant combination of suitably averaged bond spherical harmonics,

$$Q_6 = \left\{\frac{4\pi}{13} \sum_{m=-6}^{6} |\langle Q_{6m}\rangle|^2\right\}^{1/2}, \tag{4.32}$$

where

$$\langle Q_{6m}\rangle = \frac{1}{N_{\text{bonds}}} \sum_{\text{bonds}} Q_{6m}(\mathbf{r}). \tag{4.33}$$

Q_6 is an overall measure of icosahedral symmetry. It vanishes identically in an isotropic liquid. To quantify spatial correlation between neighboring icosahedra, Nelson and co-workers defined a bond angle correlation function G_6,

$$G_6(r) = \frac{1}{13} \sum_{m=-6}^{6} \langle Q_{6m}^*(\mathbf{r})\, Q_{6m}(\mathbf{0})\rangle \langle Q_{00}^*(\mathbf{r})\, Q_{00}(\mathbf{0})\rangle^{-1}, \tag{4.34}$$

where * denotes complex conjugation, and angular brackets denote averaging over all pairs of bonds separated by a distance $r = |\mathbf{r}|$ [not to be confused with the bond vector of Equations (4.31)–(4.33)]. The Q_{6m} are to be computed according to the orientation of each bond with respect to the external coordinate system. The denominator, a normalization factor, is such that $G_6(r)$ equals unity if all bonds have the same spherical harmonics. The range of G_6 is a measure of the distance over which icosahedral correlations persist. In an isotropic liquid, G_6 decays exponentially to zero [$G_6 \approx \exp(-r/\xi_6)$], with ξ_6, a temperature- and density-dependent orientational correlation length, being of the order of the mean interparticle separation (Nelson, 1983).

Steinhardt et al. (1981, 1983) studied the supercooled Lennard-Jones fluid via molecular dynamics, and used a nearest-neighbor cutoff equal to the separation at the potential minimum, $2^{1/6}\sigma$. Upon isochoric supercooling at $\rho^* = N\sigma^3/V = 0.973,$[15] the bond-orientational order parameter Q_6 became nonzero at $T^* = kT/\varepsilon = 0.63$, 10% below the melting temperature of the Lennard-Jones solid at the given density, and increased sharply below the melting temperature of the Lennard-Jones solid at the given density. Q_6 increased sharply below this temperature. Likewise, the orientational correlation length ξ_6 dropped sharply for $T^* > 0.63$ (Figure 4.11). No concurrent increase in the translational correlation length was found upon supercooling. Only partial confirmation of these findings has subsequently been reported in molecular dynamics studies by Mountain and co-workers. Mountain and Brown (1984) found Q_6 values intermediate between lack of icosahedral order in the Lennard-Jones liquid (< 0.07) and orientational order in the corresponding fcc crystal (> 0.3) only for amorphous solids, the presence of which was easily detected from a saturation in particle displacements without a concurrent evolution of latent heat. In their study of the supercooled soft sphere ($n = 12$) fluid, Amar and Mountain (1987) found very small increases in Q_6 upon supercooling, but a pronounced increase in the lifetime of the weak orientational correlations, as measured by the rate of decay of the autocorrelation function $\langle Q_6(t)Q_6(0) \rangle$. For a binary mixture of unequal-sized soft spheres, Mountain and Thirumalai (1987) found neither Q_6 nor its autocorrelation to be useful indicators of the onset of vitrification. Instead, they defined a correlation function $\varphi(t)$,

$$\varphi(t) = N^{-1} \sum_{i,j} \langle \cos\theta_{ij}(t) \rangle, \qquad (4.35)$$

where N is the number of atoms in the simulation; the sum is over all central atoms i, and all nearest neighbors j of atom i; and $\theta_{ij}(t)$ is the angle between bond ij at times 0 and t. Thirumalai and Mountain (1987) found a marked increase in the decay time of $\varphi(t)$ upon supercooling. This suggests the growth of regions where individual configurational exploration becomes gradually more hindered.

Jonsson and Andersen (1988) found icosahedral ordering in supercooled binary Lennard-Jones systems.[16] They defined two atoms as neighbors when they are closer than the first minimum in the appropriate radial distribution func-

---

[15]For comparison, the liquid density at the triple point for this fluid is $\rho^* = 0.85 \pm 0.01$ (Hansen and Verlet, 1969).

[16]Mean-field calculations by Haymet (1983) also support the notion of icosahedral symmetry in cooperatively rearranging regions.

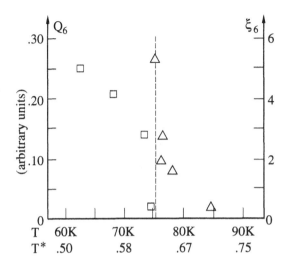

Figure 4.11. Temperature dependence of the icosahedral order parameter Q_6 ($\square$) [Equation (4.32)]; and of the icosahedral orientational correlation length ξ_6 ($\triangle$) for the supercooled Lennard-Jones fluid [$\rho^* = 0.973$; kT_m/ε (at $\rho^* = 0.973$) $= 0.701$]. The vertical dashed line denotes the temperature below which Q_6 becomes nonzero. The conversion between T^* and T (K) is T (K) $= 120T^*$ ($\varepsilon/k = 120$ K, the Lennard-Jones energy parameter for Ar). [Reprinted, with permission, from Steinhardt et al., *Phys. Rev. Lett.* 47: 1297 (1981)]

tion. In a molecular dynamics study of a mixture with [$\sigma_{11} = 1$, $\sigma_{22} = 0.8$, $\sigma_{12} = 0.9$; $\varepsilon_{11} = \varepsilon_{22} = \varepsilon_{12}$; $N_2/N_1 = 4$] they observed a substantial increase in the number of neighbor pairs having five common neighbors forming a pentagon of neighbor contacts, and of non-neighbor pairs sharing three common neighbors that form a triangle of neighbor contacts. Both types of pairs are signatures of icosahedral ordering: they occur in the thirteen-atom icosahedron shown in Figure 4.10. Their population was found to increase sharply upon isobaric cooling below 72 K at 1 bar (Ar units). In contrast to the findings of Steinhardt et al. (1981, 1983), Jonsson and Andersen (1988) found no relation between Q_6 and icosahedral ordering in a noncrystalline phase: increases in Q_6 consistent with the values reported by Steinhardt et al. (1981, 1983) were found only in partially crystalline samples. Jonsson and Andersen (1988) also found a marked system-size dependence in the extent of icosahedral ordering, with a system-spanning cluster of 138 interpenetrating or face-sharing icosahedra formed in a 1500-atom simulation (Figure 4.12), but a relatively smaller proportion (only 36 icosahedra) formed in a 500-atom system. This raises the

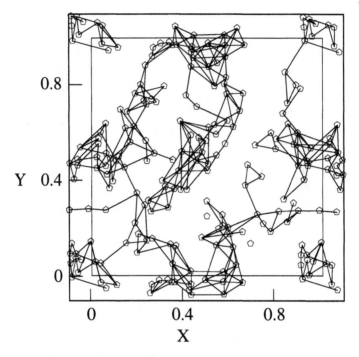

Figure 4.12. Icosahedral ordering in a 1500-atom supercooled binary Lennard-Jones system ($\sigma_{11} = 1$, $\sigma_{12} = 0.8$, $\sigma_{22} = 0.9$; $\varepsilon_{11} = \varepsilon_{12} = \varepsilon_{22}$; $N_2/N_1 = 4$) at $kT/\varepsilon = 0.2$. Solid lines connect centers of icosahedra (pentagons). The inner square is the boundary of the simulation cell. Two-dimensional projection. [Reprinted, with permission from Jonsson and Andersen, *Phys. Rev. Lett.* 60: 2295 (1988)]

serious question, first asked by Hoare (1976), of whether computer simulations of systems composed of 10^2–10^3 atoms are capable of providing statistically meaningful information on the formation of large (> 100 atoms) amorphons upon supercooling. In fact, as shown by Andersen and co-workers, severely system-size-dependent results are the norm in simulations of nucleation and ordering in deeply supercooled liquids. This important and often unrecognized limitation of simulations of supercooled liquids is discussed in detail in Section 4.7.

The above-described simulation studies have provided only qualitative information about cooperativity in supercooled liquids. Quantitative results (e.g., Figure 4.11) must be interpreted with care because of the discrepancies[17] with

---

[17]The discrepancy between the increase in Q_6 for the supercooled Lennard-Jones liquid reported by Steinhardt et al. (1981, 1983), and the observation by Mountain and Brown (1984) that only upon vitrification are such increases observed in noncrystalline systems, is noteworthy. Since Steinhardt et al. did not calculate mean squared displacements or other equivalent indicators of equilibration,

subsequent studies (Ernst et al., 1991). A promising recent development due to Mountain (1995) is the identification of a growing length scale in a model super-cooled liquid. Mountain found that the largest-wavelength propagating shear wave that a glass-forming binary mixture of soft spheres can support grows sharply upon supercooling. This suggests that the cooperatively rearranging region consists of clusters that exhibit some degree of rigidity over intervals long enough to allow the propagation of shear waves. Further work is needed to relate this length scale to the size of regions with icosahedral order.

An interesting perspective into cooperativity in supercooled liquids and glasses has resulted from the tiling model proposed originally by Stillinger and Weber (1986), and subsequently revisited by Stillinger and co-workers (Weber et al., 1986; Bhattacharjee and Helfand, 1987; Weber and Stillinger, 1987; Harris and Stillinger, 1989, 1990). The tiling model is based on the Adam-Gibbs concept of a cooperatively rearranging region. It describes the equilibrium statistics of tessellations (tilings) involving a collection of vari-ously sized domains, the cooperatively rearranging regions; and the kinetics of the interconversion processes allowed to such domains. Specifically, the model assumes a periodic two-dimensional system[18] tessellated without gaps or overlaps by square domains of any size. The potential energy associated with a particular tiling is given by

$$U = 2\lambda \sum_{j \geq 1} j \, n_j, \tag{4.36}$$

where n_j is the number of $(j \times j)$ squares in the tiling; and λ is an energy penalty per unit interdomain length. The physical picture is one of internally coherent domains whose boundaries are locations of high energy due to strained bonds or space-filling constraints such as arise, for example, when the coopera-tively rearranging regions have icosahedral symmetry (Stillinger, 1988b). The absence of gaps or overlaps imposes the constraint

$$\sum_{j \geq 1} j^2 n_j = N, \tag{4.37}$$

where N is the total area. The energy is minimized when all domain boundaries are eliminated in favor of a single system-spanning tile; it is maximized when the system is tessellated with unit (1×1) squares. Accordingly,

$$2N^{-1/2} \leq U/\lambda N \leq 2. \tag{4.38}$$

The above prescriptions define the equilibrium aspects of the tiling model. Clearly, the system will have coarse-textured, low-temperature equilibrium

---

the possibility of vitrification in their simulations cannot be ruled out. On the other hand, Jonsson and Anderson (1988) found that Q_6 increased only upon partial crystallization.

[18] See Harris and Stillinger (1989, 1990) for a generalization to three dimensions.

states with the minimal amount of domain boundaries, and high-temperature equilibrium states in which the energy penalty due to the proliferation of domain boundaries is compensated by the increase in entropy brought about by fine-textured tessellations with smaller domains.

Domain interconversion occurs through a set of allowed transitions. The ones prescribed by Stillinger and co-workers are of two types: minimal aggregation (or fragmentation), and boundary shift. In the former, domains of size $(pq) \times (pq)$ can fragment into p^2 squares of size $(q \times q)$ if and only if p is the smallest prime factor of pq. Conversely, a square arrangement of p^2 domains of size $(q \times q)$ can aggregate into a $(pq) \times (pq)$ domain if and only if p is the smallest prime factor of pq. In a boundary shift event, a $(p + 1) \times (p + 1)$ square fragments into a $(p \times p)$ domain and an L-shaped array of $(2p + 1)$ unit squares; conversely, the L-shaped array of unit squares coalesces with the partially enclosed $(p \times p)$ square to yield a $(p + 1) \times (p + 1)$ domain. Within any one class of moves, every configuration can be accessed from every other one with a finite number of moves. The rate of a minimal aggregation event was chosen as

$$r_a[p^2(q \times q) \to (pq) \times (pq)] = v_0 \, \alpha^{2pq(p-1)} \qquad (4.39)$$

and, for the inverse fragmentation event,

$$r_f[(pq) \times (pq) \to p^2(q \times q)] = v_0 \, \alpha^{2pq(p-1)} \exp[-2\beta \, \lambda pq(p-1)], \quad (4.40)$$

where v_0 is the attempt frequency, the reciprocal of which imposes the time scale of domain interconversion dynamics; α $(0 < \alpha < 1)$ is a rate parameter; $2pq(p-1)$ is the length of interdomain boundary that disappears (is created) in an aggregation (fragmentation) event; $2\lambda pq \, (p-1)$ is the potential energy decrease (increase) caused by an aggregation (fragmentation) event; $\beta = 1/kT$; and the exponential factor in (4.40) follows from detailed balancing. Similarly, for boundary shift moves, the rates satisfy

$$r_a[(p \times p) + (2p + 1)_L(1 \times 1) \to (p + 1) \times (p + 1)] = v_0 \, \alpha^{4p}, \quad (4.41)$$

$$r_f[(p + 1) \times (p + 1) \to (p \times p) + (2p + 1)_L(1 \times 1)] = v_0 \, \alpha^{4p} \exp(-4\beta \lambda p). \qquad (4.42)$$

It follows from the form of the potential energy that the equilibrium size of the average domain increases as the temperature is lowered. However, the aggregation rates for both types of moves also decrease as the domain size increases [recall that $\alpha < 1$; then see (4.39) and (4.41)]. This, in addition to the increased difficulty of finding ever larger domains properly arranged so as to allow aggregation, imposes a kinetic barrier on the evolution towards a coarsely textured equilibrium state of low energy. The ratio of aggregation to fragmentation rates increases as the temperature decreases and as the domain size increases. This introduces sluggish dynamics, and, eventually, glassy behavior.

The model exhibits a first-order phase transition between a condensed state of zero entropy involving a single macroscopic tile, and a heterogeneously tiled high-temperature phase. The phase transition has been variously reported to occur at $\beta\lambda = 0.323$ (using simple statistics to estimate the system's entropy: Stillinger and Weber, 1986); $\beta\lambda = 0.271$ (fitting Monte Carlo results for the energy, followed by thermodynamic integration to obtain the free energy: Weber and Stillinger, 1987; Weber et al., 1986); $\beta\lambda = 0.27$ (using a transfer-matrix-based method to compute the partition function: Bhattacharjee and Helfand, 1987); and $\beta\lambda = 0.264$ (using improved statistics to compute the system's entropy: Harris and Stillinger, 1990). Since the low-temperature phase has zero entropy and is the lowest-energy "amorphous" configuration that can result from the model, the transition can be thought of as an ideal glass transition. Interestingly, adding an energy penalty to account for the space-filling inability of amorphous packings in the system's Euclidean dimension smooths the first-order transition into a continuous transition (Harris and Stillinger, 1989, 1990).

The relaxation behavior of the tiling model has been studied via Monte Carlo simulations (Weber et al., 1986; Weber and Stillinger, 1987). As the temperature is lowered, the attainment of equilibrium involving ever larger tiles is kinetically impeded. Consequently, the system falls out of equilibrium and exhibits glassy behavior even above the thermodynamic transition. This is evidenced, for example, in cooling-rate-dependent energies and hysteresis upon reheating (Weber et al., 1986). The dynamics were investigated via the energy autocorrelation function

$$\phi(t) = \frac{\langle U(t)U(0)\rangle - \langle U\rangle^2}{\langle U^2\rangle - \langle U\rangle^2}, \tag{4.43}$$

where, as explained above, time is measured in units of $(\nu_0)^{-1}$. The numerical data were fitted in Kohlrausch-Williams-Watts (KWW) form (Kohlrausch, 1854; Williams and Watts, 1970),

$$\phi(t) = \exp[-(t/\tau_U)^{1-n}], \tag{4.44}$$

with τ_U and $1 > n > 0$ adjustable parameters. KWW dynamics, commonly referred to as stretched exponential behavior, occur when a system has a spectrum of relaxation times. Relaxation processes in supercooled liquids and glasses often exhibit KWW dynamics,[19] with τ_U a strongly temperature-dependent and

---

[19]Though the KWW function is generally used as a convenient empiricism for fitting data, its ubiquity suggests a more fundamental significance. Palmer et al. (1984), for example, proposed a class of relaxation models that gives rise to KWW dynamics. Stillinger (1985) explored the topology of the many-body potential energy surface that is consistent with KWW relaxation.

generally non-Arrhenius relaxation time. Weber et al. (1986) found it convenient to study the temperature dependence of the relaxation time not through τ_U, but through an average relaxation time $\langle \tau \rangle$,

$$\langle \tau \rangle = \int_0^\infty \phi(t)\, dt. \tag{4.45}$$

They found small deviations from both Arrhenius and Adam-Gibbs behavior (ln $\langle \tau \rangle$ not strictly linear either in $\beta\lambda$ or in $\beta\lambda k N/S$, with S, the system's entropy); and faster relaxation rates for boundary shift than for minimal aggregation kinetics.

The tiling model illustrates the insights into complex problems that can result from idealized models that capture the essence of the underlying physics. In this case, a plausible prescription for the energy cost of creating a domain boundary is all that is required to generate equilibrium behavior characterized by growth of internally coherent domains (cooperatively rearranging regions) at low temperatures. Although in two dimensions the resulting phase transition is implausibly first order, it is significant that the transition is smeared by introducing an energy penalty to account for the space-filling inability of amorphous domains in three dimensions. An important feature of the tiling model is that it provides an equilibrium length scale for the domain size.[20] In the language of the Adam-Gibbs theory, this corresponds to the size of the cooperatively rearranging region. This is illustrated in Figure 4.13.

Another interesting model based on the idea of cooperative relaxations is the kinetic Ising model of Fredrickson and Andersen (Fredrickson and Andersen, 1984, 1985; Fredrickson and Brawer, 1986; Fredrickson, 1986). Each site on a lattice can be in one of two states: spin up or spin down. An external field favors the alignment of all spins in the down configuration at equilibrium. The rate at which an up spin flips down is proportional to $m(m-1)$, where m is the number of the spin's nearest neighbors that are in the up state. Therefore a given spin cannot flip unless it has two or more neighboring spins in the up state. One can view an up spin as representing a high-energy, low-density, high-compressibility region in the liquid; and a down spin as representative of low-energy, high-density, low-compressibility regions that are favored at low temperature. A spin flip, therefore, corresponds to a local structural rearrangement. Then, requiring that an up spin flip only if two or more of its neighbors are also up simply says that structural change requires a compressible, low-density local environment. At low temperature, when the system is composed largely of low-energy, high-density regions, relaxations become

---

[20]Specifically, above the transition temperature, the equilibrium concentration of ($j \times j$) domains satisfies $\ln(n_j/N) = -Kj^2 - Lj - M$, where n_j is the number of ($j \times j$) domains, N the system's area, and K, L, and M temperature-dependent coefficients (Weber and Stillinger, 1987). K is positive at high temperature and becomes negative at the transition temperature: in the thermodynamic limit this implies the appearance of a single system-spanning tile at equilibrium.

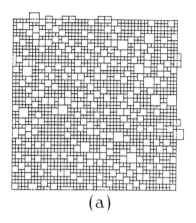

(a)

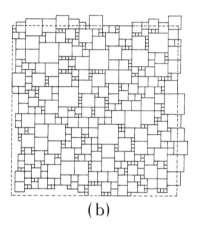

(b)

Figure 4.13. Configurations of the two-dimensional tiling model. (a): $\beta\lambda = -0.2$; domain boundary creation is energetically favored. (b): $\beta\lambda = 3.33$; nonequilibrated glassy state. [Reprinted, with permission, from Weber et al., *Phys. Rev. B*. 34: 7641 (1986)]

progressively cooperative and sluggish. Monte Carlo simulations of this model in two dimensions exhibited KWW stretched exponential relaxation dynamics (Fredrickson and Brawer, 1986), with relaxation times in excellent agreement with the Adam-Gibbs prediction as to their temperature dependence. These simulations suggest that the analytical results for this model (Fredrickson and Andersen, 1984, 1985) that predict a purely dynamic transition are incorrect.

An insightful way of visualizing structural relaxations in supercooled liquids and glasses is to study the spatial distribution of relaxation times (Harrowell, 1993; Foley and Harrowell, 1993). Figure 4.14 is a plot of the time elapsed before a spin's first flip in the two-dimensional Fredrickson-Andersen model (Foley and Harrowell, 1993). The salient features are the spatial correlation of relaxation rates, and the coarsening of the structure at low temperatures. The correlation length associated with the second moment of the distribution of first flip times bears a simple power-law relationship with the average relaxation time.[21] This correlation length is thus a good measure of the size of a cooperatively rearranging region, an important quantity about which the Adam-Gibbs theory provides no information.

4.3.3 Free Volume Theory

According to the free volume approach, vitrification occurs when the volume available for translational molecular motion falls below a critical value (Turnbull and Cohen, 1958, 1961, 1970; Cohen and Turnbull, 1959, 1964). A relationship between free volume and vitrification was first proposed by Fox and Flory (1950, 1951, 1954). A more recent version of the theory views the experimentally observable glass transition as a manifestation of an underlying first-order percolation transition occurring below T_g (Cohen and Grest, 1979, 1980, 1981; Grest and Cohen, 1980, 1981).

Consider a glass. Let the volume per molecule be v, and let v_0 denote the volume per molecule excluded to all other molecules. Upon heating, the excess volume, $v - v_0$, increases. At very low temperatures, packing constraints impose a severe energy penalty upon any nonuniform redistribution of the excess volume. Consequently, thermal expansion results mainly in increased anharmonicity in the vibrational motion of molecules about essentially fixed positions, and the excess volume is uniformly distributed. Furthermore, the entropy increase due to thermal expansion is small, since the volume change is not accompanied by randomization of molecular motion. Under these conditions, the thermal expansion coefficient is similar to that of a crystal. Eventually, $v - v_0$ reaches a critical value, beyond which any additional volume can be redistributed at random without an energy penalty. We refer to that part of the

---

[21] Consider dividing an $L \times L$ lattice into $(l \times l)$-sized cells. The second moment of relaxation times is defined as

$$m_2(l) = \left[n_l^{-1} \sum_{j=1}^{n_l} \left(\tau_{j,l} - \langle \tau \rangle \right)^2 \right] \times \left[N^{-1} \sum_{j=1}^{N} \left(\tau_{j,1} - \langle \tau \rangle \right)^2 \right]^{-1},$$

where n_l is the number of $l \times l$ cells, $\tau_{j,l}$ is the mean relaxation time (first flip time) in the jth $l \times l$ cell, $\langle \tau \rangle$ is the overall mean relaxation time (first flip time), and $N = L \times L$. The correlation length associated with the second moment of the distribution is given by $\xi_2 = 1 + \int_1^L m_2(l) \, dl$. Foley and Harrowell (1993) found $\langle \tau \rangle \propto \xi_2^{12.6}$. The very large exponent means that glassy structural arrest in the Fredrickson-Andersen model is not associated with diverging length scales.

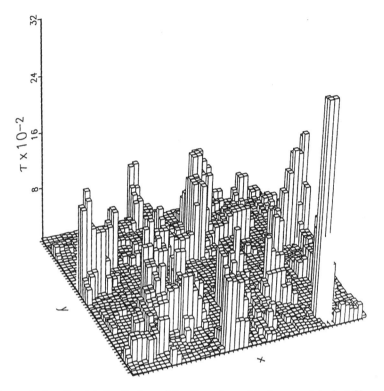

Figure 4.14a. Spatial distribution of the time elapsed before a spin's first flip on a 50×50 lattice, for the Fredrickson-Andersen kinetic Ising model at $kT/B = 1.67$. B is the magnetic field. One unit of τ corresponds to 2,500 spin-flip attempts. [Reprinted, with permission, from Foley and Harrowell, *J. Chem. Phys.* 98: 5069 (1993)]

excess volume that can be distributed randomly without energy cost as the free volume v_f,

$$v - v_0 = v_f + \Delta v_c, \tag{4.46}$$

where Δv_c is that part of the thermal expansion that cannot be randomly distributed.[22] The randomization of molecular motion arising from the redistribution of free volume contributes significantly to the entropy. Under these conditions, the thermal expansion coefficient is liquidlike. The ideal glass transition occurs when $v_f = 0$, and therefore $v - v_0 = \Delta v_c$ (Turnbull and Cohen, 1961). For the laboratory transition, $v_f > 0$.

---

[22] A more common way of defining free volume (e.g., Doolittle, 1951) is to equate it to the excess volume, that is to say, not to distinguish between excess volume that is randomly distributed and that which is uniformly distributed.

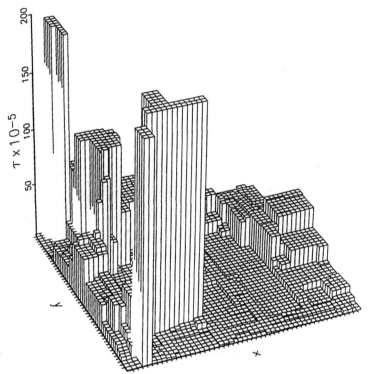

Figure 4.14b. Spatial distribution of the time elapsed before a spin's first flip on a 50×50 lattice, for the Fredrickson-Andersen kinetic Ising model at $kT/B = 0.83$. B is the magnetic field. One unit of τ corresponds to 2,500 spin-flip attempts. [Reprinted, with permission, from Foley and Harrowell, *J. Chem. Phys.* **98**: 5069 (1993)]

Molecules in a dense liquid are confined to a cage formed by their immediate neighbors. Spontaneous density fluctuations occasionally result in the formation of holes within a cage that are large enough to allow an appreciable displacement of the trapped central molecule. This can give rise to diffusive motion only if another molecule occupies the hole before the first one can return to its original position. Figure 4.15 shows a two-dimensional cage in a fluid composed of impenetrable disks. For the given positions of its immediate neighbors, the center of molecule A can move freely within the shaded area. Also shown in Figure 4.15 is a schematic representation of the work required to remove a central molecule from its cage, as a function of the cage radius. Typically, this function has a minimum at R_0, and increases very steeply for $R < R_0$. For $R > R_0$, the rise is less steep, and $U(R)$ is linear over a range of R. In the Turnbull-Cohen theory (Cohen and Turnbull, 1959; Turnbull and Cohen,

1961), the liquidlike regime of random free volume redistribution corresponds to this linear portion of the $U(R)$ curve.

The main result of the Cohen-Turnbull theory is a relationship between translational diffusion coefficient and free volume. Denoting the free volume associated with a given molecule as v', the translational diffusion coefficient is written as a sum over all possible contributions from molecules having different available free volumes:

$$D = \int_{v*}^{\infty} D(v')p(v')\,dv' = g\,u \int_{v*}^{\infty} a(v')p(v')\,dv', \qquad (4.47)$$

where $D(v')$ is the contribution to the translational diffusion coefficient due to molecules having free volume v'; $p(v')$ is the probability density of finding a free volume element between v' and $v' + dv'$; $a(v')$ is a length scale indicative of the cage diameter corresponding to a molecule having free volume v'; u is the thermal velocity $[u \approx (kT/m)^{1/2}]$; g is a geometric factor; and $v*$ is the minimum free volume capable of accommodating another molecule after the original displacement in the cage. Since free volume is distributed at random, the function $p(v')$ can be easily calculated. Specifically, we wish to maximize

$$\Omega = N! / \prod_i N_i! \qquad (4.48)$$

subject to

$$\gamma \sum_i N_i\, v_i = N\, v_f, \qquad (4.49)$$

$$\sum_i N_i = N, \qquad (4.50)$$

where Ω is the number of ways of redistributing the total available free volume, N is the number of molecules, N_i is the number of molecules having free volume v_i, v_f is the average free volume per molecule, and γ is a geometric factor that corrects for the overlap of free volume. For given N and v_f, and upon making the free volume distribution continuous, this constrained maximization yields

$$p(v') = \frac{\gamma}{v_f} \exp\left(-\frac{\gamma v'}{v_f}\right). \qquad (4.51)$$

Since $D(v')$ varies more slowly than $p(v')$, and in general $v* \gg v_f$ (Cohen and Turnbull, 1961), we can write

$$\begin{aligned} D &= \int_{v*}^{\infty} D(v')p(v')\,dv' \\ &\approx D(v*) \int_{v*}^{\infty} p(v')\,dv' = ga*u \exp\left(-\frac{\gamma v*}{v_f}\right), \end{aligned} \qquad (4.52)$$

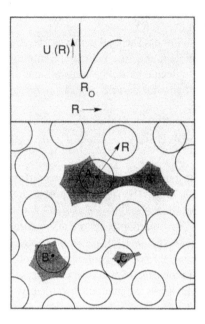

Figure 4.15. Two-dimensional cages in a fluid of hard disks. The shaded areas are the regions accessible to the centers of molecules A, B, and C. Molecule A can undergo a diffusive motion but B and C are trapped in their respective cages. The inset shows schematically the work U required to remove a molecule from its cage, as a function of the cage's radius R. (Adapted from Zallen, 1983)

where, consistently with the interpretation given to v^*, we take a^* to be of the order of the molecular diameter. The free volume expression predicts that D, and hence the possibility for translational motion, will vanish when the liquid has no more free volume available for random redistribution ($v_f = 0$).

Equation (4.52) is related to the empirical Doolittle equation, proposed originally to describe the viscosity of hydrocarbon liquids (Doolittle, 1951),

$$\eta^{-1} = A \, \exp\left(-\frac{b \, v_0}{v_f}\right) \tag{4.53}$$

where A and b are constants (the latter of order unity), v_0 and v_f are the occupied and free volume per molecule, respectively;[23] and the implied inverse relationship between translational diffusion and viscosity follows from

---

[23]The quantity v_f in the Doolittle equation is the total excess volume $v - v_0$. It therefore includes the term Δv_c of (4.46).

the Stokes-Einstein equation,

$$D = \frac{kT}{6\pi a \eta} \quad (4.54)$$

in which a is the molecular diameter.

Starting from the Doolittle equation, and noting that for a highly supercooled liquid $v_0 \gg v_f$ [hence $v_f/(v_f + v_0) \approx v_f/v_0$], we can write

$$\ln \eta = \text{const} + \frac{b}{f}, \quad (4.55)$$

where f is the Doolittle fractional free volume, v_f/v_0. Therefore,

$$\ln \frac{\eta(T)}{\eta(T_g)} = b \left(f^{-1} - f_g^{-1} \right), \quad (4.56)$$

where f_g is the fractional free volume at the glass transition. Assuming, somewhat arbitrarily, that the fractional free volume satisfies

$$f = f_g + \Delta\alpha_p \left(T - T_g \right), \quad (4.57)$$

where $\Delta\alpha_p$ is the difference between the thermal expansion coefficients of the liquid and the glass,[24] we obtain

$$\log_{10} \frac{\eta(T)}{\eta(T_g)} = \frac{b}{2.303} \left(f^{-1} - f_g^{-1} \right)$$

$$= \frac{-(b/2.303 f_g) (T - T_g)}{(T - T_g) + f_g/\Delta\alpha_p} = \frac{-C_1(T - T_g)}{C_2 + T - T_g}. \quad (4.58)$$

Equation (4.58) is the Williams-Landel-Ferry (WLF) viscosity equation (Williams et al., 1955). Taking b to be unity, and using the "universal" WLF parameters $C_1 = 17.44$ and $C_2 = 51.6°C$ (Williams et al., 1955), we obtain $f_g = 0.025$ and $\Delta\alpha_p = (4.8 \times 10^{-4})°C^{-1}$ (see Table 4.4).

The Cohen-Turnbull free-volume theory was revisited by Cohen and Grest (Cohen and Grest, 1979, 1980, 1981; Grest and Cohen, 1980, 1981), who extended the model to include thermodynamics explicitly. Using the term "cell" to denote a molecule and its surrounding cage, Cohen and Grest assumed that the relationship between a cell's local free energy a and its volume v is qualitatively similar to the $U(R)$ curve in Figure 4.15. Of particular relevance to

---

[24]To justify (4.57), we write $v(T) = v(T_g) + \alpha_l v(T_g)(T - T_g)$ and $v_0(T) = v_0(T_g) + \alpha_g v_0(T_g)(T - T_g)$, where v is the specific volume, and the subscripts l and g denote the liquid and the glass. Subtracting, and assuming $\alpha_l v(T_g) - \alpha_g v_0(T_g) \approx v(T_g)\Delta\alpha_p$, we obtain $v_f(T) = v_f(T_g) + \Delta\alpha_p v(T_g)(T - T_g)$ [note that this is the Doolittle free volume, which includes the Δv_c term in (4.46)]. Finally, assuming $v_f(T)/v(T_g) \approx v_f(T)/v(T)$, we obtain (4.57).

the theory is the difference between the quadratic portion close to the minimum v_0, and the linear portion at larger volumes ($v > v_c$), which allows one to write a schematic local Helmholtz energy

$$a(v) = a_0 + \frac{1}{2}\kappa(v - v_0)^2, \quad v < v_c,$$

$$a(v) = a_0 + \frac{1}{2}\kappa(v_c - v_0)^2 + \zeta(v - v_c), \quad v > v_c, \qquad (4.59)$$

where a_0, K, and ζ are temperature- and density-dependent parameters. Cells with $v > v_c$ are termed liquidlike; those with $v < v_c$, solidlike. Only the former possess free volume, which is given by $v_f = v - v_c$. Free exchange and redistribution of free volume can only occur between neighboring cells, provided they have a sufficient number ($\geq z$, say) of liquidlike nearest neighbors (Cohen and Grest, 1979). If a liquidlike cell has z or more liquidlike neighbors, it belongs to a liquidlike cluster. In analogy with percolation theory (see, e.g., Ziman, 1979), the glass transition is viewed as occurring when the fraction of liquidlike cells drops below a critical value, causing clusters to be isolated from one another. When the critical fraction of liquidlike cells is exceeded, an infinite connected cluster is formed, and the material becomes a liquid.

The difference between liquid- and solidlike clusters is that molecules in the former can move freely throughout the cluster. The entropy difference between a solid cluster, in which molecules are translationally localized, and a liquidlike cluster, in which they are not, is often referred to as communal entropy (Hill, 1956). In the Cohen-Grest theory, the Helmholtz energy A of a dense supercooled liquid to which the cage picture applies is written as

$$A = N \int dv \, p(v) \, [a(v) + kT \ln p(v)] \; - \; T \, S_{comm}, \qquad (4.60)$$

where $p(v)$ is the probability that a cell has volume v, N is the number of cells (and hence molecules), and S_{comm} is the communal entropy. The key quantities in the theory are the cell volume distribution function $p(v)$ and a cluster distribution function, the communal entropy being an explicit function of the latter. With f denoting the fraction of liquidlike cells,

$$f = \int_{v_c}^{\infty} p(v) \, dv, \qquad (4.61)$$

the $f(T)$ and $A(f)$ relationships that result from solving for the volume and cluster distribution functions are shown schematically in Figure 4.16. These are the key results of the theory. The fraction of liquidlike cells changes discontinuously at a temperature T_p. Correspondingly, the $A(f)$ curve shows one liquid minimum at high temperature, two equally deep minima representing the glass and liquid states at T_p, and two unequal minima thereafter, the deeper one corresponding to the glass. The metastable liquid minimum at $f > f_c$ is predicted

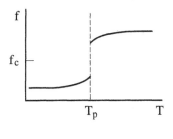

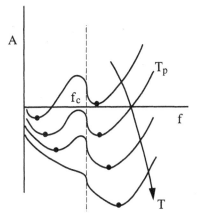

Figure 4.16. Prediction of a first-order glass transition in the Cohen-Grest free-volume theory. The upper curve shows the temperature dependence of the fraction of liquidlike cells, which changes discontinuously at T_p. The lower curve shows the relationship between the Helmholtz energy A and the fraction of liquidlike cells, at several temperatures. The arrow indicates the direction of increasing temperatures, and f_c is the fraction of liquidlike cells at the predicted first-order transition. At T_p, the two minima are equally deep. The free energy minimum at $f > f_c$ corresponds to the liquid; that at $f < f_c$ to the glass. (Adapted from Cohen and Grest, 1979)

to persist down to $T = 0$. The Cohen-Grest free volume theory predicts that the true liquid-glass transition is first order, and is therefore accompanied by volume, enthalpy, and entropy discontinuities. This is in contrast with experimental evidence. Cohen and Grest argued that observable glass transitions (in which $T_g > T_p$) mask the true underlying first-order transition, the supercooled liquid falling out of equilibrium before the equilibrium thermodynamic event at T_p can be observed.

The Cohen-Grest theory, being thermodynamic, makes predictions about Δc_p and $\Delta \alpha_p$. Direct comparison with experiment is difficult because of the number of parameters involved. Superimposed on the specific heat discontinuity is a δ-function spike at T_p due to the latent heat involved in the first-order

transition (see the crystal line in Figure 4.4). In addition, the treatment yields
an expression for the free volume,

$$\bar{v}_f = \frac{k}{2\zeta_0} \left\{ T - T_0 + \left[(T - T_0)^2 + 4 v_a \zeta_0 T k^{-1} \right]^{12} \right\}, \tag{4.62}$$

$$\zeta = \zeta_0 + \frac{k T_1}{v_a + \bar{v}_f}, \tag{4.63}$$

$$kT_0 = kT_1 + \zeta_0 v_a, \tag{4.64}$$

where $\bar{v}_f$ is the average free volume, ζ is defined in Equation (4.59), and ζ_0,
v_a, and T_1 are parameters. Note that the free volume only vanishes at $T = 0$.
Starting from $p(v)$, an expression for the diffusion coefficient can be obtained
in strict analogy with the Cohen-Turnbull treatment [see Equation (4.47)]; this
yields a Doolittle-type equation (Cohen and Grest, 1979); substitution of Equation (4.62) for the free volume results in the following viscosity expression:

$$\log_{10} \eta = C + D \left\{ T - T_0 + \left[(T - T_0)^2 + 4 v_a \zeta_0 T \right]^{1/2} \right\}^{-1}, \tag{4.65}$$

in which C, D, T_0, and $4v_a\zeta_0$ are adjustable parameters. The large number of constants detracts from the equation's fundamental significance, but
gives it great flexibility for fitting experimental data. Figure 4.17 (Cohen
and Grest, 1979) shows the temperature dependence of the viscosity of tri-
α-naphthylbenzene at atmospheric pressure covering sixteen decades (η) and
250 degrees (T), and the excellent fit provided by Equation (4.65).

Figure 4.18 shows the diffusion coefficient of supercooled hard spheres,
calculated by molecular dynamics simulation (Woodcock and Angell, 1981).
The line through the data is a fit to the Doolittle form, assuming an inverse
relation between viscosity and diffusion coefficient. The volume terms used in
this Doolittle plot were v_{fcc} for v_0, and $(v - v_K)$ for v_f. v_{fcc} is the specific volume
corresponding to close packing in a face-centered cubic lattice (fcc), with hard
spheres occupying 74% of the available space; and v_K is the Kauzmann volume,
that is to say, the volume at which equation-of-state calculations predict the
vanishing of the entropy difference between supercooled and fcc hard spheres
at the same temperature and pressure (Angell et al., 1981; Woodcock, 1981).
Figure 4.18 shows that the diffusion coefficient of supercooled hard spheres
obeys a functional form that is not inconsistent with free volume predictions
over a volume fraction range $0.495 < \phi < 0.562$.[25]

The apparent success of free volume-based interpretations of vitrification
and supercooled liquid behavior suggested by Figures 4.17 and 4.18 should be
viewed with caution. Among the glass-forming systems whose $T_g(P)$ relation

---

[25] $\phi = (\pi \sqrt{2}/6) (1.163 + 1/x)$; x (see Figure 4.18) $= 1/(v/v_{\text{fcc}} - 1.163)$.

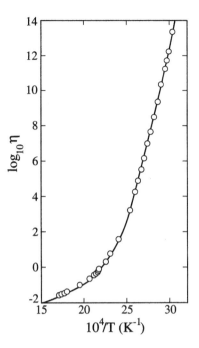

Figure 4.17. Temperature dependence of the viscosity of tri-α-naphthylbenzene at atmospheric pressure. Points are experimental data; full line is the fit to Equation (4.65). (Adapted from Cohen and Grest, 1979)

is shown in Figure 4.9, $CH_3CO_2Li \cdot 10H_2O$ is noteworthy, in that its glass transition temperature decreases with increasing pressure. According to Equation (4.28), this would imply $\Delta \alpha_p < 0$, since $\Delta K_T > 0$ always. However, (4.28) is never satisfied experimentally, and we rely on the measurements of Williams and Angell (1977), who report $\Delta \alpha_p = -0.4 \ (\pm 0.4) \times 10^{-4} \ °C^{-1}$ for this substance. This small difference is almost on the limit of detectability (Williams and Angell, 1977). Nevertheless, it is very difficult to reconcile negative or vanishing thermal expansion coefficient differences between a supercooled liquid and a glass with free volume theory. This follows from Equation (4.46), which when differentiated reads

$$\frac{dv_f}{dT} = \frac{dv}{dT} - \frac{d(\Delta v_c + v_0)}{dT} = \alpha_l \, v_l(T) - \alpha_g \, v_g(T) \approx \frac{dv}{dT} - \frac{d \Delta v_c}{dT} , \quad (4.66)$$

where we have assumed that v_0 is temperature independent. Applying (4.66) at the glass transition,

$$\left. \frac{d v_f}{d T} \right|_{T_g} \approx \Delta \alpha_p \, v \, (T_g). \quad (4.67)$$

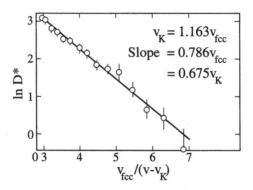

Figure 4.18. Density dependence of the diffusion coefficient of supercooled hard spheres. Points are molecular dynamics results; the line is a fit to the Doolittle equation, with v_{fcc} the specific volume at face-centered-cubic close packing, and v_K the volume at which equation-of-state calculations (Woodcock, 1981; Angell et al., 1981) predict the vanishing of the entropy difference between liquid and crystalline hard spheres. D^* is the dimensionless diffusion coefficient, $D(m/kT)^{1/2}\sigma^{-1}$, with m the mass and σ the hard-sphere diameter [Reprinted, with permission, from Woodcock and Angell, *Phys. Rev. Lett.* 47: 1129 (1981)]

The very basis of free volume theories is that increased mobility is due to a free volume increase, which is clearly incompatible with $\Delta\alpha_p < 0$ at the glass transition. This, coupled with the consistent failure of the volume-based, Ehrenfest-type equation for $T_g(P)$ [i.e., Equation (4.28)], means that the free volume viewpoint, though often useful for data correlations (Herbst et al., 1993), is not rigorous.

4.4 DYNAMIC VIEWPOINT OF THE GLASS TRANSITION: MODE COUPLING

Mode-coupling theory, in the form in which it was originally applied to supercooled liquids (Leutheusser, 1984; Bengtzelius et al., 1984), views vitrification as a transition from ergodic to nonergodic behavior in the relaxation dynamics of density fluctuations. In the former case, all microscopic configurations are accessible; in the latter, entire regions of phase space become inaccessible due to structural arrest. Insofar as structural arrest arises solely from the solution of an evolution equation for density fluctuations, and is not accompanied by singularities in any thermodynamic quantity, this is a dynamic viewpoint. More recently (e.g., Götze and Sjörgen, 1992), the emphasis has been on applying this theory to describe the dynamics of relaxation processes well above the laboratory glass transition. Both viewpoints will be described in this section.

In most of the mode-coupling literature on glass formation, heavy emphasis is placed on mathematical formalism, rather than on the underlying physics. An exception is the insightful early work of Geszti (1983), a good introduction to the topic. Three facts make up the basic picture: (i) shear stress relaxation occurs primarily through diffusive motion; (ii) diffusivity is inversely related to viscosity; (iii) viscosity is proportional to the shear stress relaxation time. Combined, (i)–(iii) lead to a viscosity feedback mechanism, which can be described (Geszti, 1983) by first writing the viscosity η as a product,

$$\eta = G_\infty \tau, \tag{4.68}$$

where τ is the shear relaxation time, and G_∞ is the instantaneous (elastic) shear modulus. Writing the shear relaxation time as a sum of vibrational and structural contributions, (4.68) reads

$$\eta \approx G_\infty \tau_{\text{vib}} + G_\infty c(T) D^{-1} = \eta_0(T) + b(T) D^{-1}, \tag{4.69}$$

where it is assumed that the structural relaxation time is inversely related to the diffusion coefficient D, and where b and c are to be obtained from a detailed microscopic theory. Invoking the Stokes-Einstein equation (4.54), we obtain

$$\eta = \eta_0(T) + [6\pi \, a \, b(T) / kT]\eta = \eta_0(T) + B(T)\eta \tag{4.70}$$

and, therefore,

$$\eta = \eta_0(T)/[1 - B(T)]. \tag{4.71}$$

Equation (4.70) implies a feedback mechanism whereby the viscosity controls the shear relaxation time, and hence the viscosity itself. Since the structural relaxation time does not increase if the temperature is raised while keeping D constant, and the elastic modulus decreases with increasing temperature, $b(T)$ is a nonincreasing function of temperature, and hence $B(T)$ increases monotonically as the temperature is decreased. Thus Equation (4.71) predicts a viscosity increase upon cooling; the resulting divergence is identified with vitrification. The mode-coupling approach to the glass transition recasts this phenomenological picture on a more rigorous basis.

The basic quantity in the theory, F, is the Fourier transform of the van Hove density-density correlation function G (Hansen and McDonald, 1986):

$$F_\mathbf{k}(t) = N^{-1}\langle \rho_\mathbf{k}(t)\, \rho_{-\mathbf{k}}(0)\rangle = \int G(\mathbf{r}, t)\exp(-i\,\mathbf{k}\cdot\mathbf{r})\, d\mathbf{r}, \tag{4.72}$$

$$
\begin{aligned}
G(\mathbf{r}, t) &= N^{-1}\left\langle \sum_i \sum_j \int \delta[\mathbf{r}' + \mathbf{r} - \mathbf{r}_i(t)]\delta[\mathbf{r}' - \mathbf{r}_j(0)]\, d\mathbf{r}'\right\rangle \\
&= N^{-1}\left\langle \sum_i \sum_j \delta\left[\mathbf{r} + \mathbf{r}_j(0) - \mathbf{r}_i(t)\right]\right\rangle,
\end{aligned}
\tag{4.73}
$$

or, equivalently,

$$G(\mathbf{r}, t) = \rho^{-1}\langle \rho(\mathbf{r}, t)\, \rho(\mathbf{0}, 0)\rangle, \tag{4.74}$$

where

$$\rho_{\mathbf{k}} = \int \rho(\mathbf{r})\, \exp\left(-i\mathbf{k} \cdot \mathbf{r}\right) d\mathbf{r} = \sum_i \exp\left(-i\mathbf{k} \cdot \mathbf{r}_i\right). \tag{4.75}$$

In the above equations, angular brackets denote thermodynamic averaging; ρ is the bulk number density; $\rho(\mathbf{r})$ the number density at $\mathbf{r}$; N the number of molecules in the system; δ the Dirac delta function; summations over i and j run from 1 to N (i and j identify molecules); $\mathbf{k}$ identifies a Fourier component; $\int \cdots d\mathbf{r} = \int \cdots dV$, where V is the system's volume; and 0 and t are time labels. The van Hove function is defined so that $G(\mathbf{r}, t)d\mathbf{r}$ is proportional to the probability of observing a particle i at $\mathbf{r} \pm d\mathbf{r}$ at time t, given that there was a particle j at the origin at time 0 (Hansen and McDonald, 1986). The zero-time value of F is the static structure factor S

$$S_{\mathbf{k}} = N^{-1}\langle \rho_{\mathbf{k}}(0)\, \rho_{-\mathbf{k}}(0)\rangle \tag{4.76}$$

(Hansen and McDonald, 1986). Central to the theory is a differential equation for F whose solution yields the time evolution of the decay of density fluctuations. It is found that, for certain values of density and temperature, F decays to zero, while for other values it decays to a finite, nonzero number. The former condition is identified with the liquid (ergodic behavior); the latter, in which density fluctuations cannot be fully relaxed due to structural arrest, is identified with the glass (non-ergodic behavior),[26]

$$\lim_{t \to \infty} \Phi_{\mathbf{k}}(t) = 0, \text{ liquid,}$$

$$\lim_{t \to \infty} \Phi_{\mathbf{k}}(t) \neq 0, \text{ glass,} \tag{4.77}$$

where

$$\Phi_{\mathbf{k}}(t) = F_{\mathbf{k}}(t)\, S_{\mathbf{k}}^{-1}. \tag{4.78}$$

The differential equation for Φ reads

$$\ddot{\Phi}_{\mathbf{k}}(t) + \nu_0 \dot{\Phi}_{\mathbf{k}}(t) + \Omega_{\mathbf{k}}^2\, \Phi_{\mathbf{k}}(t) + \Omega_{\mathbf{k}}^2 \int_0^t \Gamma_{\mathbf{k}}(t - t')\dot{\Phi}_{\mathbf{k}}\left(t'\right) dt' = 0, \tag{4.79}$$

(Bengtzelius et al., 1984; De Raedt and Götze, 1986; Götze, 1990; Götze and Sjörgen, 1992), with initial conditions

$$\Phi_{\mathbf{k}}(0) = 1, \quad \dot{\Phi}_{\mathbf{k}}(0) = 0. \tag{4.80}$$

---

[26]The ergodic behavior of supercooled liquids has been investigated by Mountain and co-workers by molecular dynamics (Thirumalai et al., 1989; Mountain and Thirumalai, 1989).

In Equation (4.79), single and double dots denote first and second time derivatives, v_0 is a damping constant, and Ω_k is a characteristic frequency, given by

$$\Omega_k{}^2 = (k\,v_0)^2\,S_k^{-1}; \; v_0 = (kT/m)^{1/2}\,, \tag{4.81}$$

where m is the mass of a molecule, and $\Gamma_k(t)$ is a memory function.

The feedback mechanism outlined in (4.68)–(4.71) enters the theory via the yet to be specified Φ dependence of Γ, the introduction of which causes the instantaneous rate of relaxation of density fluctuations to depend on its own history through the integral term in (4.79).[27] Introducing the Laplace transform of Φ (Götze, 1985; Götze and Sjörgen, 1992),

$$\Phi_k(z) = i \int_0^\infty \Phi_k(t) \exp(izt)\,dt, \quad \mathrm{Im}(z) \geq 0, \tag{4.82}$$

and taking into account (4.80), Equation (4.79) becomes

$$\Phi_k(z) = \frac{-1}{z - \dfrac{\Omega_k^2}{z + M_k(z)}}\,, \tag{4.83}$$

$$M_k(z) = i v_0 + \Omega_k^2 \Gamma_k(z) \tag{4.84}$$

(Götze,1990; Götze and Sjörgen, 1992), where $\Gamma_k(z)$ is the transform of $\Gamma_k(t)$ [see (4.82)].

The Φ dependence of the memory function can be written in general form as

$$\Gamma_k(t) = \sum_{m=1}^{m_0} (1/m!) \sum_{k_1,\ldots,k_m} V^{(m)}(k, k_1, \ldots, k_m)\, \Phi_{k_1}(t) \cdots \Phi_{k_m}(t), \tag{4.85}$$

(Götze, 1990) where $V^{(m)}$ is a so-called vertex function. Truncated versions of (4.85) are used in calculations: the two-mode ($m = 2$) approximation of Bengtzelius et al. (1984) is an example,

$$\Gamma_k(t) = (2\pi)^{-3} \int V^{(2)}(k, k')\, \Phi_{k'}(t)\, \Phi_{|k-k'|}(t)\,dk'. \tag{4.86}$$

A kinetic theory of dense atomic liquids yields the following expression for the two-mode vertex (Bengtzelius et al., 1984; Sjörgen, 1980a; Sjörgen and Sjölander, 1979):

$$V^{(2)}\left(k, k'\right) = (\rho kT/m)\left(\hat{k} \cdot k\right) c\left(k'\right) S\left(k'\right) S\left(|k - k'|\right)$$

$$\times \left\{\left(\hat{k} \cdot k'\right) c\left(k'\right) + \left[\hat{k} \cdot (k - k')\right] c\left(|k - k'|\right)\right\}, \tag{4.87}$$

---

[27] For detailed discussions of the memory function treatment of correlation functions see Zwanzig (1960), Mori (1965a,b), Boon and Yip (1980), and Hansen and McDonald (1986).

with $\hat{\mathbf{k}} = \mathbf{k} \, |\mathbf{k}|^{-1}$, and where $c(\mathbf{k})$ is the Fourier transform of the direct correlation function,

$$\rho \, c(\mathbf{k}) = [S(\mathbf{k}) - 1]/S(\mathbf{k}). \tag{4.88}$$

The static structure factor can be calculated from

$$S(k) = 1 + 4\pi \rho \int r^2 g(r) \, (kr)^{-1} \sin(kr) \, dr, \tag{4.89}$$

where $g(r)$, the pair correlation function, gives the average ratio of local to bulk density a distance r away from a central molecule. It can be obtained from knowledge of the pairwise interaction potential between molecules using integral equation theories (see, e.g., McQuarrie, 1976).

Equations (4.79), (4.80), (4.86), and (4.87) constitute a (two-)mode-coupling model of the dynamics of supercooled atomic liquids relevant to the glass transition. The approach is predictive, requiring as input only the static structure factor at the given density and temperature, a quantity that can be calculated from knowledge of the interaction potential between molecules. Recalling (4.77), we write the vitrification condition

$$\lim_{t \to \infty} \Phi_{\mathbf{k}}(t) = f_{\mathbf{k}} \neq 0 \Leftrightarrow \lim_{z \to 0} \Phi_{\mathbf{k}}(z) = -z^{-1} f_{\mathbf{k}} \tag{4.90}$$

(Götze and Sörgen, 1992), and, taking into account (4.83) and (4.84),

$$\frac{f_{\mathbf{k}}}{1 - f_{\mathbf{k}}} = \Gamma_{\mathbf{k}}(t \to \infty), \tag{4.91}$$

where the long-time limit of the memory function is obtained when $f_{\mathbf{k}}$ is used instead of $\Phi_{\mathbf{k}}$ in (4.86). The quantity $f_{\mathbf{k}}$ is called the non-ergodicity parameter, the Debye-Waller factor, or the Edwards-Anderson parameter (Edwards and Anderson, 1975; Götze and Sjörgen, 1992). Temperature and density combinations that lead to nonzero solutions for $f_{\mathbf{k}}$ in (4.91) correspond to glassy states; zero solutions pertain to the supercooled liquid.

The theory therefore yields the (ρ, T) locus of a sharp change from ergodic to nonergodic behavior, which is identified with the glass transition. Also predicted by the theory is the time-dependent behavior of the supercooled liquid, through numerical, asymptotic, and theoretical analysis of the governing dynamic equation, (4.79) or (4.83). We now review mode-coupling predictions on the location of the glass transition, the behavior of the shear viscosity and the self-diffusion coefficient of highly supercooled liquids, and the relaxation dynamics of supercooled liquids.

Mode-coupling predictions for the glass transition of the hard sphere fluid are listed in Table 4.7. In this case, there is only one variable of interest: the density at which structural arrest occurs.

These values are somewhat low compared to most other estimates. Using a linear diffusivity-volume extrapolation of stable liquid hard sphere diffusivities,

Table 4.7. Mode-Coupling Predictions for the Hard Sphere Glass Transition

Reference	ϕ^a	$Method^b$
Bengtzelius et al. (1984)	0.516	WTPY^c
Bengtzelius (1986a)	0.525	VWWTPY^d
Barrat et al. (1989)	0.525	VWWTPY^d
Barrat el al. (1989)	0.510	VWWTPY^{d,e}

^{a}Volume fraction ($= \rho\pi\sigma^3/6$) at the glass transition; σ is the hard sphere diameter.

^{b}Method for the calculation of the static structure factor.

^{c}Wertheim-Thiele (Wertheim, 1963, 1964; Thiele, 1963) solution of the Percus-Yevick equation for hard spheres (Percus and Yevick, 1958).

^{d}Verlet-Weis (Verlet and Weis, 1972) correction to the Wertheim-Thiele solution of the Percus-Yevick equation for hard spheres.

^{e}Calculation includes triplet correlations.

Woodcock and Angell (1981) obtained a value of 0.536 for the volume fraction at which structural arrest occurs due to the vanishing of the self-diffusion coefficient. Hudson and Andersen (1978) calculated volume fractions of $\phi = 0.506$ and 0.546 for the hard sphere glass transition by locating the intersection of liquid and amorphous solid equations of state (PV/NkT vs. ϕ). They used the Carnahan-Starling equation for the supercooled liquid (Carnahan and Starling, 1969), and both truncated and self-consistent free volume–based expressions for the solid. This gives rise to the two different values of ϕ. Finally, we recall from Section 4.3.1 that Speedy (1994) obtained a value of $\phi = 0.562$ using molecular dynamics to locate the intersection of the liquid and vitreous branches of the equation of state; and that Woodcock (1981) reported $\phi = 0.565$ as the homogeneous nucleation limit of the hard sphere fluid.

Bengtzelius (1986a) solved Equation (4.91) for the Lennard-Jones fluid. His results are shown in Figure 4.19, which also includes comparisons with computer simulations. These comparisons should be interpreted with caution. As will be discussed in detail in Section 4.7, simulations yield a very broad transition that is shifted towards higher temperatures due to the inevitably high simulated cooling rates ($> 10^{12}$ K sec^{-1}; Angell et al., 1981; Fox and Andersen, 1984). In this type of simulation, the glass transition is determined by the intersection of straight-line fits to the high- and low-temperature branches of isobaric density vs. temperature or enthalpy vs. temperature curves. The transition region where the isobars exhibit appreciable curvature can be as wide as $0.36\varepsilon/k$ (Fox and Andersen, 1984; equivalent to 63.5°C for the ε/k value used by these authors). Furthermore, at these conditions, simulated diffusion coefficients are of the order of 10^{-8}–10^{-7} cm^2 sec^{-1} (Fox and Andersen, 1984), whereas in laboratory glass transitions this value is much lower, typically

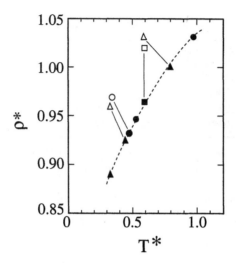

Figure 4.19. Comparison between mode-coupling predictions for the location of the glass transition (full symbols) and the corresponding molecular dynamics calculations (open symbols). Circles refer to the Lennard-Jones potential (Clarke, 1979; isobaric simulation at zero pressure); triangles to a shifted and truncated Lennard-Jones potential (Fox and Andersen, 1984; two isobaric simulations at $P\sigma^3/\varepsilon = 3$ and 10, pressure increasing from left to right); squares to a purely repulsive Lennard-Jones potential (Ullo and Yip, 1985; isothermal simulation at $kT/\varepsilon = 0.6$). The thin lines connect corresponding theoretical and simulation points. $\rho^* = \rho\sigma^3$; $T^* = kT/\varepsilon$. (Adapted from Bengtzelius, 1986a)

10^{-18} cm^2 sec^{-1} (Angell et al., 1981). Thus, although the simulated system has "vitrified" in one sense, it is far from a condition of structural arrest, and it would be more meaningful to compare the mode-coupling glass transition with simulation-based extrapolations to a condition of vanishing diffusion. The problem with this approach, however, is that extrapolations based on data at $D \approx 10^{-7}$cm^2 sec^{-1} are inaccurate, because a single functional form, $D(\rho)$ or $D(T)$, is unlikely to apply throughout the entire extrapolated range.

Mode-coupling theory also predicts the functionality that describes the vanishing of the self-diffusion coefficient (and therefore the viscosity divergence) as structural arrest is approached. The mode-coupling prediction reads

$$D(\text{or } \eta^{-1}) \propto \Delta^{[1/(1-x)+1/(y-1)]} = \Delta^y,$$

$$\Delta = (\rho_x - \rho)/\rho_x \text{ or } (T - T_x)/T_x \qquad (4.92)$$

(Götze, 1985), where ρ_x and T_x are the density and temperature at which structural arrest occurs; Δ describes the proximity to the glass transition along isothermal or isobaric paths; and x (< 1) and y (> 1) are the two roots of the

Table 4.8. Mode-Coupling Predictions on the Exponent Characterizing Structural Arrest

Reference	γ	Memory Function
Leutheusser (1984)[a]	1.765	Schematic M2
Bengtzelius et al. (1984)[b]	1.765	Equation (4.86)[d]
Kirkpatrick (1985)[b]	1.890	Kadanoff and Swift (1968)
Bengtzelius (1986b)[c]	2.368	Sjörgen (1980b)[e]
Barrat et al. (1989)[c]	2.580	Equation (4.68)[f]

[a] Model with no $\mathbf{k}$ dependence.

[b] Simplified model with all $\mathbf{k}$ dependence lumped into $\mathbf{k}_0$, corresponding to first $S(\mathbf{k})$ peak.

[c] Full $\mathbf{k}$ dependence.

[d] Hard sphere fluid. Wertheim-Thiele Percus-Yevick structure factor (see Table 4.7).

[e] Lennard-Jones fluid. Structure factor calculated via random phase approximation (Weeks et al., 1971; Andersen et al., 1972; Bengtzelius, 1986a).

[f] Hard sphere fluid. Verlet-Weis, Wertheim-Thiele, Percus-Yevick structure factor (see Table 4.7).

equation

$$\lambda = \Gamma \left(\frac{1+\xi}{2} \right)^2 / \Gamma(\xi) \qquad (4.93)$$

with Γ denoting the gamma function. For two-mode models, the exponent parameter λ is given by

$$\lambda = (1/2) \sum_{\mathbf{k},\mathbf{q},\mathbf{p}} \hat{e}_{\mathbf{k}}^{gl} \left[\frac{\partial^2 \Gamma_{\mathbf{k}}}{\partial \Phi_{\mathbf{q}} \partial \Phi_{\mathbf{p}}} (1 - \Phi_{\mathbf{q}})^2 (1 - \Phi_{\mathbf{p}})^2 \right]_{gl} e_{\mathbf{q}}^{gl} e_{\mathbf{p}}^{gl}, \qquad (4.94)$$

where the superscript and subscript gl (glass) denote quantities evaluated at the transition, and $\hat{e}_{\mathbf{q}}$ and $e_{\mathbf{k}}$ denote left and right eigenvectors of the so-called stability matrix

$$\sum_{\mathbf{q}} C_{\mathbf{kq}}^{gl} e_{\mathbf{q}}^{gl} = e_{\mathbf{k}}^{gl}, \quad \sum_{\mathbf{k}} \hat{e}_{\mathbf{k}}^{gl} C_{\mathbf{kq}}^{gl} = \hat{e}_{\mathbf{q}}^{gl}, \quad C_{\mathbf{kq}} = \frac{\partial \Gamma_{\mathbf{k}}}{\partial \Phi_{\mathbf{q}}} (1 - \Phi_{\mathbf{q}})^2 \qquad (4.95)$$

(Götze, 1985). Table 4.8 lists the mode-coupling predictions for the exponent γ of Equation (4.92).

The terminology "schematic M2" in Table 4.8 refers to memory function expressions of the type

$$\Gamma(t) = \sum_{m=1}^{m_0} V^{(m)} \Phi^m(t), \qquad (4.96)$$

where the $V^{(m)}$ are pure numbers (Götze, 1990). An M2 model corresponds to

$V^{(1)} = 0$, $V^{(2)} \neq 0$, $V^{(j)} = 0$ $(j > 2)$; an M12 model to $V^{(1)} \neq 0$, $V^{(2)} \neq 0$, $V^{(j)} = 0$ $(j > 2)$.[28]

Figure 4.20 shows mode-coupling predictions for the self-diffusion coefficient of the supercooled Lennard-Jones fluid by Bengtzelius (1986b).[29] The points represent calculations at $kT/\varepsilon = 0.6$, the straight line is a power-law fit with $\gamma = 1.66$, and the curved line a Doolittle fit with singular density $\rho\sigma^3 = 0.965$. At $kT/\varepsilon = 0.6$, Bengtzelius calculated vitrification to occur at $0.9601 < \rho\sigma^3 < 0.9602$ (the triple point of the Lennard-Jones fluid occurs at $kT/\varepsilon = 0.68 \pm 0.02$; Hansen and Verlet, 1969). It is impossible to distinguish power-law from Doolittle behavior in these calculations. There is an obvious inconsistency in Figure 4.20 between the exponent of the power-law fit ($\gamma = 1.66$) and the corresponding theoretical prediction ($\gamma = 2.368$; Table 4.8).

Figure 4.21 shows a mode-coupling power-law fit (Barrat et al., 1989) of molecular dynamics calculations of supercooled hard sphere diffusivities (Woodcock and Angell, 1981). The full line is the equation

$$D^* = D_0(\phi_{gl} - \phi)^{2.58} \tag{4.97}$$

with D_0 (= 0.716) and the glass volume fraction (= 0.6) as fitting parameters, and $D^* = D\left(m\sigma^{-2}/kT\right)^{1/2}$. Since the fitted diffusivities span only one decade, Figure 4.21 cannot be considered as a rigorous test of the mode-coupling exponent $\gamma = 2.58$. Furthermore, using the glass volume fraction as fitting parameter detracts from the fundamental significance of a plot such as Figure 4.21, especially in light of the large discrepancy between the optimum fit value and the theoretical predictions (Table 4.7).

The viscosities of many moderately supercooled liquids exhibit an isobaric temperature dependence that is not inconsistent with the mode-coupling prediction

$$\eta \propto (T - T_x)^{-\gamma}, \tag{4.98}$$

where, typically, $1.5 < \gamma < 2.3$ (Angell, 1988b; Taborek et al., 1986). This behavior is illustrated in Figure 4.22, where the straight line corresponds to $\gamma = 2$ and T_x is obtained as the temperature intercept of this line. Note that the viscosity does not diverge at T_x. Rather, this temperature is indicative of a

---

[28]Such models are useful in the mathematical study of mode-coupling theory (Götze, 1990), as well as in the analysis of the dynamics of supercooled liquids. From this point of view, the model of Bengtzelius et al. (1984) [i.e., Equation (4.86)] is an M2 model.

[29]This quantity can be obtained by combining the viscosity expression due to Geszti (1983) with the Stokes-Einstein equation, to arrive at

$$D = \frac{10\pi}{\sigma} \left[\int_0^\infty dk\, k^4\, [dS(k)/dk]^2/[S(k)]^4 \int_0^\infty dt\, [F_k(t)]^2 \right]^{-1},$$

which shows how the condition $\lim_{t \to \infty} F_k(t) \neq 0$ causes structural arrest.

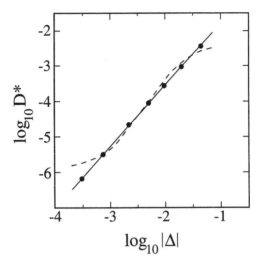

Figure 4.20. Mode-coupling calculations for the self-diffusion coefficient of the supercooled Lennard-Jones fluid at $kT/\varepsilon = 0.6$ (circles) $[D^* = D\sigma^{-1}(m/\varepsilon)^{1/2}; \Delta = (\rho_x - \rho)/\rho_x]$. Subscript x denotes structural arrest. The straight line corresponds to power-law behavior, $D^* \propto \Delta^\gamma$, with $\gamma = 1.66$; the dashed line to Dolittle behavior with a singular specific volume $v\sigma^{-3} = 1.036$. [Reprinted, with permission, from Bengtzelius, *Phys. Rev. A.* 34: 5059 (1986)]

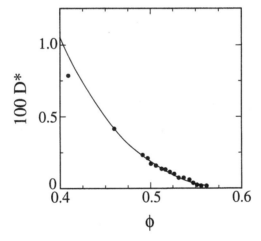

Figure 4.21. Volume fraction dependence of the self-diffusion coefficient of the supercooled hard sphere fluid $[D^* = D\left(m/\sigma^2 kT\right)^{1/2}]$. The points are molecular dynamics calculations (Woodcock and Angell, 1981). The curve is Equation (4.97), with $D_0 = 0.716$ and $\phi_{gl} = 0.6$ used as fitting parameters. [Reprinted, with permission, from Barrat et al., *J. Phys. Cond. Matt.* 1: 7163 (1989)]

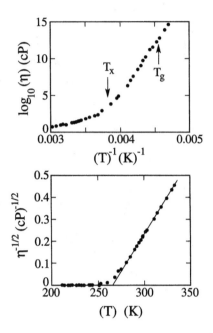

Figure 4.22. Temperature dependence of the viscosity of salol (phenyl salicylate) at atmospheric pressure. The straight line is consistent with Equation (4.98), with $\gamma = 2$. Note the breakdown of power-law, mode-coupling-type behavior around 275 K. (Adapted from Taborek et al., 1986)

transition from power-law behavior at $T > T_x$ to a different low-temperature regime in which the behavior is either Arrhenius or VTF. The power-law behavior illustrated in Figure 4.22 is quite general: the paper by Taborek et al. (1986) lists values of γ for seventeen liquids, including water, alcohols, aromatic and aliphatic hydrocarbons, halogenated hydrocarbons, and noble gases. Viscosities at T_x fall typically in the 10^2–10^3 P range, far short of the 10^{13} P normally associated with the glass transition.

It is largely as a result of the generality of behavior such as is shown in Figure 4.22 that more recent work (e.g., Barrat et al., 1989; Ullo and Yip, 1989; Götze, 1990; Rössler, 1990; Götze and Sjörgen, 1992) places emphasis on mode coupling as a theory of the $T > T_x$ regime, whereas the original literature (e.g., Leutheusser, 1984; Bengtzelius et al., 1984) aimed at describing loss of ergodicity. Recent extensions of the theory (Götze and Sjörgen, 1987; Sjörgen and Götze, 1991; Fuchs et al., 1992) incorporate a coupling between density and momentum fluctuations. These terms allow for activated diffusion, and restore ergodicity at low temperatures. This extended mode-coupling theory has not been studied in as much detail as the original formulation, which predicts a sharp loss of ergodicity.

Another important prediction of mode-coupling theory concerns the relaxation dynamics of supercooled liquids in the $T > T_x$ regime. Defining

$$a = \frac{1-x}{2}, \tag{4.99}$$

$$b = \frac{y-1}{2}, \tag{4.100}$$

hence $\gamma = (1/2a) + (1/2b)$ [see (4.92) and (4.93)], there are two characteristic times,

$$\tau_{\mathrm{I}} = t_0 \Delta^{-\gamma}, \tag{4.101}$$

$$\tau_{\mathrm{II}} = t_0 \Delta^{-1/2a}, \tag{4.102}$$

$$\lim_{\Delta \to 0} \frac{\tau_{\mathrm{I}}}{\tau_{\mathrm{II}}} = \Delta^{-1/2b} = \infty, \tag{4.103}$$

where t_0 is a microscopic time scale, say $d(m/kT)^{1/2}$, with d of atomic dimensions. For $\tau_{\mathrm{I}} \gg t \gg t_0$, density fluctuations relax according to

$$\Phi_{\mathbf{k}}(t) = f_{\mathbf{k}} + A_{\mathbf{k}} \Delta^{1/2} g\left(t/\tau_{\mathrm{II}}\right) \tag{4.104}$$

(Götze and Sjörgen, 1992), where $f_{\mathbf{k}}$ is the non-ergodicity parameter (see Götze, 1985, for $A_{\mathbf{k}}$ expressions for two-mode theories). The scaling function g has the following short- and long-time limiting forms:

$$g\left(t/\tau_{\mathrm{II}}\right) = \left(t/\tau_{\mathrm{II}}\right)^{-a}, \qquad t/\tau_{\mathrm{II}} \ll 1, \tag{4.105}$$

$$g\left(t/\tau_{\mathrm{II}}\right) = -B\left(t/\tau_{\mathrm{II}}\right)^{b}, \qquad t/\tau_{\mathrm{II}} \gg 1, \tag{4.106}$$

where $B > 0$. In general, g is the solution of the equation

$$-1 + \lambda g\left(\hat{t}\right)^2 + \frac{d}{d\hat{t}} \int_0^{\hat{t}} d\hat{t}' g\left(\hat{t} - \hat{t}'\right) g\left(\hat{t}'\right) = 0, \tag{4.107}$$

where $\hat{t} = t/\tau_{\mathrm{II}}$ and λ has been defined before [see (4.93) and (4.94)]. In particular, (4.107) means that B is a function of λ.

For $t \gg \tau_{\mathrm{II}}$, the dynamics obeys the scaling law

$$\Phi_{\mathbf{k}}(t) = F_{\mathbf{k}}(t/\tau_{\mathrm{I}}), \tag{4.108}$$

(Götze and Sjörgen, 1992), where $F_{\mathbf{k}}$ is the solution of the equation

$$F_{\mathbf{k}}\left(\tilde{t}\right) = \Gamma_{\mathbf{k}}\left(\tilde{t}\right) - \frac{d}{d\tilde{t}} \int_0^{\tilde{t}} d\tilde{t}' \, \Gamma_{\mathbf{k}}\left(\tilde{t} - \tilde{t}'\right) F_{\mathbf{k}}\left(\tilde{t}'\right) \tag{4.109}$$

where $\tilde{t} = (t/\tau_1)$, and the memory function $\Gamma_{\mathbf{k}}$ is to be evaluated at the condition of structural arrest, $\Delta = 0$. The short-time asymptote of $F_{\mathbf{k}}$ satisfies

$$\Phi_{\mathbf{k}} \left(t/\tau_1\right) = f_{\mathbf{k}} - A_{\mathbf{k}} B \left(t/\tau_1\right)^b \tag{4.110}$$

(Götze and Sjörgen, 1992). Equations (4.104), (4.105), (4.106), (4.108), and (4.110) apply also to correlators between variables X and Y, Φ_{XY}, for which $\langle \delta X \delta \rho \rangle$ and $\langle \delta Y \delta \rho \rangle$ do not vanish (δ denotes a fluctuation with respect to equilibrium). In this case, $f_{\mathbf{k}}$ and $A_{\mathbf{k}}$ depend on X and Y. Equations (4.106) and (4.110) are sometimes called the von Schweidler law (Götze and Sjörgen, 1992).

Mode-coupling theory thus makes detailed predictions about the relaxation dynamics of supercooled liquids for times that are intermediate between the microscopic and hydrodynamic limits. The exponents a and b are predicted to be independent of density, temperature, wave number, and the particular correlation function. The exponents are related to the system's long-time behavior, since $\tau_1 \propto \eta \propto D^{-1} \propto \Delta^{-\gamma}$ and $\gamma = (1/2a) + (1/2b)$ [see (4.92) and (4.99)]. Relaxation occurs in two steps: $\Phi_{\mathbf{k}}$ approaches a condition of apparent structural arrest [$\Phi_{\mathbf{k}} \rightarrow f_{\mathbf{k}}$ according to (4.105)]; subsequently, $\Phi_{\mathbf{k}}$ decays to zero and away from $f_{\mathbf{k}}$. The initial decay occurs according to the von Schweidler law [see (4.106) and (4.110)].

Figure 4.23(a) (Kob and Andersen, 1994) shows the evolution of $\Phi_s(\mathbf{k}, t)$ in a supercooled binary Lennard-Jones mixture, calculated by molecular dynamics. The subscript s stands for "self," $\Phi_s(\mathbf{k}, t)$ being the Fourier transform of the normalized van Hove function [Equation (4.73)] with $i = j$,

$$G_s \left(\mathbf{r}, t\right) = N^{-1} \left\langle \sum_{i=1}^{N} \delta \left[\mathbf{r} + \mathbf{r}_i \left(0\right) - \mathbf{r}_i \left(t\right)\right] \right\rangle, \tag{4.111}$$

$$\Phi_s \left(\mathbf{k}, t\right) = \frac{\int G_s \left(\mathbf{r}, t\right) \exp\left(-i\mathbf{k} \cdot \mathbf{r}\right) d\mathbf{r}}{\int G_s \left(\mathbf{r}, 0\right) \exp\left(-i\mathbf{k} \cdot \mathbf{r}\right) d\mathbf{r}}. \tag{4.112}$$

$G_s(\mathbf{r}, t)d\mathbf{r}$ is proportional to the probability of observing particle i at $\mathbf{r} \pm d\mathbf{r}$ at time t, given that particle i was at the origin at time 0. $\Phi_s(\mathbf{k}, t)$ is called the self-intermediate scattering function. At sufficiently low temperatures, $\Phi_s(\mathbf{k}, t)$ exhibits a plateau; the approach to this plateau is described by (4.104) and (4.105). The initial departure from the plateau is described by (4.104) and (4.106). Figure 4.23(b) (Kob and Andersen, 1994) was obtained from 4.23(a) by scaling the time with τ such that $\Phi_s(\tau) = e^{-1}$. The dashed line is a master curve that can be fitted with $\Phi_s = 0.783 - A \left(t/\tau\right)^{0.488 \pm 0.015}$. Thus, von Schweidler power-law behavior with temperature-independent f and exponent b is obeyed over almost three decades of rescaled time. Kob and Anderson (1994) also showed that the von Schweidler exponent b is independent of the wave vector $\mathbf{k}$ and is quite similar for both mixture species; that the singular temperature T_x obtained from fitting the diffusion coefficient to $D \propto (T - T_x)^\gamma$ is the same for the two mixture components; and that the exponent γ obtained from such a fit differs for the two components. While the last prediction is at odds with the theory, the remaining findings represent a significant confirmation of theoretical predictions. They show that the system attains a condition of apparent structural arrest, the plateau in Φ_s, whose value is the **k**-dependent non-ergodicity parameter, and whose duration increases with proximity to the singularity, according to (4.102). Kob and Andersen fitted the scaling time τ to $\tau \propto (T - T_x)^{-\gamma}$, and obtained a γ value which agreed with the independently determined a and b exponents. This is a stringent test of the theory, because a and b can not only be determined independently from fits to Φ according to (4.105) and (4.106), but they are also related mathematically by Equations (4.93), (4.99), and (4.100); furthermore, one has the relationship $\gamma = (1/2a) + (1/2b)$. Note that the von Schweidler law does not apply to the very-long-time limit, $\Phi \to 0$. The simulations of Kob and Andersen span a time of ca. 10^{-8} sec, using Lennard-Jones Ar parameters to convert to dimensional time. This amounts to two orders of magnitude longer than previous computational studies of mode-coupling predictions (Ullo and Yip, 1985).

Figure 4.24 shows the relaxation dynamics of a supercooled Lennard-Jones liquid at $kT/\varepsilon = 0.6$, as predicted by numerical solution of the mode-coupling equations (Bengtzelius, 1986b). Using Ar parameters to convert to dimensional time, the x axis is found to span the range 10^{-14}–10^{-6} sec. These numerical calculations span a longer time than both the molecular dynamics study of the supercooled Lennard-Jones fluid at the same temperature $t \le 10^{-10}$ sec (Ullo and Yip, 1985), and the binary Lennard-Jones simulations of Kob and Andersen ($t \le 3 \times 10^{-8}$ sec; see Figure 4.23). Note the similarity with Figure 4.23(a), including short-time relaxation on a time scale $\sigma \, (m/\varepsilon)^{1/2}$, an intermediate period of apparent structural arrest, and final relaxation away from the plateau. Bengtzelius found that the evolution of Φ_k after the plateau could be described by the KWW function, with **k**-dependent exponent. The stretched exponential behavior reported by Bengtzelius (1986b) has an important unrealistic feature: the exponent n [see Equation (4.44)] approaches 0 at long times. There is no experimental evidence of this return to exponential relaxation at long times.

The two-step relaxation predicted by mode-coupling theory and shown in Figures 4.23 and 4.24 is often incorrectly confused with the so-called β and α processes of supercooled and glassy dynamics, which we discuss briefly below.

The dynamics of deeply supercooled liquids can be described in terms of two well-separated processes: slow collective motions associated with the exploration of deep configurational energy minima; and faster, noncollective motions associated with the exploration of local minima in the vicinity of a given deep configurational minimum (Stillinger, 1995; see also Section 4.3.2). The relaxation behavior due to collective motions is called α relaxation; that due to faster,

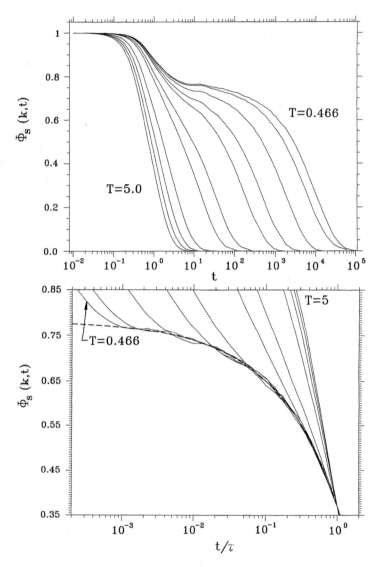

Figure 4.23. (a, top): Evolution of the self-intermediate scattering function for A-type atoms, $\Phi_s(k, t)$, in a supercooled binary Lennard Jones mixture ($\varepsilon_A = 1$, $\varepsilon_B = 0.5$, $\varepsilon_{AB} = 1.5$, $\sigma_A = 1$, $\sigma_B = 0.88$, $\sigma_{AB} = 0.8$, $N_A = 800$, $N_B = 200$), at $k\sigma_A = 7.251$, corresponding to the first peak in the static structure factor of species A. Time is in units of $\sigma_A(m/48\varepsilon_A)^{1/2}$, and temperature in units of ε_A/k_b, with k_b Boltzmann's constant. Temperatures from left to right: 5, 4, 3, 2, 1, 0.8, 0.6, 0.55, 0.5, 0.475, and 0.466. (b, bottom): Scaling behavior of the self-intermediate scattering function. Same curves as in (a), but with time scaled by τ, such that $\Phi_s(k, \tau) = e^{-1}$. The dashed line is the von Schweidler fit $\Phi_s = 0.783 - A(t/\tau)^{0.488}$. (Adapted from Kob and Andersen, 1994)

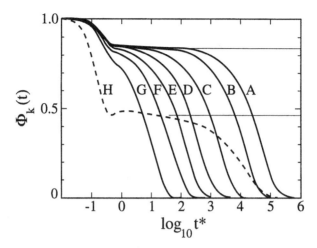

Figure 4.24. Numerical mode-coupling calculations of the relaxation of the supercooled Lennard-Jones fluid at $kT/\varepsilon = 0.6$. Time is in units of $\sigma (m/\varepsilon)^{1/2}$. Solid curves are for $k\sigma = 7$; dashed curve, $k\sigma = 6$. Curve A, $\Delta [= (\rho_x - \rho)/\rho_x] = 0.00031$; B, $\Delta = 0.00073$; C, $\Delta = 0.0023$; D, $\Delta = 0.0054$; E, $\Delta = 0.0106$; F, $\Delta = 0.021$; G, $\Delta = 0.052$; H, $\Delta = 0.00031$; x denotes structural arrest. The straight lines show the values of f_k [see Equation (4.90)]. [Reprinted, with permission, from Bengtzelius, *Phys. Rev. A.* 34: 5059 (1986)]

noncollective motion, β relaxation (Johari, 1973, 1976; Angell, 1990b; Stillinger, 1995). Figure 4.25 (Johari, 1973) shows the frequency dependence of the imaginary component of the dielectric permittivity (dielectric loss spectrum) slightly above T_g. The time scales associated with the α and β relaxations at the given conditions are 1–10 sec and 10^{-5} sec, respectively. The corresponding temperature dependence of the frequency of maximum loss is shown in Figure 4.26 (Johari, 1973). The β curve shows Arrhenius behavior; the α curve becomes progressively non-Arrhenius in the vicinity of T_g, and can be fitted with the VTF equation (Johari, 1976). These two figures suggest a picture of relaxation in supercooled liquids that can be summarized as follows (Johari, 1976). Each isolated and deep configurational energy minimum (an α minimum) contains multiple local minima separated by small energy barriers. The configurational changes associated with the exploration of deep, separated minima determine the rate of the α process, while those involving contiguous minima determine the rate of the β process. At the glass transition α relaxation is arrested, but β motion persists. As a glass is heated, α minima become progressively mutually accessible, while β minima associated with a given configuration are largely unaffected. At a certain $T > T_g$, the distinction between α and β motion disappears, a condition that would correspond to the extrapolated merging of the α and β curves in Figure 4.26. Above this temper-

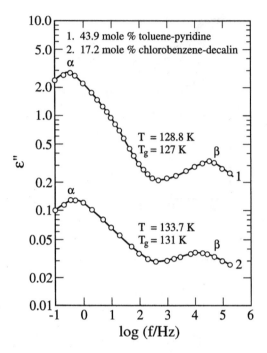

Figure 4.25. Frequency dependence of the imaginary component of dielectric constant for two glass-forming liquids slightly above T_g. Note the α and β peaks. [Reprinted, with permission, from Johari, *J. Chem. Phys.* 58: 1766 (1973)]

ature, accessible configurations are numerous and separated by shallow energy barriers; the exploration of such configurations is therefore indistinguishable from a β process.

In a classic paper, Goldstein (1969) argued that when a liquid is supercooled, normal diffusive motion gradually becomes activated, and relaxation increasingly occurs via large-amplitude, space- and time-localized, and highly infrequent events (Barrat and Klein, 1991). Goldstein argued that when the shear relaxation time exceeds 10^{-9} sec (or, equivalently, when the viscosity exceeds 10 P), relaxation is dominated by the infrequent visitation of deep potential energy minima (α relaxations). This crossover from continuous to activated behavior is illustrated in Figure 4.27 (Signorini et al., 1990), which shows the rotational motion of nitrate ions in a molecular dynamics simulation of the glass-forming system $[Ca(NO_3)_2]_{0.4}[KNO_3]_{0.6}$. θ is given by $\cos^{-1}[\mathbf{u}(t) \cdot \mathbf{u}(0)]$ with $\mathbf{u}$ denoting a unit vector perpendicular to the ion's plane. At high temperature, angular diffusion coexists with larger angular jumps; at low temperature, orientational relaxation is only possible through large and infrequent jumps of amplitude close to π.

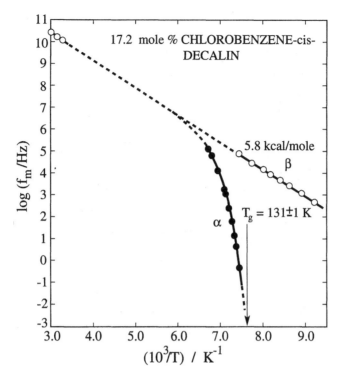

Figure 4.26. Arrhenius plot of the frequency of maximum loss for a glass-forming liquid. α relaxations become arrested at T_g. α and β relaxations become indistinguishable at high temperature. [Reprinted, with permission, from Johari, *J. Chem. Phys.* 58: 1766 (1973)]

Because of the above-mentioned behavior involving fast and slow processes with well-separated time scales, the times τ_I and τ_{II} of mode-coupling theory [see (4.101) and (4.102)] are sometimes referred to as τ_α and τ_β. The approach and early departure, and intermediate departure from the plateau, are termed the β and α processes, respectively (Götze and Sjörgen, 1992). This notation is confusing and should be avoided because in the $T > T_x$ regime where mode-coupling predictions apply there is no distinction between slow collective motions due to the exploration of deep configurational minima (α relaxation) and faster, noncollective exploration of local minima (β relaxation).

Activated relaxation is an important feature of the dynamics of highly super-cooled liquids. If we identify T_x (Figure 4.22) as the temperature below which activated processes become important, then mode coupling becomes a theory of liquid dynamics in the higher-temperature regime ($T > T_x$). Recent extensions (Götze and Sjörgen, 1987; Sjörgen and Götze, 1991; Fuchs et al., 1992) that in-corporate ergodicity-restoring activated diffusion at low temperatures appear to

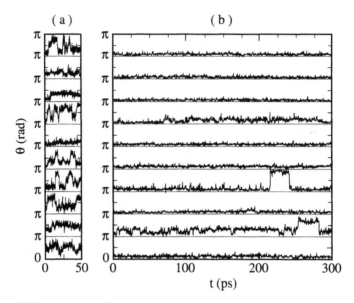

Figure 4.27. Rotation of nitrate ions in the molecular dynamics simulation of $[Ca(NO_3)_2]_{0.4}[KNO_3]_{0.6}$ at 700 K (a), and at 350 K (b). [Reprinted, with permission, from Signorini et al., *J. Chem. Phys.* 92: 1294 (1990)]

resolve the contradiction between T_x and the ideal glass transition. This contradiction, in other words, is between what the original formulation of the theory (Leutheusser, 1984; Bengtzelius et al., 1984) attempted to describe, structural arrest at the glass transition; and what it actually describes, the dynamics of liquids well above the glass transition.

A nonsharp glass transition is also predicted in the hydrodynamic treatment of Mazenko and co-workers (Das and Mazenko, 1986; Das, 1987, 1990; Kim and Mazenko, 1991). Starting from generalized Langevin equations for the evolution of the mass and momentum density in the presence of thermal fluctuations (Das et al., 1985), these authors find a pronounced slowing down in the decay of mass and momentum density fluctuations at low enough temperature or high enough density, but no structural arrest.

At conditions such that diffusion is not activated (i.e., high temperature), mode coupling predicts that the decay of correlation functions with nonvanishing density overlap exhibits universal scaling. The form of the scaling laws, as well as some of the predicted relationships between the scaling exponents, have been confirmed by detailed molecular dynamics studies of atomic systems (e.g., Kob and Andersen, 1994, 1995). Although the theory was developed for such simple liquids exclusively, it appears to be more generally applicable. Behavior that is consistent with at least some aspects of the theory has been reported for polymers (Frick et al., 1990; Sjörgen, 1991; Sidebottom et al., 1992), ionic and oxide glass formers (Mezei et al., 1987; Sidebottom et al., 1993), and molecular liquids (Elmroth et al., 1992; Sokolov et al., 1995). Agreement with mode-coupling predictions has also been reported for colloidal suspensions (van Megen and Pusey, 1991; van Megen et al., 1991; Götze and Sjörgen, 1991); this is less surprising than in the other cases mentioned above, given the similarity with hard sphere systems. At present, it is not well understood why some aspects of the theory can be used to describe complex fluids such as polymers, while at the same time simple lattice-gas models show important disagreements (Kob and Andersen, 1993a,b). This is one of the most interesting open questions in this field; computer simulations of well-defined model liquids are ideally suited to explore the limits of applicability of mode coupling to supercooled liquids.

Mode coupling, in sum, describes many aspects of the relaxation dynamics of supercooled liquids accurately. It is indeed an important theoretical development. This is so in spite of a literature often concerned more with formal rigor than with providing physical insight; in spite, too, of the appreciable difference between what the theory describes accurately, relaxations at $T > T_x$, and what it sought to describe when it was formulated, structural arrest at the glass transition. Recent extensions of the theory (e.g., Sjörgen and Götze, 1991; Fuchs et al., 1992) which restore ergodicity below T_x have narrowed this conceptual gap.

4.5 STRONG AND FRAGILE LIQUIDS

In 1985, Angell proposed a useful classification scheme for supercooled liquids (Angell, 1985; see also Angell, 1988a, 1990a, 1991a,b). Angell plotted the temperature dependence of the viscosities of a wide variety of supercooled liquids at atmospheric pressure in scaled Arrhenius fashion (Figure 4.28). Following an idea first proposed by Laughlin and Uhlmann (1972), the scaling temperature was chosen to be that at which the viscosity reaches 10^{13} P.[30] Liquids that exhibit Arrhenius behavior over the entire temperature spectrum (straight lines in Figure 4.28) were termed strong by Angell; those which exhibit marked deviations from Arrhenius behavior were termed fragile. Laughlin and Uhlmann (1972) were actually the first to note that a scaled Arrhenius viscosity plot brings forth these two characteristic types of behavior.

---

[30] Alba et al. (1990) have pointed out that it is more appropriate to use the calorimetrically determined T_g as the scaling temperature. This is normally a higher temperature than that at which the viscosity reaches 10^{13} P [i.e., the viscosity at the calorimetric T_g is lower than 10^{13} P; for example, $10^{9.5}$ P in some fragile liquids (Angell, 1990a)].

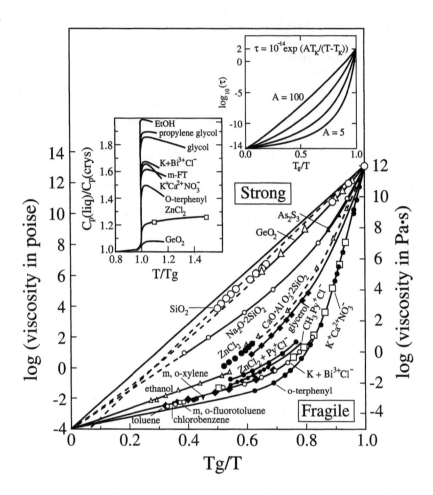

Figure 4.28. Dependence of viscosity on scaled temperature showing strong and fragile patterns of behavior. T_g is the temperature at which the viscosity reaches 10^{13} P. Insets show how the temperature dependence can be reproduced by varying the VTF A parameter [Equations (4.22) and (4.113)], and changes in the heat capacity across the glass transition for several liquids [Reprinted, with permission, from Angell, *J. Non-Cryst. Sol.* 102: 205 (1988)]

Strong liquids, such as the network oxides SiO_2 and GeO_2, have tetrahedrally coordinated structures that resist thermal degradation. When heated across the glass transition, their short- and intermediate-range order[31] tends to persist.

---

[31] Short-range order denotes the mutual arrangement of nearest neighbors and the regularities

This structural stability is reflected in the small heat capacity and thermal expansion coefficient changes that accompany the vitrification of strong liquids. In contrast, any remnants of the structure in which a fragile liquid is trapped below T_g disappear rapidly upon heating above T_g. This is reflected in relatively large changes in heat capacity and thermal expansion coefficient at T_g. Ionic and simple molecular liquids tend to be fragile. A validation of the underlying idea whereby strong and fragile behavior reflect the stability (or lack thereof) of short- and intermediate-range order to thermal degradation can be found in the observation that strong liquids (GeO_2, SiO_2) can be made fragile by introducing additional components that disrupt the network structure (Angell, 1990a; Jeong et al., 1986).

There exists an important relationship between Angell's strong-fragile classification and the Adam-Gibbs theory. To see this, we first note that the stability of short- and intermediate-range structure in strong liquids translates into a relatively temperature-insensitive configurational entropy. In contrast, the rapidly degrading structure in fragile liquids gives rise to a strongly temperature-dependent configurational entropy. According to the Adam-Gibbs theory [Equation (4.21)], this gives rise to Arrhenius behavior for strong liquids, and to non-Arrhenius behavior for fragile liquids, in agreement with the trends shown in Figure 4.28. If we further assume an inverse relationship between Δc_p, the difference between liquid and crystal heat capacities, and temperature (i.e., $\Delta c_p = K/T$; Alba et al., 1990; see Section 4.3.2), we obtain a VTF-type viscosity expression [Equation (4.22)]. The constant A in the VTF expression, $\eta \propto \exp[A/(T - T_K)]$, reads

$$A = \frac{T_K \, s_{conf}^* \, \delta\mu \, k^{-1}}{K}. \tag{4.113}$$

Angell has described the numerator as kinetic in nature, and the denominator as thermodynamic. This distinction follows from the fact that the chemical potential difference in the numerator represents a free energy barrier to cooperative rearrangement, and hence plays the role of an activation energy for the relaxation rate. The denominator, on the other hand, is indicative of the magnitude of the heat capacity jump at the glass transition. Fragile liquids have small A values; strong liquids, large A values. This is illustrated in the inset to Figure 4.28, where the entire strong-fragile spectrum is reproduced by varying A.[32] Also shown as an inset to Figure 4.28 is the heat capacity change at T_g for several liquids. In general, there is a consistent trend towards larger

---

resulting therefrom over distances up to 6–8 Å. Intermediate-range order describes regularities occurring in the range from 6–8 to 30–50 Å (Zarzycki, 1990.)

[32]The quantity A in Figure 4.28 (inset) is dimensionless, and is related to the quantity defined in (4.113) by A (Figure 4.28) $= (A$ [Equation (4.113)])$/T_K$. The relationship between T_K and T_g needed to generate the curves in the inset follows from setting $\log \tau = 2$ at T_g, whence $\left(T_g/T_K\right) = [1 + A$ (Figure 4.28)]$/(\ln 2 + 14 \ln 10)$.

heat capacity changes at T_g as the spectrum of liquid behavior is traversed in the direction of increasing fragility. The one notable exception is the alcohols, which exhibit the largest heat capacity change, yet fall in the middle of the strong-fragile spectrum. A qualitative interpretation of this anomaly was proposed by Angell (1990a, 1991b). It involves the potential energy hypersurface of a macroscopic system shown schematically in two dimensions in Figure 4.29, where the x axis represents a generic collective coordinate, and the y axis potential energy barriers. Angell conjectured that strong liquids have a low density of potential energy minima, reflecting the configurational constraints imposed by their underlying rigid structure. At the other extreme, fragile liquids should have a much higher density of potential energy minima, reflecting the lack of an underlying rigid structure, and hence the proliferation of possible configurations. The alcohols are thought to have a fragilelike density of local minima, but stronglike (that is to say, large) energy barriers between the minima. Alcohols, in other words, are thermodynamically fragile, but kinetically strong. This kinetic strength is due to the cost of breaking hydrogen bonds in order to sample different configurations, a feature not encountered in non-hydrogen-bonded liquids. In terms of the VTF A parameter, the alcohols would have both a high K (fragile behavior) and a high $\delta\mu$ (strong behavior). Clearly, more work is required to substantiate these intriguing but qualitative ideas. One obvious need is for viscosity measurements for supercooled liquids spanning a spectrum of systematically varied hydrogen-bond strengths. Some results along these lines have been reported by Angell (1991b), but their interpretation is inconclusive.

When the height of the typical potential energy barrier in Figure 4.29 is appreciably greater than kT, relaxation becomes activated and occurs by the infrequent visitation of deep potential energy minima. Hence, the mode-coupling ideas discussed in Section 4.4 are applicable to fragile liquids above the glass transition. The potential energy hypersurface of a fragile liquid has features on two distinct length scales. Each deep potential energy minimum is an inherent structure or α minimum (Stillinger, 1995); β minima are contiguous local minima separated by small ridges. α minima define the salient topography of the hypersurface, its large craters. β minima define the ruggedness of the terrain.

Recent experiments (Fujara et al., 1992; Cicerone and Ediger, 1993; Cicerone et al., 1995) have shown that the Stokes-Einstein equation, (4.54), breaks down in several deeply supercooled fragile liquids (Rössler, 1990). Translational diffusion is roughly two orders of magnitude faster than predicted from the measured viscosity. At the same time, rotational diffusion remains coupled to viscosity ($D_{rot}\eta T^{-1}$ is approximately constant). The extent to which this interesting phenomenon represents generic fragile behavior (e.g., Stillinger and Hodgdon, 1994) is not well understood at present.

The strong-fragile classification is an intriguing and useful concept. Its implications are sufficiently broad as to suggest a possible unifying viewpoint of the physics of supercooled liquids. For example, a correlation between

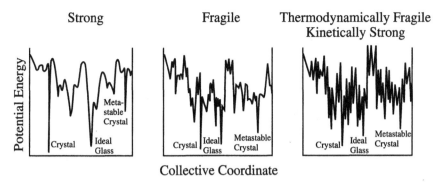

Figure 4.29. Schematic two-dimensional representation of the potential energy hyper-surface of strong (left), fragile (center), and hydrogen-bonded (right) liquids. Strong liquids have a low density of potential energy minima, separated by very high potential energy barriers. Fragile liquids have a high density of potential energy minima. Hydrogen-bonded liquids such as alcohols have a high density of potential energy minima (thermodynamic fragility), separated by high potential energy barriers (kinetic strength). [Reprinted, with permission, from Angell, in *NATO ASI E*, 188: 133 (1990). Copyright 1990, Kluwer Academic Publishers]

fragility and nonexponential relaxation has been found to exist for a wide variety of supercooled liquids, ionic melts, amorphous polymers, and covalent glass formers (Böhmer et al., 1993). At present, however, the concept of strength and fragility remains largely qualitative. Additional work is required in order to bring quantitative and predictive considerations to bear on this promising idea.

4.6 SUPERCOOLED AND GLASSY WATER

In Section 2.2.8 data were presented on the isobaric heat capacity, isothermal compressibility, and thermal expansion coefficient of supercooled water (Figures 2.12, 2.13, 2.14, and 2.15). The unusual behavior of supercooled water, exemplified by the increase in its response functions at low temperatures, has generated continued interest since the early 1970s. There is now a large body of data on the thermodynamic, transport, and structural properties of liquid water below its freezing point. Two excellent reviews by Angell (1982, 1983) are the standard reference on experimental work on supercooled water. Here we focus primarily on theoretical work aimed at interpreting the measurements. In Section 4.6.1 we summarize the experimental studies on the thermophysical properties of supercooled water; theoretical interpretations are discussed in Section 4.6.2; the different forms of glassy water and their relation to supercooled water are addressed in Section 4.6.3. Recent theoretical advances notwithstanding, it will be shown that understanding supercooled water remains a major scientific challenge. To date, there is no theory that can explain all the known facts.

Before discussing experiments and their interpretation, it is convenient to summarize water's salient structural and geometric properties. A fundamental characteristic of water is its ability to form strong, directional hydrogen bonds (Stillinger, 1980). The average strength of a hydrogen bond is approximately 23 kJ mol^{-1}.[33] Figure 4.30 shows the structure of ordinary hexagonal ice (ice Ih), the stable form of solid water at 1 bar and $T < 273$ K. Each water molecule has four nearest neighbors and acts as a hydrogen donor to two of them and as hydrogen acceptor to the other two. These four nearest neighbors are located at the vertices of a regular tetrahedron surrounding the central oxygen atom, and each O-O distance is 2.77 Å. The HOH bond angle of the water molecule (104.5°) is in fact very close to the tetrahedral angle (109.5°). Hexagonal ice forms a three-dimensional network held together by hydrogen bonds. Because of the strong directionality of the bonds, the network is open: when ice melts at atmospheric pressure, loss of long-range order is accompanied by a 9% density increase. The heat of melting of ice Ih is only 13% of its heat of sublimation (Stillinger, 1980). This implies that the majority of hydrogen bonds are unbroken upon melting. Thus, in liquid water close to the melting point and even more so in supercooled water, local tetrahedral symmetry persists, although this order is transient and short ranged (e.g., Geiger et al., 1979; Mezei and Beveridge, 1981; Bosio et al., 1983; Blumberg et al., 1984; Mezei and Speedy, 1984; Speedy and Mezei, 1985; Geiger et al., 1986; Speedy et al., 1987; Bellissent-Funel et al., 1989; Kowall et al., 1990; Sciortino et al., 1990a,b, 1991, 1992; Ruocco et al., 1993).

4.6.1 Experiments

Table 4.9 lists several studies of the equilibrium thermophysical properties of supercooled water (see Sato et al., 1991, for a critical compilation).[34] The most important result of these studies is the increase in isothermal compressibility, isobaric heat capacity, and the magnitude of the thermal expansion coefficient upon cooling (recall Figures 2.12–2.15). As a further illustration of this behavior, Figure 4.31 shows the isothermal compressibility of supercooled H_2O and D_2O between 100 and 1900 bar (Kanno and Angell, 1979). The isochoric heat capacity, calculated from c_p, K_T, and α_p measurements by means of Equations (2.43) and (2.44), remains approximately constant (Angell, 1982). Interesting questions persist about the behavior of the sound velocity in supercooled water (Magazú et al., 1989; Sciortino and Sastry, 1994). The apparent minimum with respect to temperature in the low-frequency sound velocity close to −25°C

---

[33]This estimate is obtained from knowledge of the cohesive energy of ordinary ice, using the fact that in this phase there are, on average, two hydrogen bonds per molecule (Stillinger, 1980).

[34]For reviews including nineteenth and early twentieth century studies, see Angell (1982, 1983).

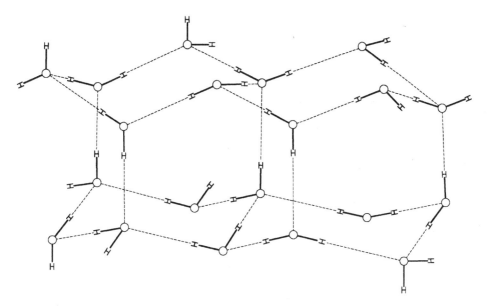

Figure 4.30. The network structure of ice Ih. Hydrogen bonds are shown dashed. [Reprinted, with permission, from "Water Revisited," by F. H. Stillinger, *Science*, 209: 451 (1980). Copyright 1980, American Association for the Advancement of Science]

(Trinh and Apfel, 1978) is not well understood, and deserves further study. On the other hand, this question concerns the adiabatic compressibility [see (2.75) and (2.76)]; hence it is not important for understanding the stability of the liquid, for which the relevant quantity is the isothermal compressibility.

Table 4.10 lists several studies of transport properties and relaxation in supercooled water. The data show a sharp increase in relaxation times upon supercooling, and marked deviations from Arrhenius behavior characterized by an increase in the effective activation energy at low temperatures (Pruppacher, 1972; Gillen et al., 1972). This is shown in Figure 4.32 for the viscosity and the diffusivity. The pressure dependence of the transport properties, illustrated in Figure 4.33 for the viscosity, is quite interesting. At sufficiently low temperatures, pressure increases cause an increase in the diffusion coefficient (Angell et al., 1976) and a corresponding decrease in viscosity (DeFries and Jonas, 1977) and in the rotational and dielectric relaxation times (Prielmeier et al., 1987, 1988). Hence the molecular mobility of supercooled water is enhanced by the application of pressure. This is a consequence of the increase in tetrahedral coordination at low temperatures, which hinders the translational and rotational mobility of water molecules. Upon application of pressure, the tetrahedral network is disrupted, and the ability of molecules to translate and rotate increases correspondingly (Sciortino et al., 1990b, 1991, 1992). The work described in the following section aims at explaining the increase of water's

Table 4.9. Experimental Studies of the Equilibrium Thermophysical Properties of Supercooled Water[a]

Property	Technique	Range	Reference
Isothermal compressibility	Capillary	1 bar; −26/45°C	Speedy and Angell (1976)
	Capillary	1/1900 bar; −30/25°C	Kanno and Angell (1979)
	Small-angle x-ray scattering	1 bar; −34/25°C	Xie et al. (1993)
Density, thermal expansion coefficient	Capillary	1 bar; −23/4°C	Schufle (1965)
	Capillary	1 bar; −40/24°C	Schufle and Venugopalan (1967)
	Capillary	1 bar; −34(H), −29(D)/4°C	Zheleznyi (1969)
	Dilatometry on emulsions	1 bar; −35(H), −30(D)/5°C	Rasmussen and MacKenzie (1973)
	Capillary	1/1400 bar; −30/20°C (D)	Kanno and Angell (1980)
	Capillary	1 bar; −34(H), −19(D)/40°C	Hare and Sorensen (1986)
	Capillary	1 bar; −33/−5°C	Hare and Sorensen (1987)
Isobaric heat capacity	N/A	1 bar; −7.5/31°C	Anisimov et al. (1972)
	Calorimetry on emulsions	1 bar; −35(H), −30(D)/5°C	Rasmussen and MacKenzie (1973)
	Calorimetry on bulk and emulsions	1 bar; −38(H), −34(D)/5°C	Rasmussen et al. (1973)
	Calorimetry on bulk and emulsions	1 bar; −38(H), −33(D)/4°C	Angell et al. (1973)
	Calorimetry on bulk and emulsions	1 bar; −35/17°C	Oguni and Angell (1980)[b]
	Calorimetry on bulk and emulsions	1 bar; −37(H), −33(D)/17°C	Angell et al. (1982b)

Table 4.9. *Continued*

Property	Technique	Range	Reference
Speed of sound	Brillouin scattering	1 bar; −9/100°C	Rouch et al. (1976)
	Ultrasonic laser diffraction	1 bar; −17/177°C	Trinh and Apfel (1978)
	Brillouin scattering	1 bar; −20/20°C	Teixeira and Leblond (1978)
	Acoustic levitation	1 bar; −33/−15°C	Trinh and Apfel (1980)
	Brillouin scattering	1 bar; −20(H), −17(D)/80°C	Conde et al. (1982)
	Echo overlap	1/4600 bar; −20/22°C	Petitet et al. (1983)
	Brillouin scattering	1 bar; −27/20°C	Magazú et al. (1989)
Vapor pressure	Capacitance transducer, droplets	−14(H), −12(D)/−2(H), 2°C (D)	Bottomley (1978)
	Capacitance transducer, droplets	−22(H), −15(D)/0(H), 4°C(D)	Kraus and Greer (1984)
Dielectric constant	Capacitance	1 bar; −36/20°C	Hasted and Shahidi (1976)
		1 bar; −35/0°C	Hodge and Angell (1978)
		1 bar; −17/20°C	Das et al. (1982)

[a]H = H_2O; D = D_2O. Results are for H_2O unless otherwise indicated.
[b]Study on $H_2O + H_2O_2$ and $H_2O + N_2H_4$ mixtures. Contains also data for pure water.

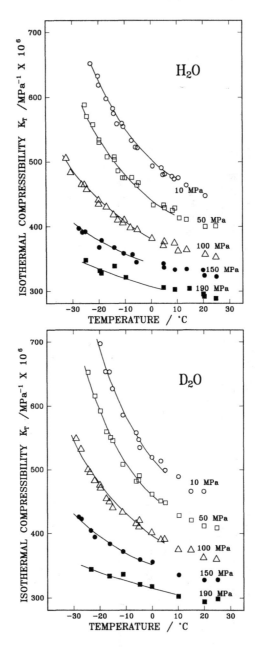

Figure 4.31. Isothermal compressibility of supercooled H_2O (a, top) and D_2O (b, bottom). Lines are fits of the experimental data to power-law divergences of the form of Equation (4.114), with exponent $\gamma = 0.373$ (a) and 0.415 (b). (Adapted from Kanno and Angell, 1979)

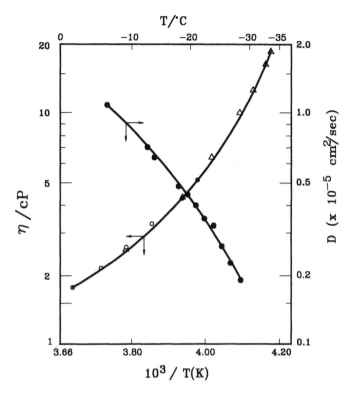

Figure 4.32. Arrhenius plot of the diffusivity and shear viscosity of supercooled water at atmospheric pressure. The diffusivity data are from Gillen et al. (1972); the viscosity data from Osipov et al. (1977).

response functions upon supercooling. Much less attention has been given to understanding transport and relaxation in supercooled water.

In addition to the thermophysical properties listed in Tables 4.9 and 4.10, much has been learned from structural studies of atomic and molecular correlation functions in supercooled and glassy water by x-ray and neutron diffraction. This work is discussed in Sections 4.6.2 and 4.6.3.

4.6.2 Interpretation

The stability limit conjecture, already discussed in Section 2.2.8, is an important attempt at understanding the properties of supercooled water. The conjecture is due originally to Speedy (1982a,b), and was later generalized by D'Antonio and Debenedetti (1987) and Debenedetti and D'Antonio (1988). The main features of this theory, which is based entirely on thermodynamic consistency arguments, are reiterated in Figure 4.34. According to this interpretation, water

Table 4.10. Experimental Studies of the Transport Properties and Relaxation Behavior of Supercooled Water[a]

Property	Technique	Range	Reference
Self-diffusion coefficient	Tritium tracer	1 bar; −25/30°C	Pruppacher (1972)
	Spin-echo NMR	1 bar; −31/25°C	Gillen et al. (1972)
	Spin-echo NMR	1/2380 bar; −20/2°C	Angell et al. (1976)
	Spin-echo NMR	1/3000 bar; −30/0°C	Prielmeier et al. (1987)
	Spin-echo NMR	1/4000 bar; −70/0°C (H)	Prielmeier et al. (1988)
		1/2000 bar; −30/0°C (D)	
Viscosity	Capillary flow	1 bar; −24/25°C	Hallett (1963)
	Rolling ball	1/6000 bar; −15/10°C	DeFries and Jonas (1977)
	Moving capillary column	1 bar; −35(H), −31(D)/4°C	Osipov et al. (1977)
Spin-lattice relaxation time	Spin-echo NMR	1 bar; −37/15°C (D)	Hindman and Svirmickas (1973)
	Spin-echo NMR	1 bar; −16/145°C	Hindman et al. (1973)
	Spin-echo NMR	1/6000 bar; −15/10°C	DeFries and Jonas (1977)
	Spin-echo NMR	50/2500 bar; −87/20°C	Lang and Lüdemann (1977)
	Spin-echo NMR	1/3000 bar; −85/10°C (D)	Lang and Lüdemann (1980)
	Spin-echo NMR	50/2500 bar; −35/184°C[b]	Lang and Lüdemann (1981)
	Spin-echo NMR	50/3000 bar; −30/184°C[c]	

[a] H = H_2O; D = D_2O. Results are for H_2O unless otherwise indicated.
[b] Spin-lattice relaxation of ^{17}O in $H_2^{17}O$.
[c] Spin-lattice relaxation of ^{17}O in $D_2^{17}O$.

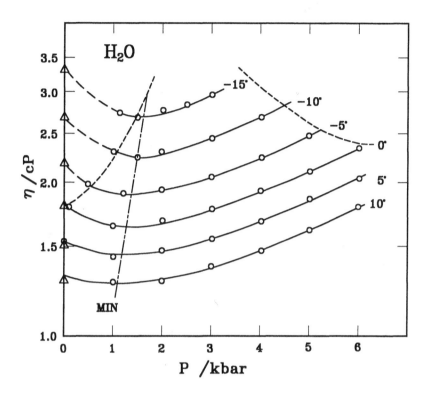

Figure 4.33. Pressure dependence of the shear viscosity of H_2O. The dotted lines are phase boundaries (ice I, left; ice V, right). Triangles are values from Hindman (1974). (Adapted from DeFries and Jonas, 1977)

has a continuous spinodal curve bounding the superheated, supercooled, and stretched metastable states. The observed increases in the response functions upon supercooling are viewed as being caused by the underlying spinodal, where K_T, c_p, and $-\alpha_p$ diverge. A diverging compressibility indicates loss of stability with respect to a fluid phase. The theory in its original form, therefore, does not take into account that supercooled water is metastable with respect to an anisotropic, crystalline phase. In two of the more recent theories to be discussed below, however (Poole et al., 1994; Borick et al., 1995), a continuous, Speedy-like retracing spinodal is present but so too is an isotropic phase, with respect to which supercooled water becomes metastable at low temperatures.

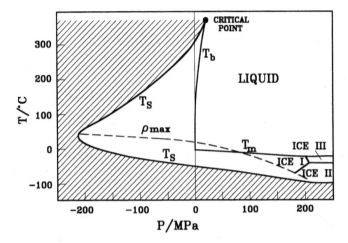

Figure 4.34. Phase diagram of water, including the postulated continuous, retracing spinodal according to Speedy's stability limit conjecture. (Adapted from Angell, 1988c)

Within the framework of the stability limit conjecture, data for deeply supercooled water should be plotted according to

$$X = A \left(\frac{T}{T_s} - 1 \right)^{-\gamma} + X_0, \tag{4.114}$$

where X is a quantity that diverges at the spinodal, A is a constant, T_s is the singular (spinodal) temperature, and X_0 is the background, nondiverging component. If the Helmholtz energy is an analytic function of temperature and density at the spinodal, $\gamma = 1/2$ should be observed when a spinodal is approached at constant pressure (see Section 2.2.4), the usual cooling path in experimental studies. It is indeed common practice to plot data for supercooled water according to (4.114). Since the very existence of the singularity, let alone its location, is uncertain, it is not surprising that this procedure yields ambiguous results. For example, Speedy and Angell (1976) found that the compressibility data shown in Figure 2.12 could be best fitted with an expression of the form of (4.114), without the background term, and with $\gamma = 0.349$ and $T_s(1 \text{ bar})$ = 228 K. Kanno and Angell (1979) used $\gamma = 0.37$ to obtain a $T_s(P)$ locus from compressibility measurements up to 1900 bar. Oguni and Angell (1980) measured c_p for binary $H_2O + H_2O_2$ and $H_2O + N_2H_4$ solutions, and plotted the data against solute mole fraction. Extrapolation to zero solute content allowed these authors to calculate a background contribution, and to fit the anomalous component, $X - X_0$, to a power-law divergence. For the H_2O_2-based extrapolation, the data were best fitted with $\gamma = 1.35$ and $T_s = 224.9$ K; for the N_2H_4-based extrapolation, they found $\gamma = 0.933$ and $T_s = 226.4$ K. Clearly, $\gamma \geq 1$ is unphysical, because it implies that the enthalpy and the volume diverge at the spinodal. Finally, use of (4.114) to analyze the α_p data of Hare and Sorensen (1986), Figure 2.14, proved inconclusive, the values of T_s and γ being sensitive to the background term. Some of the different (γ, T_s) pairs obtained with different background choices (Hare and Sorensen, 1986, 1987) include (1.35, 216 K), (0.95, 225 K), and (0.89, 228 K).

The viscosity, inverse diffusivity, and dielectric and spin-lattice relaxation times of supercooled water can be correlated with power laws such as Equation (4.114) without the background term (e.g., Speedy and Angell, 1976; Lang and Lüdemann, 1980; Prielmeier et al., 1987). Such results imply structural arrest at T_s. The experimental verification of this interesting possibility requires data at temperatures closer to the assumed singularity than it is possible to attain: at atmospheric pressure the closest attainable approach to the assumed singularity temperature, $T_s = 228$ K (Speedy and Angell, 1976) is 233 K, the homogeneous nucleation temperature. This corresponds to a fractional approach $(T/T_s - 1)$ of 0.02, compared to 10^{-5} for critical-point experiments (Sengers and Levelt-Sengers, 1978). Nevertheless, it is remarkable that the pronounced slowing down of relaxation processes takes place far above the glass transition.[35]

The other interpretations to be discussed here are based not on macroscopic thermodynamic arguments but on microscopic models or computer simulations. Sastry et al. (1993) have proposed an interesting waterlike lattice model, summarized in Figure 4.35. They divide a simple cubic lattice into two interpenetrating sublattices. The interaction energy between two nearest-neighbor sites on the same sublattice is -2ε if both sites are occupied, and 0 otherwise. The interaction energy between two neighboring sites on different sublattices is -2ε if both sites are occupied, $-2J$ if one site is occupied with a correctly oriented molecule while the other site is empty, and 0 otherwise. A molecule can be oriented in q ways, only one of which leads to bonding. J is thus an orientation-dependent bonding energy. For $J > \varepsilon$, the state of minimum energy (ground state) is a half-occupied lattice: one sublattice is empty, the other one full, and all molecules are correctly oriented. Thus the model has an open, low-density, fully bonded ground state. These simple interactions define a system in which states of low density are energetically favored $(J > \varepsilon)$, but entropically disfavored (only 1 out of q orientations leads to bonding), whereas the opposite is true for high-density configurations. The key feature of this model is the fact that bonding requires that molecules be not only properly oriented but also sufficiently separated. Figure 4.36 shows the phase diagram obtained by solving the model in the mean-field approximation. Several waterlike features are apparent. First, the melting curve is negatively sloped, indicating that the solid phase is less dense than the liquid.[36] Second, there is a locus

---

[35] Water undergoes a glass transition around 130 K, but it is not known if the high-temperature phase is liquid. See Section 4.6.3.

[36] The solid phase is obtained by maximizing the partition function with respect to an order parameter that quantifies the preferential occupation of one of the sublattices.

of density maxima, to the left of which the stable or supercooled liquid has a negative thermal expansion coefficient, and to the right of which it expands when heated isobarically (positive thermal expansion coefficient). The locus of density maxima is negatively sloped in the (P, T) plane, just as water's is in the range of pressures where it has been measured (-200 to 1200 bar; Henderson and Speedy, 1987; Angell and Kanno, 1976). In agreement with the thermodynamic consistency arguments of Speedy (1982a) and of Debenedetti and D'Antonio (1988), the intersection of the locus of density maxima and the liquid spinodal causes the spinodal to retrace towards positive pressures. This is an interesting, microscopically based validation of a key feature of the stability limit conjecture. In addition, the model corrects the important limitation of the conjecture: the spinodal is superseded by a locus of limits of stability of the liquid with respect to the solid. The instability is caused by fluctuations in sub-lattice occupancy at fixed total density. Along the locus of liquid-solid limits of stability, the compressibility remains finite. Sasai (1993) reached very similar conclusions using a lattice-gas implementation of a random graph model of water's hydrogen-bonded network.[37]

According to the waterlike model of Sastry et al. (1993), the response functions of the supercooled liquid increase sharply (Sastry, 1993) but do not diverge. The spinodal retraces, as predicted by the stability limit conjecture, but it disappears at negative pressure, and the relevant limit of stability upon supercooling is with respect to the solid phase. This is not inconsistent with available experimental evidence because it is not known whether the observed increases in response functions are true divergences. Extrema in the response functions have not been observed in the range of temperatures so far explored.

The elegant model of Sastry is descriptive rather than predictive: its geometry, for example, is not waterlike. Nevertheless, it illustrates the usefulness of simple models that capture the basic physics of a complicated phenomenon. In this case, the model incorporates in a most satisfying way the competition between icelike configurations of low energy, entropy, and density, and denser configurations with higher energy and entropy.

An entirely different interpretation has been proposed by Poole et al. (1992). In molecular dynamics studies of ST2 (Poole et al., 1992, 1993a; Stanley et al., 1994) and TIP4P water (Poole et al., 1993a; Stanley et al., 1994),[38] these authors did not observe a Speedy-like retracing spinodal. Their interpretation of supercooled water's behavior is shown in Figure 4.37. As the pressure is reduced, the locus of density maxima, which is negatively sloped at high pressure, becomes

---

[37]See, however, Sasai (1990) for lattice model predictions of a low-temperature liquid-solid instability with diverging response functions.

[38]The ST2 (Stillinger and Rahman, 1974) and TIP4P (Jorgensen et al., 1983) are commonly used pair potentials in which the water molecule is represented as a rigid body interacting with other water molecules via a Lennard-Jones potential between the oxygen sites, plus Coulombic forces between localized fractional charges.

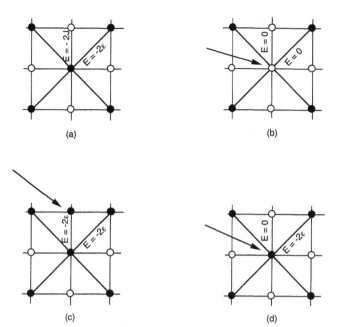

Figure 4.35. Two-dimensional representation of elementary interactions in the waterlike lattice model of Sastry et al. (1993). The central site and its four diagonal neighbors are on sublattice A, and the remaining four sites are on sublattice B. Energies correspond to an AB and an AA interaction between the central site and two neighbor sites. In (a), sublattice A is full, sublattice B is empty, and the central molecule is properly oriented for hydrogen bonding. This is the state of lowest energy. In (b), the molecule in the central site has been removed (arrow). In (c), a molecule has been added to sublattice B. In (d), the central molecule is improperly oriented for hydrogen bonding. [Reprinted, with permission, from Sastry et al., *J. Chem. Phys.* 98: 9863 (1993)]

infinitely sloped, and then positively sloped under high tension. Because of this retracing, the locus of density maxima does not intersect the spinodal, which therefore does not retrace towards positive pressures. In their thermodynamic analysis, Debenedetti and D'Antonio (1988) assumed that the locus of density maxima has a finite slope, hence they did not consider the case shown in Figure 4.37. At any given pressure, the locus of density maxima represents the highest temperature at which the liquid's thermal expansion coefficient is negative. Thus, in the interpretation of Poole et al. (1992), the thermal expansion coefficient again becomes positive provided sufficient tension is applied. Stretched water, in other words, should be a "normal" liquid.

In the absence of a retracing spinodal, what causes the anomalous increase in supercooled water's response functions? Poole et al.'s intriguing answer to this

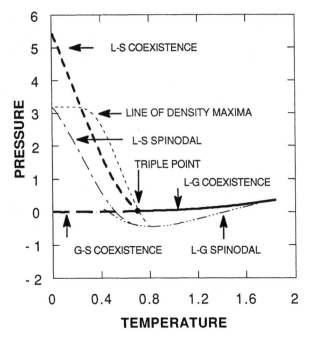

Figure 4.36. Phase equilibrium, stability limits, and locus of density maxima obtained by solving the waterlike lattice model of Sastry et al. in the mean-field approximation. The pressure is in units of J/v_0 (v_0 is the volume of a lattice site), and the temperature, in units of J/k (k is Boltzmann's constant). [Reprinted, with permission, from Sastry et al., *J. Chem. Phys.* 98: 9863 (1993)]

question is that such anomalies are caused by a metastable critical point, the low-pressure termination of a first-order liquid-liquid transition (see Figure 4.37). They arrived at this conclusion based on the observation that low-temperature isotherms of ST2 water exhibit an inflection point at high pressure; the shape and temperature dependence of the inflection region are similar to what would be observed if a critical point were approached from above in temperature (Poole et al., 1992). Nonequilibrium isothermal compression of low-temperature, low-density amorphous solid ST2 in fact yielded sharp inflections at which the volume changed abruptly, suggesting the existence of a first-order phase transition between two amorphous forms of ST2. The inflections of the supercooled liquid isotherms and the isothermal compression simulations of the low-density amorphous solid are shown in Figure 4.38. The response functions of ST2 water increase in the supercooled region, in agreement with experimental observations (Poole et al., 1992). Qualitatively identical behavior was also found in TIP4P water (Poole et al., 1993a; Stanley et al., 1994).

The interpretation of supercooled water's behavior shown in Figure 4.37 can be summarized as follows: in addition to the vapor-liquid transition, there is a

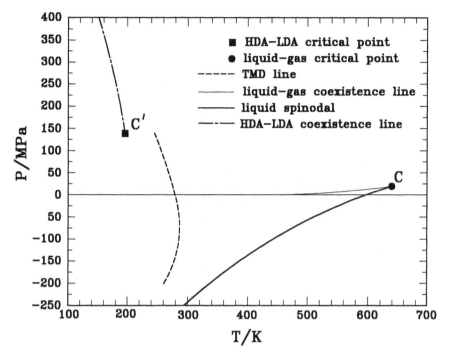

Figure 4.37. The two-critical point interpretation of supercooled water's phase behavior. The locus of density maxima (temperature of maximum density, TMD) does not intersect the liquid-vapor spinodal. The source of anomalies is the critical point C', above which two different forms of amorphous ice (high- and low-density amorphs, HDA and LDA) become indistinguishable. [Reprinted, with permission, from Poole et al., *Nature*, 360: 324 (1992). Copyright 1992, Macmillan Magazines Limited]

second phase transition between two amorphous forms of water. Because of the low temperatures involved, the transition occurs between two structurally arrested glassy phases. As interpreted by Poole et al., this second phase transition is first order, the two amorphous phases differing in density and entropy. The slope of the transition line indicates that the high-density phase has the higher entropy. The low-pressure, high-temperature termination of this line of first-order transitions is the critical point that is responsible for the observed anomalies of supercooled water. The liquid spinodal plays no role in these anomalies, because the retracing of the locus of density maxima under high tension prevents it from intersecting the spinodal, and therefore the spinodal does not retrace. The simulations of Poole et al. (1992, 1993a) show clearly that the locus of density maxima of ST2 and TIP4P water retraces. Though the inflection of the supercooled liquid isotherms is suggestive, the existence and location of the critical point remain to be established. There is, on the

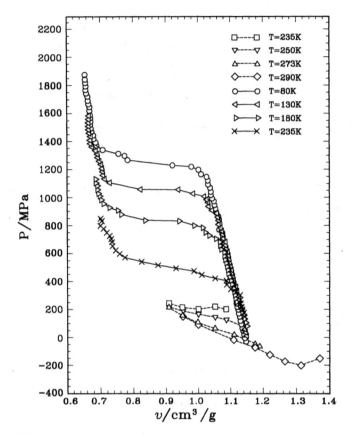

Figure 4.38. Isothermal compression curves for LDA obtained by molecular dynamics simulation of ST2 water (full lines), and low-temperature portion of liquid isotherms showing inflection points (dotted lines). Note the abrupt decrease in volume upon compression into an apparently different amorphous solid phase. The compression curves were generated by increasing the pressure by 25 MPa every 10 psec. The solid curves represent nonequilibrium processes, as can be seen by comparing the liquid and solid curves at 235 K. [Reprinted, with permission, from Poole et al., *Nature* 360: 324 (1992). Copyright 1992, Macmillan Magazines Limited]

other hand, abundant experimental evidence in support of the existence of two forms of vitreous water (e.g., Mishima, 1994). These are commonly referred as low-density amorphous ice (LDA) and high-density amorphous ice (HDA). Glassy water is discussed in Section 4.6.3.

Theoretical work aimed at understanding the notion of a second critical point followed quite rapidly the publication of the original computer simulation study by Poole et al. (1992). In an illuminating paper, Poole et al. (1994)

proposed a modification of the van der Waals equation for hydrogen-bonding liquids. These authors partitioned the range of hydrogen-bond energies into two categories: weakly bonded pairs, for which the energy is zero, and strongly bonded pairs, for which the energy is $-J$.[39] They also noted that a strongly hydrogen-bonded environment in the vicinity of a given molecule requires that adjacent molecules be arranged in an open tetrahedral structure similar to that found in ordinary hexagonal ice. This geometric constraint is incorporated by assuming that a bond can exist in $(q^2 - 1) \gg 1$ configurations with zero energy, corresponding to weak bonds, and only one strongly bonded configuration with energy $-J$. These configurations differ by virtue of the mutual orientation of the two molecules participating in bond formation. The other important assumption is that strong hydrogen bonds are most likely to occur when the bulk density has a value such that every molecule has the optimal local volume required to bond strongly to its neighbors. Away from this optimal density, a fraction f of the bonds is strong, and the remaining fraction, $(1 - f)$, is weak. The density dependence of f was assumed to be a Gaussian centered at the optimal bulk density. These assumptions lead to the following expression for the Helmholtz energy per bond:

$$A_b = -kT \left\{ f \ln \left[q^2 - 1 + \exp(\beta J) \right] + (1 - f) \ln q^2 \right\} \qquad (4.115)$$

(Poole et al., 1994), and the Helmholtz energy per molecule is given by

$$A = A_{\text{vdW}} + 2A_b, \qquad (4.116)$$

where A_{vdW} is the usual Helmholtz energy of the van der Waals fluid, and the factor 2 accounts for the fact that there are two hydrogen bonds per molecule. The equation of state is obtained by differentiation of the Helmholtz energy [see Equation (2.26)].

Figure 4.39 shows the resulting phase behavior. For hydrogen bond strengths greater than 16.5 kJ mol^{-1}, the equation predicts the retracing of the locus of density maxima and the appearance of a second fluid-fluid phase transition terminating at a critical point [Figure 4.39(a)]. Below this critical temperature, the liquid phase splits into a low-density and a high-density phase. The hydrogen bond term gives rise to a rather steep minimum in the energy-volume isotherms, centered around the optimal bulk density for hydrogen bonding. This minimum is superimposed upon a broad nonbonded minimum. Thus the energy-volume isotherms become double welled. Since $A = U - TS$, at low enough temperatures the Helmholtz energy becomes double welled, and a second phase transition occurs. The effect of hydrogen bonds is to impose a second region of stability at low temperatures, centered around the optimal bulk density for

---

[39] This notation differs from that of Poole et al. (1994), but is consistent with that used here to discuss the lattice model of Sastry et al. (1993).

bonding. For hydrogen-bond strengths less than 16.5 kJ mol^{-1}, the phase behavior is that shown in Figure 4.39(b). There is now a triple point between fluid phases; in a (T, ρ) projection, the low-density liquid phase is completely enclosed within a region of states that would otherwise be unstable (Poole et al., 1994). Remarkably, the locus of density maxima does not retrace; instead, it intersects the spinodal along which the superheated liquid loses stability with respect to the vapor, causing it to retrace. Thus, for low enough hydrogen-bond strengths, the phase behavior includes the key prediction of the stability limit conjecture. Note, however, that the phase with respect to which the liquid loses stability is now isotropic (the intermediate-density liquid), so that a locus of diverging compressibility is plausible.

The most interesting aspect of the ingenious modification of the van der Waals equation proposed by Poole et al. (1994) is the fact that it reproduces both the stability limit conjecture and the two-critical-point scenario. It makes a persuasive case for considering these interpretations as fundamentally related, rather than mutually exclusive, as they had been hitherto regarded. Of course, both these scenarios cannot be valid for a given liquid. However, the model's generality strongly suggests that similar behavior in the supercooled region may occur in other network-forming fluids, in which case molecular parameters such as the bond strength should determine which of these scenarios applies to a specific substance.

A phase diagram topologically identical to that shown in Figure 4.39(a) [two fluid-fluid transitions, with the one occurring at high pressure having a negatively-sloped binodal in the (P, T) plane that terminates at a second critical point] is predicted by a two-state model in which liquid water is treated as a mixture of two types of clusters (Ponyatovsky et al., 1994). This idea is related to the bonded and nonbonded states of Poole et al.'s van der Waals model. The phenomenological two-state mixture model is the simplest theory that predicts a second critical point for water.

The work of Borick et al. (1995) lends further theoretical support to a fundamental relation between the different proposed scenarios for the phase behavior of supercooled water. These authors studied a lattice model of network-forming fluids. In this model, there are two interpenetrating simple cubic lattices; the basic interactions are shown in Figure 4.40. Two molecules occupying nearest-neighbor positions experience a weak attraction, $-\varepsilon$, regardless of their mutual orientation; a pair of next-nearest-neighbor molecules that are correctly mutually oriented experience a strong attraction, $-J$. There are q^2 possible mutual orientations, of which one leads to bonding. The presence of a molecule in any one of the four centers of cells that share the edge along which the bond forms weakens the bond by an amount $cJ/4$ $(c > 0)$. The ground state of this model consists of one fully bonded, occupied sublattice, with the other one empty. As in the work of Sastry et al. (1993), to which this model is closely related, bonding requires that molecules be correctly mutually oriented and sufficiently

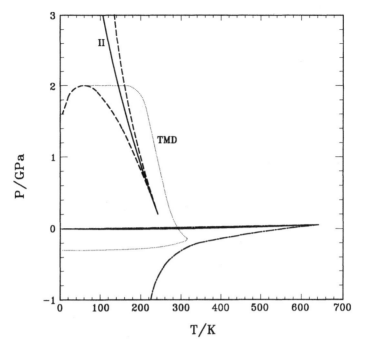

Figure 4.39a. Phase equilibrium (—), stability limits (– – – and — · —), and locus of density maxima (⋯) generated by the van der Waals equation of state, modified to account for hydrogen bonding. Note the appearance of a second phase transition at low temperatures (II). $v_{HB} = 1.087$ cm³ g⁻¹ (optimal specific volume for hydrogen bonding); $b = 12.01$ cm³ mol⁻¹ (van der Waals excluded-volume parameter); $a = 0.218$ Pa m⁶ mol⁻² (van der Waals attractive parameter); $q^2 - 1 = 5.02 \times 10^4$ (number of nonbonded orientations per bond). For $J = 22$ kJ mol⁻¹, the second fluid-fluid transition terminates at a critical point, and the locus of density maxima retraces without intersecting the spinodal. [Reprinted, with permission, from Poole et al., *Phys. Rev. Lett.* 73: 1632 (1994)]

separated. The treatment of bonds is more detailed in the model of Borick et al. (1995) because three-body effects are incorporated explicitly. The relevance to water comes from the existence of an open ground state and the competition between bonded states of low density, energy, and entropy, and nonbonded states of high density, energy, and entropy. However, the ground-state network, which is sixfold coordinated, does not have waterlike geometry.

The mean-field solution of the model yields behavior topologically identical to that shown in Figure 4.36: negatively sloped melting line (the liquid expands when it freezes), negatively sloped locus of density maxima, spinodal retracing,

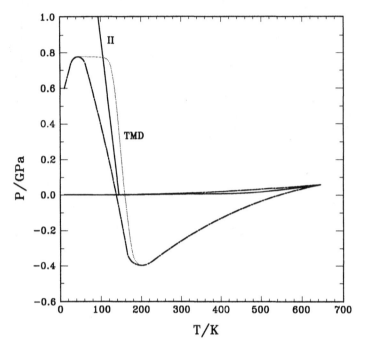

Figure 4.39b. Phase equilibrium (—), stability limits (– – – and — - —), and locus of density maxima (⋯) generated by the van der Waals equation of state, modified to account for hydrogen bonding. Note the appearance of a second phase transition at low temperatures (II). $v_{HB} = 1.087$ cm^3 g^{-1} (optimal specific volume for hydrogen bonding); $b = 12.01$ cm^3 mol^{-1} (van der Waals excluded-volume parameter); $a = 0.218$ Pa m^6 mol^{-2} (van der Waals attractive parameter); $q^2 - 1 = 5.02 \times 10^4$ (number of nonbonded orientations per bond). For $J = 14$ kJ mol^{-1}, there is a three-fluid triple point, and a Speedy-like retracing spinodal. [Reprinted, with permission, from Poole et al., *Phys. Rev. Lett.* 73: 1632 (1994)]

and liquid-solid instability line superseding the spinodal. Interestingly, when the bonding strength is sufficiently high, the behavior shown in Figure 4.41 results. The locus of density maxima retraces, as in the simulations of Poole et al. (1992), and does not intersect the spinodal; however, no additional phase transition appears at low temperatures. In the absence of a retracing spinodal or a critical point, the response functions remain finite at all pressures when the liquid is cooled isobarically below its equilibrium freezing temperature. The increase in the response functions at low temperatures is due exclusively to the density anomalies. In Monte Carlo simulations of the two-dimensional version

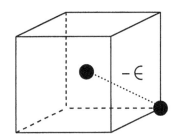

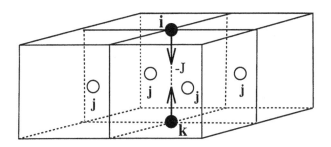

Figure 4.40. Elementary interactions in the network-forming lattice model of Borick et al (1995). The vertices of each cubic cell define sublattice A, and the centers define sublattice B. Two nearest neighbors on different sublattices (top) experience a weak attraction, $-\varepsilon$, regardless of their mutual orientation. Next nearest neighbors on the same sublattice (i, k) experience a strong attraction, $-J$, if their mutual orientation is correct for bonding (bottom). The presence of a molecule j in any one of the four centers of cells that share the edge along which the bond forms weakens the bond by an amount $cJ/4$. [Reprinted, with permission, from Borick et al., *J. Phys. Chem.* 99: 3781 (1995). Copyright 1995, American Chemical Society]

of their model, Borick et al. (1995) noted that bonds form cooperatively, via low-density regions consisting of cyclic bonded clusters, rather than randomly, as assumed in the mean-field calculations. This effect is more pronounced close to the freezing density. To capture this behavior, they incorporated a Gaussian density dependence into the effective bond strength. The resulting phase behavior is topologically identical to that shown in Figures 4.37 and 4.39(a): retracing of the locus of density maxima; appearance of a second fluid-fluid transition terminating at a critical point; negatively sloped binodal for the second fluid-fluid transition. Upon increasing the orientational entropy (larger q) and decreasing the width of the Gaussian, the phase behavior becomes topo-

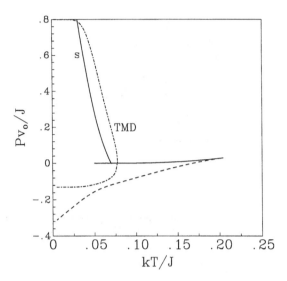

Figure 4.41. Phase equilibrium, stability limits, and density anomalies in the lattice model of Borick et al. for the strong, density-independent bonding case. Parameter values $J/\varepsilon = 10$ (ratio of bonding to nonbonding interaction strength), $c = 0.7$ (fraction of $J/4$ by which each one of four surrounding molecules can weaken a bond), and $q = 8$ (number of distinguishable orientations per molecule). v_0 is the hard-core volume per molecule. s is the liquid-solid binodal; TMD is the locus of density maxima. Note that the locus of density maxima retraces; hence the liquid spinodal does not retrace, and there is only one phase transition between isotropic phases. [Reprinted, with permission, from Borick (1995)]

logically identical to that of Figure 4.39(b): three-fluid triple point; negatively sloped, nonretracing locus of density maxima; retracing spinodal.

An earlier, related model was formulated by Stanley (1979), and discussed in detail by Stanley and Teixeira (1980). Each water molecule i is assigned to one of j species based on the number of bonds that it can form. A bond is formed when the interaction between two molecules is stronger than an arbitrary cutoff. When implemented on a lattice, the model views each occupied site as an oxygen atom, and each bond as a hydrogen bond between neighboring pairs of oxygen atoms. In any given configuration, molecule i on a lattice with coordination number 4 can form 0, 1, 2, 3, or 4 bonds ($j = 5$). For a given probability that two neighbor molecules are bonded, the overall concentration of molecules belonging to one of the j species is determined by random statistics; the species' connectivity, however, is highly nonrandom. For example, on a lattice with coordination number 4, an empty site cannot be nearest neighbor to a species-4 molecule; similarly, if the four nearest neighbors of a given site belong to species 4, then the site must also belong to species 4 (Stanley, 1979). In particular, the positions of tetrahedrally coordinated (species-4) oxygen atoms are strongly correlated. If the volume per site depends on the number of bonds emanating from the site, with $V_4 > V_3 > V_2 > V_1 > V_0$, then an "icelike" patch consisting predominantly of spatially correlated, fully bonded, species-4 molecules will have a lower local density than, say, a patch containing the same number of nonbonded molecules. The spatial correlation among molecules of a given species introduces corresponding spatial correlations in density fluctuations because the local density is related to the type of species present. Consequently, as the temperature decreases, the characteristic size of "icelike" patches increases, and so do density fluctuations. Using these appealingly simple ideas, Stanley and Teixeira (1980) made a number of qualitative predictions that agree with trends observed in supercooled water. These include the increase in the compressibility and isobaric heat capacity, and the decrease in the density at low temperatures. The model predicts that the compressibility reaches a maximum at low temperature and remains finite upon further supercooling.

Table 4.11 summarizes the main theoretical interpretations of the thermodynamics of supercooled water.

The singularity-free scenario illustrated in Figure 4.41 and mentioned in Table 4.11 is consistent with the x-ray scattering experiments of Xie et al. (1993). These investigators measured the static structure factor $S(k)$ down to $-34°C$ and over the wave number range $0.05 < k < 0.3$ Å^{-1}, and calculated the isothermal compressibility K_T by extrapolation to zero wave number:

$$S(k = 0) = \rho k T K_T = 1 + 4\pi\rho \int [g - 1] r^2 \, dr, \qquad (4.117)$$

where ρ is the number density, and g, the pair correlation function. Xie et al. (1993) used a Lorentzian function to fit their $S(k)$ data, from which they obtained the correlation length [see (2.61) and (2.62)]. Interestingly, they found compressibility increases upon supercooling but no significant increase of the correlation length. In light of the second equality in (4.117) this result means that the compressibility increase is not due to long ranged spatial correlations, but to an increase in the height of the $g(r)$ peak (clustering). This also means that the compressibility remains finite, since K_T can only diverge if g becomes long ranged, as it does at the critical point. This is contrary to the existence of a Speedy-like retracing spinodal, and it also implies that a second critical point, if it exists, is far removed from the conditions studied by Xie et al. (1993). Sharp increases in the response functions upon supercooling are not inconsistent with a singularity-free scenario because the response function maxima may be inaccessible due to homogeneous nucleation.

Accurate x-ray scattering data in supercooled water are difficult to obtain because the scattering from water is weak. This calls for high incident-beam intensity and long accumulation times at each data point; in addition, back-

Table 4.11. Interpretations of the Thermodynamics of Supercooled Water

Supercooled Liquid Spinodal

Experimental Evidence

Many properties fit $X = \left(T - T_{sp}\right)^{-\gamma}$.

Theoretical or Computational Support

* Thermodynamic consistency based solely on stability with respect to an isotropic phase (Speedy, 1982a; Debenedetti and D'Antonio, 1988).
* Retracing spinodal in lattice models (Sastry et al., 1993; Sasai, 1993; Borick et al., 1995), but spinodal exists only at negative pressure.
* Second fluid-fluid transition with three-fluid triple point shows a retracing spinodal in modified van der Waals treatment (Poole et al., 1994; low-hydrogen-bond-strength case), in two-state model of Ponyatovsky et al. (1994), and in lattice model of Borick et al. (1995) modified to make bonding density dependent (scenario appears at high orientational entropy and lower width of the bonding strength Gaussian).
* Diverging response functions upon supercooling in lattice model of Sasai (1990). Implausibly, this divergence happens along a liquid-solid instability.

Two Critical Points

Experimental Evidence

Apparently first-order transition between LDA and HDA (e.g., Mishima, 1994; see Section 4.6.3).

Theoretical or Computational Support

* Low-temperature inflection point in liquid isotherms, but no certainty about critical point (Poole et al., 1992, 1993a; Stanley et al., 1994).
* Isothermal compression simulations yield two apparently distinct amorphous solid phases (Poole et al., 1992).
* Second fluid-fluid transition with critical point in modified van der Waals treatment (Poole et al., 1994; high-hydrogen-bond-strength case), in two-state model of Ponyatovsky et al. (1994), and in lattice model of Borick et al. (1995) modified to make bonding density dependent.

No Singularity Upon Supercooling

Experimental Evidence

Small-angle x-ray scattering shows no anomalous growth in the correlation length (Xie et al., 1993).

Theoretical or Computational Support

* Stanley-Teixeira site correlation model shows response function extrema but no divergence upon supercooling (Stanley and Teixeira, 1980).
* Lattice model of Borick et al. (1995) shows a retracing density maxima locus, no spinodal retracing, and no additional phase transition (high and density-independent bond strength).

ground subtraction and cell geometry are important sources of error (Xie et al., 1993). Accordingly, it is not surprising that conflicting results have been reported in several x-ray scattering studies of supercooled water: increase in $S(k)$ with decreasing k at low temperature, with a correlation length that increases up to 8 Å at $-20°C$ (Bosio et al., 1981); no k dependence in S (Michielsen et al., 1988; Dings et al., 1992).

Understanding the properties of supercooled water remains one of the most interesting problems in the physics of liquids. A fundamental question is whether there exists a mechanism for diverging density fluctuations. If so, is this mechanism a second critical point [Figure 4.39(a)] or a retracing spinodal [Figure 4.39(b)]? If indeed the properties of supercooled water reflect the transition between LDA and HDA (see Section 4.6.3), is density the only order parameter for this transition, as has hitherto been assumed (e.g., Mishima, 1994), or does density couple with another order parameter? Neutron and x-ray scattering structural studies of supercooled water at low temperature and high pressure (e.g., Bellissent-Funel and Bosio, 1995) can shed light on this question. Improved understanding of phase transitions between different glassy phases of the same substance (vitreous polyamorphism: Grimsditch, 1984; Itie et al., 1989; Brazhkin et al., 1992; Ponyatovsky and Barkalov, 1992; Angell, 1995; Smith et al., 1995) is clearly needed.

More studies of structure and bulk behavior at lower temperatures are needed. Measuring the compressibility in capillaries below $-26°C$ (Speedy and Angell, 1976), and pushing the limits of state-of-the-art neutron and x-ray scattering capabilities (e.g., Bellissent-Funel et al., 1989; Dore, 1986, 1990; Xie et al.,

1993) towards still greater supercooling are especially important goals. Also important is the information on spatial bond correlations provided by Raman spectroscopy (Green et al., 1986, 1987), and its extension to lower temperatures (< 248 K). The most interesting theoretical development to date is the notion that there exists a fundamental relation between the various interpretations of water's behavior. Although only one scenario can apply to a given substance, the work of Poole et al. (1994) and of Borick et al. (1995) suggests that metastable phase behavior of the types shown in Figures 4.36, 4.37, 4.39, and 4.41 can occur in other network-forming liquids. Further theoretical and computational studies of model network formers are needed to clarify this interesting viewpoint (Stanley et al., 1994). All of the theories and interpretations discussed in this section deal almost[40] exclusively with thermodynamics. The dynamics of supercooled water is equally deserving of theoretical attention.

4.6.3 Glassy Water

Low-temperature transitions between amorphous phases are important for understanding the thermodynamics of supercooled water. The extremely rapid cooling of liquid water ($> 10^6$ K sec^{-1}) leads to the formation of an amorphous solid (Brüggeller and Mayer, 1980; Dubochet and McDowell, 1981; Dubochet et al., 1982; Mayer and Brüggeller, 1982; Mayer, 1985a,b; Johari et al., 1987; Hallbrucker et al., 1989). The resulting material is called glassy water (GW; Hallbrucker et al., 1989). Amorphous solid water can also be made starting from the vapor and crystalline phases. The resulting materials are known as amorphous solid water (ASW; Sceats and Rice, 1982), and high-density amorphous ice (HDA; Mishima et al., 1984), respectively.

The so-called hyperquenching route to GW involves rapid cooling of liquid water droplets. In the method developed by Mayer (1985a), an aqueous aerosol is produced by ultrasonic nebulization, giving rise to droplet sizes in the 3–25 μm range. The aerosol is conveyed with gaseous nitrogen into a vacuum chamber, where the droplets are deposited on a liquid-nitrogen-cooled copper plate. In contrast to other hyperquenching techniques (Brüggeller and Mayer, 1980; Dubochet and McDowell, 1981; Dubochet et al., 1982; Mayer and Brüggeller, 1982), this approach does not require the dispersion of water droplets in a liquid cryomedium. ASW is formed by deposition of water vapor onto a cold substrate, typically Cu at 77 K (Burton and Oliver, 1935; Narten et al., 1976; Sceats and Rice, 1982). HDA can be obtained by compressing ice Ih above 10 kbar at 77 K. In this novel approach (Mishima et al., 1984, 1985; Whalley et al., 1986) ice is compressed at a temperature that is low enough (77 K) so that crystallization to more stable forms is too slow, until it reaches a

---

[40] See Stanley and Teixeira (1980) and Sasai (1993) for brief discussions of transport and relaxation in supercooled water.

pressure that would presumably correspond to the low-temperature metastable extrapolation of its melting curve (ca. 10 kbar at 77 K). The crystal then melts, but, the temperature being so low, it does so directly into a glass.

When ASW is annealed in vacuum from 77 K to ca. 113 K, it relaxes to GW (Hallbrucker et al., 1989). The identity between the structures of GW and relaxed ASW has been established by comparison of the respective pair correlation functions and structure factors determined by x-ray and neutron scattering (Bellissent-Funel et al., 1992). When HDA is heated at atmospheric pressure, it undergoes a transition to a low-density amorphous phase at ca. 120 K (Mishima et al., 1985). The resulting material is called low-density amorphous ice, LDA (Mishima, 1994). X-ray and neutron diffraction studies have shown that LDA and GW are very similar structurally (Bellissent-Funel et al., 1992). It is generally accepted that LDA and GW are in fact the same material (Whalley et al., 1989).[41] Accordingly, in what follows we refer to water's two glassy phases as LDA and HDA.

The relationship between LDA and HDA is of fundamental importance, not just for understanding the phase behavior of water, but for the physics of disordered systems in general. Since it was first reported (Mishima et al., 1985), the LDA $\leftrightarrow$ HDA transformation has attracted great attention because it constitutes the first example of what appears to be a first-order transition between two amorphous solids (Poirier, 1985).[42] The LDA $\leftrightarrow$ HDA transition has been characterized volumetrically (Mishima et al., 1985; Handa et al., 1986; Floriano et al., 1989; Mishima, 1994), thermally (Handa et al., 1986; Floriano et al., 1989) and visually (Mishima et al., 1991). The (P, T) projection of the phase diagram of solid water, including the observed transitions between LDA and HDA, is shown in Figure 4.42 (Mishima, 1994). Important facts about the transitions between LDA, HDA, and ice Ih include the following:

(i) When ice Ih is compressed to ca. 11 kbar at 77 K, it transforms to HDA. The volume contraction that accompanies this transition is 4.1 cm^{3}mol^{-1} (Mishima et al., 1984; Floriano et al., 1989). The density of HDA at 77 K and ca. 11 kbar is 1.31 ± 0.02 g cm^{-3} (Mishima et al., 1984, 1985).

(ii) When pressure is removed at 77 K, the ice Ih $\rightarrow$ HDA transition does not reverse; the density of HDA at 1 bar is 1.17 ± 0.02 g cm^{-3} (Mishima et al., 1984, 1985).

(iii) When HDA is heated at 1 bar, it transforms to LDA at ca. 120 K (Mishima et al., 1985). The density of LDA at these conditions is 0.94 g cm^{-3} (Mishima et al., 1985).

---

[41] The contrary view, namely, that LDA and GW are different materials, is advocated by Johari (1995) and Mayer (1994), based on the slightly different locations of their glass transitions.

[42] At present it is not known whether density is the sole order parameter for this transition, or whether it couples with another order parameter.

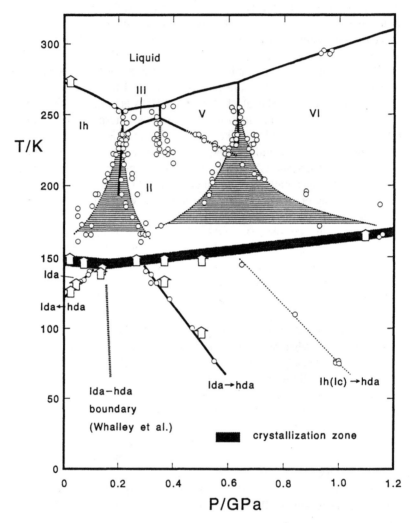

Figure 4.42. Low-temperature phase diagram of H_2O including transitions between amorphous solid phases. Open circles denote pressure-induced transitions, and arrows denote temperature-induced transitions. The line labeled lda-hda boundary is the calculated equilibrium locus of first-order transitions (Whalley et al., 1989). The lines labeled LDA → HDA and HDA → LDA are the kinetic loci along which the transition occurs. The line Ih → HDA is thought to correspond to the metastable continuation of the melting curve of ice Ih. [Reprinted, with permission, from Mishima, *J. Chem. Phys.*, 100: 5910 (1994)]

(iv) The enthalpy change that accompanies the HDA → LDA transformation at atmospheric pressure is ca. -750 J mol^{-1} (Whalley et al., 1989; Floriano et al., 1989).

(v) When LDA is compressed at 77 K, it transforms to HDA at ca. 5 kbar, with a volume change of -4 ± 0.3 cm^3 mol^{-1} (Mishima et al., 1985; Floriano et al., 1989).

(vi) At temperatures between 130 and 140 K, LDA transforms to HDA upon compression at ca. 3 kbar, and HDA reverts to LDA upon decompression at ca. 1 kbar (Mishima, 1994).

Estimates of the equilibrium pressure at 77 K between ice Ih and HDA, and between LDA and HDA, respectively, yield values of ca. 5 and 2 kbar, respectively (Floriano et al., 1989). These estimates are based on the assumption that these are indeed first-order transitions. The fact that these transitions actually happen at ca. 11 kbar (ice Ih → HDA) and ca. 5 kbar (LDA → HDA) suggests that the observed transformations are kinetically controlled manifestations of the equilibrium transitions.

When LDA is heated, it undergoes a very weak glass transition. This transition was observed at 124K in calorimetric measurements that used a heating rate of 10 K h^{-1} (Handa and Klug, 1988), and at 136 K using a heating rate of 30 K min^{-1} (Johari et al., 1987; Hallbrucker et al., 1989). The heat capacity jump at the transition is ca. 1.6 J mol^{-1}K^{-1} (Hallbrucker et al., 1989). Upon further heating, crystallization to ice Ic (cubic ice) occurs at ca. 150 K, and is accompanied by an enthalpy change of -1.33 ± 0.02 kJ mol^{-1} (Hallbrucker and Mayer, 1987). The nature of the phase that results upon heating through the glass transition is not well understood. Hallbrucker et al. (1989) and Johari et al. (1990) have assumed that it is liquid, while Jenniskens and Blake (1994) conclude from electron diffraction and direct observation that it is an amorphous solid. Clearly, further characterization studies of the amorphous phase of water immediately above the glass transition are needed. Recent isotopic exchange experiments (Fisher and Devlin, 1995) suggest that rotational degrees of freedom become active on heating through the glass transition, while translational motion remains frozen.

A proposed phase diagram for LDA and HDA that is consistent with many of the above observations is shown in Figure 4.43 (Poole et al., 1993b). In particular, according to this diagram, (a) the rapid cooling of water at atmospheric pressure leads to LDA because the region of stability of LDA is connected to the stable liquid region at 1 bar without intersecting the coexistence line or the LDA → HDA spinodal; (b) isothermal compression of LDA at 77 K will induce a transition to HDA at pressures exceeding the equilibrium pressure; (c) if the pressure is removed from HDA at 77 K, it can still be observed at 1 bar because the HDA → LDA spinodal is not crossed along this path; (d) if HDA is heated above ca. 120 K at 1 bar it transforms to LDA because the HDA → LDA spinodal is crossed.

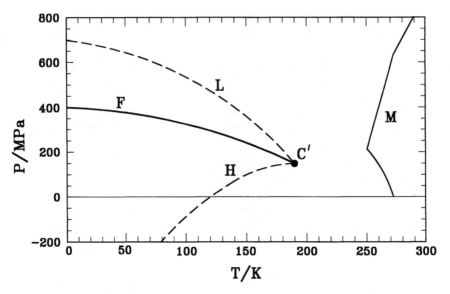

Figure 4.43. Proposed phase diagram of amorphous solid water. M is the melting curve. F is the first-order phase transition line between HDA and LDA, terminating at the critical point C'. L is the spinodal for the LDA $\rightarrow$ HDA transition. H is the spinodal for the HDA $\rightarrow$ LDA transition. [Reprinted, with permission, from Poole et al., *Phys. Rev. E.* 48: 4605 (1993)]

According to Figure 4.43, there exists a continuity of metastable states at 1 bar connecting liquid water to LDA. Structural studies by neutron and x-ray diffraction support this viewpoint (Bellissent-Funel et al., 1989; Dore, 1990; Bellissent-Funel and Bosio, 1995). Speedy (1992) has taken a contrary view. He showed, using thermodynamic integration and the known values of the entropy and enthalpy of supercooled water at 236 K and of the enthalpy of the amorphous phase at 150 K (in all cases relative to ice Ih at the given temperature and 1 bar), that continuity requires that

$$s \, (150 \text{ K}, \ 1 \text{ bar, amorph}) - s \, (150 \text{ K}, 1 \text{ bar, ice Ih}) \leq 2.9 \text{ J K}^{-1} \text{ mol}^{-1}.$$
$$(4.118)$$

Since

$$
\begin{aligned}
\Delta s \, (T) \quad &= \quad s \, (T, 1 \text{ bar, amorph}) - s \, (T, 1 \text{ bar, ice Ih}) \\[4pt]
&= \quad \Delta s \, (0) + \int_0^T \frac{\Delta c_p}{T} \, dT,
\end{aligned}
$$
$$(4.119)$$

and using a theoretical estimate for $\Delta s (0) = 6.3 \text{ J K}^{-1} \text{mol}^{-1}$ based on the random network model (Sceats et al., 1979; Sceats and Rice, 1980a,b,c; Sceats and Rice, 1982), Speedy argued that continuity is impossible $\left(\Delta c_p > 0 \right)$. The material that results from heating LDA above 136 K is then, according to Speedy (1992) (see also Angell, 1993), a distinct phase of water that cannot be connected by a thermodynamically reversible path to liquid water at 1 bar. Speedy called this phase water II. Thought provoking as this argument is, it is based on a theoretical estimate of $\Delta s(0)$. The opposite viewpoint, namely, that the theoretical estimate is wrong (Whalley et al., 1989; Johari et al., 1994) appears to be consistent with available experimental evidence, though it leaves this theoretical discrepancy unresolved. At present, the continuity issue is a matter of controversy (Speedy et al., 1995). The resolution of the argument requires ultrafast calorimetric measurements of the amorphous phase above 150 K prior to crystallization.

4.7 COMPUTER SIMULATION OF SUPERCOOLED LIQUIDS

Computer simulations have become an invaluable tool for the study of liquids (Allen and Tildesley, 1987). There exists a vast literature on simulations of supercooled liquids, nucleation, and the glass transition, including several excellent reviews (Frenkel and McTague, 1980; Angell, 1981; Angell et al., 1981; Barrat and Klein, 1991). This section is therefore not a comprehensive review of existing studies, but a discussion of peculiarities and limitations that are unique to the computational investigation of deeply supercooled liquids. There is no question but that simulations can provide important insights into the physics of supercooled liquids, as illustrated by the many examples cited in this chapter. Nevertheless, the implementation and interpretation of such simulations involves questions of time and length scales that do not arise in the case of stable liquids, and it is these questions that concern us here.

Of the two techniques, Monte Carlo and molecular dynamics, the latter has been the more frequently used to study supercooled liquids. Fox and Andersen (1984) have raised the question of whether the one-at-a-time stochastic motion of atoms in a Monte Carlo simulation (e.g., Abraham, 1980) can reproduce the collective structural rearrangements that characterize deeply supercooled liquids. This interesting question has not been given the full attention it merits. What is required is a comparative study of structures generated by Monte Carlo and molecular dynamics using techniques and structural indicators such as Voronoi tessellation (e.g., Hsu and Rahman, 1979a), inherent structure analysis (e.g., Stillinger and Weber, 1984a), and bond orientational order (e.g., Steinhardt et al., 1983; Mountain and Brown, 1984).

The basic limitation of molecular dynamics as applied to the study of supercooled liquids is one of time scales. As discussed before, Goldstein's analysis (1969) suggests that supercooled liquids begin to relax via the infrequent exploration of deep potential energy minima (α relaxations, or, equivalently, the sub-T_x regime) when the shear relaxation time becomes of the order of 10^{-9} sec.

Angell (1988a) provides evidence of low-temperature uncoupling of relaxation processes in some fragile liquids. When this occurs, shear relaxation is always faster than configurational relaxation. Thus 10^{-9} sec is the shortest time that must be probed if the characteristic features of structural relaxation of supercooled liquids are to be detected, let alone explored. More realistically, one would require sufficient supercooling (so that the α and β processes are fully separated), and observation spanning several relaxation times. Let us assume conservatively that both conditions are met when a supercooled liquid is probed during 10^{-8} sec. In molecular dynamics, the time step used for the numerical integration of Newton's equations of motion is determined by a compromise between numerical accuracy (which calls for a small time step) and the duration of a simulation (which, for a given simulated time, is inversely proportional to the size of the time step). Typically, the simulation of atomic systems is done with time steps of the order of 10^{-14} sec.[43] Simulating the dynamics of 1000 atoms over 10^{-8} sec therefore requires roughly 10 hours of computer time.[44] Because it is not the size of the system but the number of steps that is large, this is not a problem where parallel computation can be used to increase the speed of the calculations.

Implicit in the previous discussion is the assumption that crystallization does not occur during the simulation. In simulations of deeply supercooled atomic liquids, crystallization occurs readily (see, for example, the early studies of Mandell et al., 1976, 1977; Tanemura et al., 1977; Hsu and Rahman, 1979a,b). One way of preventing crystallization is to cool the system very rapidly. Because of computer time limitations, however, simulated cooling rates are always several orders of magnitude larger than their fastest attainable laboratory counterparts. The most common way of controlling the temperature in molecular dynamics is by rescaling atomic velocities, v.[45] Since $\Delta T / T = 2 \Delta v / v$, a relative velocity change of 10^{-4} implies a temperature change of 0.02 K for $T = 100$ K (the triple point of Ar occurs at 83.8 K). When even such a small velocity change takes place over one typical molecular dynamics time step, the resulting cooling rate is 2×10^{12} K sec^{-1}, which is five orders of magnitude faster than the fastest cooling rates currently attainable in the laboratory for bulk samples. The implications of these very high cooling rates as to the nature of the transition observed in simulations and its relationship to the laboratory

---

[43]In a simulation of, say, a single-component Lennard-Jones system, time is measured in units of $[m\sigma^2/\varepsilon]^{1/2}$, where m is the mass of an atom, and σ and ε are the Lennard-Jones size and energy parameters. An adequate compromise between numerical accuracy and run duration results when the integration step is of the order of $0.01[m\sigma^2/\varepsilon]^{1/2}$. Using values of mass and potential parameters appropriate to a typical atomic system (e.g., Lennard-Jones Ar, for which $m = 2.99 \times 10^{-26}$ kg, $\sigma = 3.4$ Å, $\varepsilon/k = 120$ K) gives an integration step of 1.44×10^{-14} sec.

[44]Chialvo and Debenedetti (1992) quote a value of 0.032 sec per time step for a 1372-atom Lennard-Jones system on a Cray Y-MP/432 machine.

[45]For molecular systems, both translational and rotational motion must be rescaled. Unless otherwise stated, we discuss atomic systems.

glass transition were first analyzed by Angell and co-workers (Angell, 1981; Angell et al., 1981), and are discussed below.

One way of quenching a model liquid in a simulation is to change the temperature instantaneously by an amount ΔT by rescaling atomic velocities; to keep the temperature constant during a certain equilibration interval τ_{equil}; and then to repeat this cycle. To facilitate comparisons with experiments, the system is usually quenched isobarically (e.g., Fox and Andersen, 1984). Configurations at the end of each τ_{equil} are used as the starting point for a data collection run of duration τ_{sim}. Isothermality is maintained by rescaling during both equilibration and data collection intervals. The above protocol defines an effective cooling rate

$$|\dot{q}| = \frac{\Delta T}{\tau_{equil}}. \qquad (4.120)$$

The simulated system has its own relaxation time, during which it will readjust to temperature perturbations. Assuming Arrhenius behavior, we can write

$$\tau_{relax} = \tau_0 \exp(A/T). \qquad (4.121)$$

Equilibration can only occur if τ_{equil} is substantially greater than the system's inherent relaxation time, τ_{relax}. Otherwise, structural arrest on the simulation time scale will occur when the ratio $\tau_{relax}/\tau_{equil}$ exceeds a minimum number (for example, unity, beyond which the system will still be relaxing when data collection begins). The temperature when this condition first occurs, T^*, is then given by

$$T^* = \frac{A}{\left[\ln \frac{n\Delta T}{\tau_0 |\dot{q}|}\right]}, \qquad (4.122)$$

where we have written n for the critical ratio of inherent relaxation to simulated equilibration times. Equivalently, we can write

$$\frac{d \ln |\dot{q}|}{d(1/T^*)} = -A. \qquad (4.123)$$

One notable effect of increasing the cooling rate is thus to raise the temperature at which the system falls out of equilibrium.

Intimately related to the increase in transition temperature is the smearing of the transition across a wide temperature interval. Assuming exponential relaxation, a perturbation will decay by only 1% after 0.01 relaxation times, and by 99% after 5 relaxation times (the relaxation time is here the time it takes for a perturbation to decay to $1/e$ of its initial value). Thus a temperature drop that brings about a two-order-of-magnitude increase in the relaxation time will cause the system to appear structurally arrested on the simulation time scale (Angell, 1981; Angell et al., 1981). If the relaxation time follows Arrhenius behavior, the width of the temperature interval across which the relaxation time changes by two orders of magnitude is given by

$$\delta T = T_2 - T_1 = 4.6 \, T_1 \, T_2 / A. \qquad (4.124)$$

The effective activation energy E for structural relaxation in simulated systems ($E = AR$; R is the gas constant; $A = d \ln \tau / dT^{-1}$) is low compared to laboratory systems, where marked deviations from Arrhenius behavior cause the apparent activation energy to increase markedly upon cooling. For example, for VTF behavior $d \ln \tau / dT^{-1} = A \left[T/(T - T_0) \right]^2$, and lowering the temperature from $2T_0$ to $1.2T_0$ causes a ninefold increase in the apparent activation energy. It can be seen from (4.124) that this broadens the liquid to glass transformation range with respect to what is observed in laboratory systems. For a typical laboratory activation energy for structural relaxation in supercooled liquids of 400 kJ mol^{-1} (Angell et al., 1981), δT calculated via (4.124) equals 8 K. If, on the other hand, E is only 10 kJ mol^{-1}, which is typical of molecular liquids at ambient temperature (Angell et al., 1981), δT becomes 147 K.[46] Thus, the high cooling rates used in simulations cause the threshold temperature at which the system begins to fall out of equilibrium to increase, and the transformation range across which structural arrest occurs to broaden appreciably. Figure 4.44 illustrates the widening of the transformation regime in simulations (Fox and Andersen, 1981). These authors observed that the tendency to nucleate increased with pressure. In order to avoid nucleation, faster cooling rates were used at the higher simulated pressures (ca. 10^{11} K sec^{-1} at $P\sigma^3/\varepsilon = 10$, compared to ca. 10^9 K sec^{-1} at $P\sigma^3/\varepsilon = 0.5$).[47] This resulted in a very wide transformation regime, which, at $P\sigma^3/\varepsilon = 8$, spanned 64 K (Fox and Andersen, 1984).

Figure 4.44 shows some features that are commonly associated with the laboratory glass transition, but also an unusually wide transformation range. Meaningful comparisons between computer and laboratory glass transitions, however, can only be made when the time scales being probed are similar. This was done by Angell and Torell (1983), who juxtaposed the behavior of an ionic laboratory glass former (Ca^{++}-K$^+$-NO$_3^-$; Ca^{++}/K$^+$ = 2/3) probed on the 10^{-11} sec time scale by Brillouin scattering, and that of the supercooled Lennard-Jones liquid. The transformation range of the laboratory glass former exhibited the same pronounced smearing that occurs in simulations. This is shown in Figure 4.45. One concludes that the extreme cooling rates used in simulations smear the resulting glass transition to a degree that forces comparison not with calorimetric measurements where the system's response is probed on the 10^2 sec time scale, but with high-frequency experiments, which probe dynamics thirteen orders of magnitude removed in time, i.e., over the

---

[46]In both cases we use $T_2 = 300$ K as a basis for comparison.

[47]In order to convert to dimensional quantities, we use the argon values for the Lennard-Jones parameters given by Fox and Andersen (1984), $\sigma = 3.4$ Å and $\varepsilon/k = 176.5$ K.

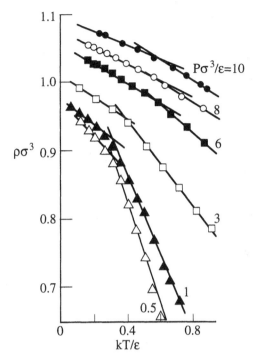

Figure 4.44. Temperature dependence of the density of the supercooled Lennard-Jones fluid at various pressures. The cooling rate at $P\sigma^3/\varepsilon = 10$ is six times higher than at 0.5, 1, and 3. Note the gradual widening of the transformation regime. [Reprinted, with permission, from Fox and Andersen, in *Ann. N.Y. Acad. Sci.* 371: 123 (1981)]

10^{-11} sec range. The smeared simulated transition is the unavoidable consequence of the fact that simple atomic systems become metastable with respect to the fully ordered crystalline state at relatively high temperatures, such that the characteristic time for configurational exploration is very short, mandating that the metastable liquid be probed on comparably short times if crystallization is to be avoided. This strategy allows the study of thermodynamic properties by simulation, but is unsuited for the adequate investigation of configurational relaxation, which occurs over time scales of at least 10^{-8} sec.

A common way of avoiding crystallization in simulations is to use a mixture of particles of different sizes. The formation of a binary critical nucleus is then, compared to the single-component case, a much more improbable event. This can be seen from the fact that an interchange of randomly chosen particles has no effect on any single-component cluster, but can destroy a binary cluster. Numerous molecular dynamics studies of supercooled and glassy binary soft sphere systems have been performed (e.g., Bernu et al., 1985, 1987; Mountain

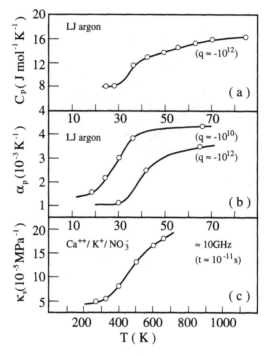

Figure 4.45. Comparison between the isobaric heat capacity (a) and the thermal expansion coefficient (b) of the Lennard-Jones fluid across the glass transition, calculated by molecular dynamics; and the adiabatic compressibility $[K_s = (\partial \ln \rho / \partial P)_s]$ of the laboratory ionic glass former Ca^{++}-K^+-NO_3^- at hypersonic frequencies (c). q denotes the quench rate ($K\ sec^{-1}$). The simulation results are made dimensional by using argon parameters for the Lennard-Jones fluid. Note that the transformation regime for the laboratory system spans more than 200 K. [Reprinted, with permission, from Angell and Torell, *J. Chem. Phys.*, 78: 937 (1983)]

and Thirumalai, 1987; Thirumalai and Mountain, 1987; Pastore et al., 1988; Barrat et al., 1988; Miyagawa et al., 1988; Miyagawa and Hiwatari, 1989; Hiwatari and Miyagawa, 1990; Kob and Andersen, 1994, 1995). This important body of work[48] has addressed the location of the glass transition (Bernu et al., 1985), scaling laws for diffusion coefficients close to the glass transition (Bernu et al., 1987), relaxation of orientational order parameters (Mountain and Thirumalai, 1987; Thirumalai and Mountain, 1987), relaxation of the density-density correlation function (Pastore et al., 1988), the behavior of the elastic constants near the glass transition (Barrat et al., 1988), jump motions (Miyagawa et al., 1988), static and dynamic structural changes upon annealing (Miyagawa and

---

[48] See Barrat and Klein (1991) for a review.

Hiwatari, 1989), self- and collective relaxation dynamics (Hiwatari and Miyagawa, 1990), the search for a correlation length in supercooled liquids (Ernst et al., 1991), and the detailed testing of mode-coupling theory (Kob and Andersen, 1994, 1995).

The fact that limited computer power makes it difficult for simulations to span more than 10^{-9} sec invites careful consideration of the meaning of structural arrest. The relaxation time for translational motion, τ_{trans}, is the characteristic time it takes for a particle (atom, molecule) to move by Brownian motion over a distance of the order of its size,

$$\tau_{trans} \approx a^2 D^{-1}, \tag{4.125}$$

where a is the particle diameter and D is the translational diffusivity. In order for the system not to appear arrested during a simulation, we must have $\tau_{sim} > \tau_{trans}$. Therefore, a system with a diffusivity smaller than $a^2/(10^{-9}$ sec) will appear structurally arrested in a simulation. Using 3 Å as a typical atomic dimension, this gives roughly 10^{-6} cm^2 sec^{-1} as the threshold diffusivity below which structural arrest on the time scale of a simulation will occur. For simple liquids, then, diffusivities much smaller than 10^{-6} cm^2 sec^{-1} cannot be computed reliably by molecular dynamics. A real liquid with such a diffusion coefficient is a viscous fluid, not a glass. This is an important difference between laboratory and computer glasses.

Whereas time-scale considerations are important in simulations of amorphous systems, the computational study of nucleation and crystallization is complicated by length-scale-related problems. Both the finiteness of the simulated system's size and the use of periodic boundary conditions[49] introduce artifacts that have frequently been overlooked. The identification of size- and periodicity-dependent artifacts in simulations of nucleation and crystallization is due to Andersen and co-workers (Honeycutt and Andersen, 1984, 1986; Swope and Andersen, 1990). Honeycutt and Andersen (1984) studied the supercooled Lennard-Jones fluid. In a molecular dynamics simulation at constant energy, crystallization is accompanied by a sharp temperature rise. If crystallization is the direct consequence of the formation of a critical-sized nucleus, the length of time elapsed before sudden crystallization should be inversely proportional to N (number of atoms in the simulation), for given density and temperature. This is because the rate of nucleation should be constant across the system, and the probability of forming a critical nucleus should thus be proportional to the system's size. Instead, Honeycutt and Andersen (1984)

---

[49]The standard way of minimizing surface effects in simulations of small systems is to impose periodic boundary conditions (see, e.g., Allen and Tildesley, 1987), whereby the computational cell is surrounded by an infinite, space-filling set of identical replicas. For every pair of particles i and j, interactions are computed only between i and the nearest image of j, chosen from among j and its periodic replicas. Furthermore, when a particle leaves the computational cell, a replica enters through the opposite face.

found that the mean time for crystallization (averaged over several simulations) actually increased with N over the range $108 \leq N \leq 1500$. Thus, these authors concluded that the crystallization event normally observed in simulations is not related to homogeneous nucleation, and is probably an artifact of periodic boundary conditions. They also showed that the nucleus present at the time of catastrophic crystal growth is much larger than a critical nucleus (it is postcritical). This was done by perturbing the system at various times before crystallization, and observing whether this affected the total time elapsed before catastrophic growth. If a system is perturbed by reassigning random velocities from a Boltzmann distribution before a critical nucleus has formed, it will have to climb again the entire free energy barrier to critical nucleus formation before crystallizing. Hence the average time for catastrophic crystallization in this case would be the sum of the average time for sudden crystallization and the time elapsed before intervention. Perturbing a system where a postcritical nucleus is present, on the other hand, should have no effect on average on the time elapsed before catastrophic crystallization. Honeycutt and Andersen periodically stored the positions and velocities of every atom in the course of a molecular dynamics simulation of a 500-particle system. The stored configurations were perturbed by changing the velocity of each particle without changing the total kinetic energy, and allowed to evolve in time until catastrophic growth occurred. Then the mean time for catastrophic growth should grow linearly with the intervention time (the time elapsed between the start of the simulation and the reassignment of atomic velocities) for precritical systems, and should be constant for postcritical systems. Honeycutt and Andersen's results (Figure 4.46) show that when the unperturbed system crystallizes catastrophically it is clearly postcritical. This suggests that the onset of catastrophic crystal formation is unrelated to nucleation, possibly an artifact of periodic boundary conditions.

In a subsequent study of the supercooled Lennard-Jones fluid, Honeycutt and Andersen (1986) used a cluster-ancestry-tracking algorithm to identify the time of critical nucleus formation and the critical nucleus size. Not only did they find that the nucleation time increased with sample size, an unphysical result, but also that the size of the critical nucleus increased with N (e.g., it contained 24 ± 12 atoms when $N = 500$, and 61 ± 5 atoms when $N = 1300$). This means that the periodic boundaries are influencing the nucleation process. The work of Honeycutt and Andersen raises serious questions about the validity of molecular dynamics studies of nucleation in systems with N of order 10^3.

Swope and Andersen (1990) addressed the question of size dependence in simulations of nucleation and crystallization. They compared the evolution of a 15,000-particle supercooled Lennard-Jones system with that of 64 subsystems having on average 15,625 particles into which the computational cell of a 10^6-particle system at identical thermodynamic conditions was divided. It was found that the 15,000-particle system evolved in time in a manner typical of

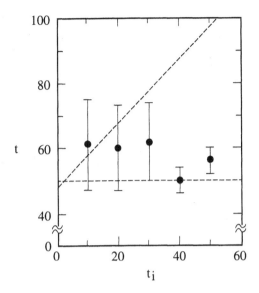

Figure 4.46. Relationship between the mean time for catastrophic crystallization, t, and the intervention time, t_i, in a five-hundred-particle supercooled Lennard-Jones system ($T^* = 0.45$; circles). The dashed line with positive slope represents the expected behavior of a precritical system, with an intercept equal to the mean time for catastrophic growth corresponding to the given size, temperature, and density. The horizontal line indicates the expected behavior of a postcritical system (the relevant time in this case is the actual time for catastrophic growth for the unperturbed system). Time is in units of $\left(m\sigma^2/\varepsilon\right)^{1/2}$, where σ and ε are Lennard-Jones parameters, and m is the mass of an atom. [Reprinted, with permission, from Honeycutt and Andersen, *Chem. Phys. Lett.* 108: 535 (1984)]

the behavior of each of the 10^6-particle-system's subregions. This implies that a 15,000-atom system is large enough not to exhibit system-size-dependent artifacts. However, it was also found that the 15,000-atom system was too small to capture the diversity of behavior seen among the 64 subsystems of the 10^6-atom simulation. Differences between results obtained in the 15,000- and 10^6-atom simulations were consequently termed statistical (as opposed to system-size artifacts) by Swope and Andersen. Thus the elimination of size-dependent artifacts and the simultaneous attainment of statistical significance in simulations of nucleation and crystallization calls for system sizes much larger than have been used in the past to study this problem. The careful consideration of both size and statistics in recent computational studies of nucleation in supercooled liquids (Báez and Clancy, 1995) reflects the importance of Swope and Andersen's work.

The computational study of structural relaxation in supercooled liquids requires very long simulations. Computational studies of nucleation and crystallization require very large systems. Massively parallel computers are ideally suited to handle large systems; their ready availability will lead to important advances in the atomistic investigation of nucleation and crystallization.

References

Abraham, F.F. 1980. "An Isothermal-Isobaric Computer Simulation of the Supercooled Liquid/Glass Transition: Is the Short-Range Order in the Amorphous Solid FCC?" *J. Chem. Phys.* 72: 359.

Adam, G., and J.H. Gibbs. 1965. "On the Temperature Dependence of Cooperative Relaxation Properties in Glass-Forming Liquids." *J. Chem. Phys.* 43: 139.

Alba, C., L.E. Busse, D.J. List, and C.A. Angell. 1990. "Thermodynamic Aspects of the Vitrification of Toluene, and Xylene Isomers, and the Fragility of Light Hydrocarbons." *J. Chem. Phys.* 92: 617.

Alder, B.J., and T.E. Wainwright. 1957. "Phase Transition for a Hard Sphere System." *J. Chem. Phys.* 27: 1208.

Alder, B.J., and T.E. Wainwright. 1962. "Phase Transition in Elastic Disks." *Phys. Rev.* 127: 359.

Allen, M.P., and D.J. Tildesley. 1987. *Computer Simulation of Liquids.* Clarendon Press: Oxford.

Amar, J.G., and R.D. Mountain. 1987. "Molecular Dynamics Study of a Supercooled Soft-Sphere Fluid." *J. Chem. Phys.* 86: 2236.

Andersen, H.C., D. Chandler, and J.D. Weeks. 1972. "Role of Repulsive Forces in Liquids: The Optimized Random Phase Approximation." *J. Chem. Phys.* 56: 3812.

Angell, C.A. 1981. "The Glass Transition: Comparison of Computer Simulation and Laboratory Studies." In *Structure and Mobility in Molecular and Atomic Glasses,* J.M. O'Reilly and M. Goldstein, eds. *Ann. N.Y. Acad. Sci.* 371: 136.

Angell, C.A. 1982. "Supercooled Water." In *Water—A Comprehensive Treatise,* F. Franks, ed. Vol. 7, Chap. 1. Plenum Press: New York.

Angell, C.A. 1983. "Supercooled Water." *Annu. Rev. Phys. Chem.* 34: 593.

Angell, C.A. 1985. "Strong and Fragile Liquids." In *Relaxations in Complex Systems,* K. Ngai and G.B. Wright, eds. National Technical Information Series. p. 1. U.S. Department of Commerce: Springfield, VA.

Angell, C.A. 1988a. "Structural Instability and Relaxation in Liquid and Glassy Phases near the Fragile Liquid Limit." *J. Non-Cryst. Sol.* 102: 205.

Angell, C.A. 1988b. "Perspective on the Glass Transition." *J. Phys. Chem. Sol.* 49: 863.

Angell, C.A. 1988c. "Supercooled Water. Approaching the Limits." *Nature.* 331: 206.

Angell, C.A. 1990a. "Relaxation, Glass Formation, Nucleation and Rupture in Normal and 'Water-Like' Liquids at Low Temperatures and/or Negative Pressures." In *Correlations and Connectivity. Geometric Aspects of Physics, Chemistry and Biology,* H.E. Stanley and N. Ostrovsky, eds., p. 133. NATO Advanced Study Institute Series E, vol. 188. Kluwer Academic Publishers: Dordrecht.

Angell, C.A. 1990b. "Dynamic Processes in Ionic Glasses." *Chem. Rev.* 90: 523.

Angell, C.A. 1991a. "Relaxation in Liquids, Polymers and Plastic Crystals—Strong/Fragile Patterns and Problems." *J. Non-Cryst. Sol.* 131–133: 13.

Angell, C.A. 1991b. "Transport Processes, Relaxation, and Glass Formation in Hydrogen-Bonded Liquids." In *Hydrogen-Bonded Liquids*, J.C. Dore and J. Teixeira, eds., p. 59. NATO Advanced Study Institute Series C, vol. 329. Kluwer Academic Publishers: Dordrecht.

Angell, C.A. 1993. "Water II is a 'Strong' Liquid." *J. Phys. Chem.* 97: 6339.

Angell, C.A. 1995. "Formation of Glasses from Liquids and Biopolymers." *Science.* 267: 1924.

Angell, C.A., and J. Donnella. 1977. "Mechanical Collapse vs. Ideal Glass Formation in Slowly Vitrified Solutions: A Plausibility Test." *J. Chem. Phys.* 67: 4560.

Angell, C.A., and H. Kanno. 1976. "Density Maxima in High-Pressure Supercooled Water and Liquid Silicon Dioxide." *Science.* 193: 1121.

Angell, C.A., and K.J. Rao. 1972. "Configurational Excitations in Condensed Matter, and the 'Bond-Lattice' Model for the Liquid-Glass Transition." *J. Chem. Phys.* 57: 470.

Angell, C.A., and W. Sichina. 1976. "Thermodynamics of the Glass Transition: Empirical Aspects." In *The Glass Transition and the Nature of the Glassy State*, M. Goldstein and R. Simha, eds. *Ann. N.Y. Acad. Sci.* 279: 53.

Angell, C.A., and D.L. Smith. 1982. "Test of the Entropy Basis of the Vogel-Tamman-Fulcher Equation. Dielectric Relaxation of Polyalcohols Near T_g." *J. Phys. Chem.* 86: 3845.

Angell, C.A., and L.M. Torell. 1983. "Short Time Structural Relaxation Processes in Liquids: Comparison of Experimental and Computer Simulation Glass Transitions on Picosecond Time Scales." *J. Chem. Phys.* 78: 937.

Angell, C.A., and J.C. Tucker. 1974. "Heat Capacities and Fusion Entropies of the Tetrahydrates of Calcium Nitrate, Cadmium Nitrate, and Magnesium Acetate. Concordance of Calorimetric and Relaxational 'Ideal' Glass Transition Temperature." *J. Phys. Chem.* 78: 278.

Angell, C.A., J. Shuppert, and J.C. Tucker. 1973. "Anomalous Properties of Supercooled Water. Heat Capacity, Expansivity, and Proton Magnetic Resonance Chemical Shift from 0 to $-38°C$." *J. Phys. Chem.* 77: 3092.

Angell, C.A., E.D. Finch, L.A. Woolf, and P. Bach. 1976. "Spin-Echo Diffusion Coefficients of Water to 2380 bar and $-20°C$."*J. Chem. Phys.* 65: 3063.

Angell, C.A., E. Williams, K.J. Rao, and J.C. Tucker. 1977. "Heat Capacity and Glass Transition Thermodynamics for Zinc Chloride. A Failure of the First Davies-Jones Relation for dT_g/dP." *J. Phys. Chem.* 81: 238.

Angell, C.A., J.H.R. Clarke, and L.V. Woodcock. 1981. "Interaction Potentials and Glass Formation: a Survey of Computer Experiments." *Adv. Chem. Phys.* 48: 397.

Angell, C.A., R.C. Stell, and W. Sichina. 1982a. "Viscosity-Temperature Function for Sorbitol from Combined Viscosity and Differential Scanning Calorimetry Studies." *J. Phys. Chem.* 86: 1540.

Angell, C.A., M. Oguni, and W.J. Sichina. 1982b. "Heat Capacities of Water at Extremes of Supercooling and Superheating." *J. Phys. Chem.* 86: 998.

Angell, C.A., D.R. MacFarlane, and M. Oguni. 1984. "The Kauzmann Paradox, Metastable Liquids, and Glasses." In *Dynamic Aspects of Structural Change in Liquids and Glasses*, C.A. Angell and M. Goldstein, eds. *Ann. N.Y. Acad. Sci.* 484: 241.

Anisimov, M.A., A.V. Voronel, N.S. Zaugolnikove, and G.I. Ovodov. 1972. "Specific Heat of Water near the Melting Point and Ornstein-Zernike Fluctuation." *JETP Lett.* 15: 317.

Avrami, M. 1939. "Kinetics of Phase Change. I. General Theory." *J. Chem. Phys.* 7: 1103.

Avrami, M. 1940. "Kinetics of Phase Change. II. Transformation-Time Relations for Random Distribution of Nuclei." *J. Chem. Phys.* 8: 212.

Avrami, M. 1941. "Granulation, Phase Change, and Microstructure. Kinetics of Phase Change. III." *J. Chem. Phys.* 9: 177.

Báez, L.A., and P. Clancy. 1995. "The Kinetics of Crystal Growth and Dissolution from the Melt in Lennard-Jones Systems." *J. Chem. Phys.* 102: 8138.

Barrat, J.-L., and M.L. Klein. 1991. "Molecular Dynamics Simulations of Supercooled Liquids near the Glass Transition." *Annu. Rev. Phys. Chem.* 42: 23.

Barrat, J.-L., J.-N. Roux, J.-P. Hansen, and M.L. Klein. 1988. "Elastic Response of a Simple Amorphous Binary Alloy near a Glass Transition." *Europhys. Lett.* 7: 707.

Barrat, J.-L., W. Götze, and A. Latz. 1989. "The Liquid-Glass Transition of the Hard Sphere System." *J. Phys. Cond. Matt.* 1: 7163.

Bellissent-Funel, M.-C., and L. Bosio. 1995. "A Neutron Scattering Study of Liquid D_2O under Pressure and at Various Temperatures." *J. Chem. Phys.* 102: 3727.

Bellissent-Funel, M.-C., J. Teixeira, L. Bosio, and J.C. Dore. 1989. "A Structural Study of Deeply Supercooled Water." *J. Phys. Cond. Matt.* 1: 7123.

Bellissent-Funel, M.-C., L. Bosio, A. Hallbrucker, E. Mayer, and R. Sridi-Dorbez. 1992. "X-Ray and Neutron Scattering Studies of the Structure of Hyperquenched Glassy Water." *J. Chem. Phys.* 97: 1282.

Bengtzelius, U. 1986a. "Theoretical Calculations on Liquid-Glass Transitions in Lennard-Jones Systems." *Phys. Rev. A.* 33: 3433.

Bengtzelius, U. 1986b. "Dynamics of a Lennard-Jones System close to the Glass Transition." *Phys. Rev. A.* 34: 5059.

Bengtzelius, U., W. Götze, and A. Sjölander. 1984. "Dynamics of Supercooled Liquids and the Glass Transition." *J. Phys. C. Sol. State Phys.* 17: 5915.

Bernu, B., Y. Hiwatari, and J.-P. Hansen. 1985. "A Molecular Dynamics Study of the Glass Transition in Binary Mixtures of Soft Spheres." *J. Phys. C. Sol. State Phys.* 18: L371.

Bernu, B., J.-P. Hansen, Y. Hiwatari, and G. Pastore. 1987. "Soft-Sphere Model for the Glass Transition in Binary Alloys: Pair Structure and Self-Diffusion." *Phys. Rev. A.* 36: 4891.

Bhattacharjee, S.M., and E.H. Helfand. 1987. "Equilibrium Behavior of the Tiling Model." *Phys. Rev. A.* 36: 3332.

Birge, N.O. 1986. "Specific Heat Spectroscopy of Glycerol and Propylene Glycol near the Glass Transition." *Phys. Rev. B.* 34: 1631.

Birge, N.O., and S.R. Nagel. 1985. "Specific Heat Spectroscopy of the Glass Transition." *Phys. Rev. Lett.* 54: 2674.

Blumberg, R.L., H.E. Stanley, A. Geiger, and P. Mausbach. 1984. "Connectivity of Hydrogen Bonds in Liquid Water." *J. Chem. Phys.* 80: 5230.

Böhmer, R., K.L. Ngai, C.A. Angell, and D.L. Plazek. 1993. "Nonexponential Relaxations in Strong and Fragile Glass Formers." *J. Chem. Phys.* 99: 4201.

Boon, J.P., and S. Yip. 1980. *Molecular Hydrodynamics.* McGraw-Hill: New York.

Borick, S.S. 1995. "The Thermodynamics of Supercooled Water: A Theoretical Study Using Lattice Models with Directional Bonding." Ph.D. thesis, Princeton University, Princeton, N.J.

Borick, S.S., P.G. Debenedetti, and S. Sastry. 1995. "A Lattice Model of Network-Forming Fluids with Orientation-Dependent Bonding: Equilibrium, Stability, and Implications for the Phase Behavior of Supercooled Water." *J. Phys. Chem.* 99: 3781.

Bosio, L., J. Teixeira, and H.E. Stanley. 1981. "Enhanced Density Fluctuations in Super-cooled H_2O, D_2O, and Ethanol-Water Solutions: Evidence from Small-Angle X-Ray Scattering." *Phys. Rev. Lett.* 46: 597.

Bosio, L., S.-H. Chen, and J. Teixeira. 1983. "Isochoric Temperature Differential of the X-Ray Structure Factor and Structural Rearrangements in Low-Temperature Heavy Water." *Phys. Rev. A.* 27: 1468.

Bottomley, G.A. 1978. "The Vapour Pressure of Supercooled Water and Heavy Water." *Aust. J. Chem.* 31: 1177.

Brazhkin, V.V., R.N. Voloshin, S.V. Popova, and A.G. Umnov. 1992. "Pressure-Temperature Phase Diagram of Solid and Liquid Te under Pressures up to 10 GPa." *J. Phys. Cond. Matt.* 4: 1419.

Brüggeller, P., and E. Mayer. 1980. "Complete Vitrification in Pure Liquid Water and Dilute Aqueous Solutions." *Nature.* 288: 569.

Brüning, R., and K. Samwer. 1992. "Glass Transition on Long Time Scales." *Phys. Rev. B.* 46: 11318.

Burton, E.F., and W.F. Oliver. 1935. "The Crystal Structure of Ice at Low Temperatures." *Proc. Roy. Soc. A.* 153: 166.

Carnahan, N.F., and K. Starling. 1969. "Equation of State for Nonattracting Rigid Spheres." *J. Chem. Phys.* 51: 635.

Chialvo, A.A., and P.G. Debenedetti. 1992. "An Automated Verlet Neighbor List Al-gorithm with a Multiple Time Step for the Simulation of Large Systems." *Comput. Phys. Commun.* 70: 467.

Cicerone, M.T., and M.D. Ediger. 1993. "Photobleaching Technique for Measuring Ultraslow Reorientation Near and Below the Glass Transition: Tetracene in o-Terphenyl." *J. Phys. Chem.* 97: 10489.

Cicerone, M.T., F.R. Blackburn, and M.D. Ediger. 1995. "How Do Molecules Move near T_g? Molecular Rotation of Six Probes in o-Terphenyl across 14 Decades in Time." *J. Chem. Phys.* 102: 471.

Clarke, J.H.R. 1979. "Molecular Dynamics Studies of Glass Formation in the Lennard-Jones Model of Argon." *J. Chem. Soc. Faraday Trans. 2.* 75: 1371.

Cohen, M.H., and G.S. Grest. 1979. "Liquid-Glass Transition, a Free-Volume Ap-proach." *Phys. Rev. B.* 20: 1077.

Cohen, M.H., and G.S. Grest. 1980. "Origin of Low-Temperature Tunneling States in Glasses." *Phys. Rev. Lett.* 45: 1271.

Cohen, M.H., and G.S. Grest. 1981. "A New Free-Volume Theory of the Glass Transi-tion." In *Structure and Mobility in Molecular and Atomic Glasses*, J.M. O'Reilly and M. Goldstein, eds. *Ann. N.Y. Acad. Sci.* 371: 199.

Cohen, M.H., and D. Turnbull. 1959. "Molecular Transport in Liquids and Glasses." *J. Chem. Phys.* 31: 1164.

Cohen, M.H., and D. Turnbull. 1964. "Metastability of Amorphous Structures." *Nature.* 203: 964.

Conde, O., J. Teixeira, and P. Papon. 1982. "Analysis of Sound Velocity in Supercooled H_2O, D_2O, and Water-Ethanol Mixtures." *J. Chem. Phys.* 76: 3747.

Courant, R., and D. Hilbert. 1989. *Methods of Mathematical Physics*. Vol. 1, chap. 7. Wiley: New York.

D'Antonio, M.C., and P.G. Debenedetti. 1987. "Loss of Tensile Strength in Liquids without Property Discontinuities: A Thermodynamic Analysis." *J. Chem. Phys.* 86: 2229.

Das, B.K., G.F. Kraus, J.T. Siewick, and S.C. Greer. 1982. "Dielectric Constant of Undercooled Water." In *Proceedings of the 8th Symposium on Thermophysical Properties*. Vol. II, p. 324. American Society of Mechanical Engineers: New York.

Das, S.P. 1987. "Effect of Structure on the Liquid-Glass Transition." *Phys. Rev. A.* 36: 211.

Das, S.P. 1990. "Glass Transition and Self-Consistent Mode-Coupling Theory." *Phys. Rev. A.* 42: 6116.

Das, S.P. and G.F. Mazenko. 1986. "Fluctuating Nonlinear Hydrodynamics and the Liquid-Glass Transition." *Phys. Rev. A.* 34: 2265.

Das, S.P., G.F. Mazenko, S. Ramaswamy, and J.J. Toner. 1985. "Hydrodynamic Theory of the Glass Transition." *Phy. Rev. Lett.* 54: 118.

Davies, R.O., and G.O. Jones. 1953a. "Thermodynamic and Kinetic Properties of Glasses." *Adv. Phys.* 2: 370.

Davies, R.O., and G.O. Jones. 1953b. "The Irreversible Approach to Equilibrium in Glasses." *Proc. Roy. Soc. A.* 217: 26.

Debenedetti, P.G., and M.C. D'Antonio. 1988. "Stability and Tensile Strength of Liquids Exhibiting Density Maxima." *Amer. Inst. Chem. Eng. J.* 34: 447.

DeFries, T., and J. Jonas. 1977. "Pressure Dependence of NMR Proton Spin-Lattice Relaxation Times and Shear Viscosity in Liquid Water in the Temperature Range -15 to $10°C$." *J. Chem. Phys.* 66: 896.

Denbigh, K. 1981. *The Principles of Chemical Equilibrium*. 4th ed., Chap. 13. Cambridge University Press: Cambridge.

De Raedt, H., and W. Götze. 1986. "Scaling Properties of Correlation Functions at the Liquid-Glass Transition." *J. Phys. C. Sol. State Phys.* 19: 2607.

Dings, J., J.C.F. Michielsen, and J. van der Elsken. 1992. "Equilibrium and Nonequilibrium Contributions to X-Ray Scattering from Supercooled Water." *Phys. Rev. A.* 45: 5731.

Doolittle, A.K. 1951. "Studies in Newtonian Flow. II. The Dependence of the Viscosity of Liquids on Free-Space." *J. Appl. Phys.* 22: 1471.

Dore, J. 1986. "Structural Studies of Amorphous Ice and Supercooled Water." In *Water and Aqueous Solutions*, G.W. Neilson and J.E. Enderby, eds., p. 89. Adam Hilger: Bristol.

Dore, J. 1990. "Hydrogen-Bond Networks in Supercooled Liquid Water and Amorphous Ices." *J. Molec. Struct.* 237: 221.

Dubochet, J., and A.W. McDowall. 1981. "Vitrification of Pure Water for Electron Microscopy." *J. Microsc.* 124: RP3.

Dubochet, J., J. Lepault, R. Freeman, J. A. Berriman, and J.-C. Homo. 1982. "Electron Microscopy of Frozen Water and Aqueous Solution." *J. Microsc.* 128: 219.

Edwards, S.F., and P.W. Anderson. 1975. "The Theory of Spin Glasses." *J. Phys. F. Met. Phys.* 5: 965.

Ehrenfest, P. 1933. "Phasenumwandlunger im Ueblichen und Erweiterten Sinn, Classifiziert nach den Entsprechenden Singularitaeten des Thermodynamischen Potentiales." *Comm. Kamerlingh Onnes Lab. Leiden*. Suppl. 75b: 628

Elmroth, M., L. Börjesson, and L.M. Torell. 1992. "Observation of a Dynamic Anomaly in the Liquid-Glass Transformation Range by Brillouin Scattering." *Phys. Rev. Lett.* 68: 79.

Ernst, R.M., S.R. Nagel, and G.S. Grest. 1991. "Search for a Correlation Length in a Simulation of the Glass Transition." *Phys. Rev. B*. 43: 8070.

Fisher, M., and J.P. Devlin. 1995. "Defect Activity in Amorphous Ice from Isotopic Exchange Data: Insight into the Glass Transition." *J. Phys. Chem.* 99: 11584.

Floriano, M.A., Y.P. Handa, D.D. Klug, and E. Whalley. 1989. "Nature of the Transformations of Ice I and Low-Density Amorphous Ice to High-Density Amorphous Ice." *J. Chem. Phys.* 91: 787.

Foley, M., and P. Harrowell. 1993. "The Spatial Distributions of Relaxation Times in a Model Glass." *J. Chem. Phys.* 98: 5069.

Fox, J.R., and H.C. Andersen. 1981. "Molecular Dynamics Simulation of the Glass Transition." In *Structure and Mobility in Molecular and Atomic Glasses*, J.M. O'Reilly and M. Goldstein, eds. *Ann. N.Y. Acad. Sci.* 371: 123.

Fox, J.R., and H.C. Andersen. 1984. "Molecular Dynamics Simulations of a Supercooled Monatomic Liquid and Glass." *J. Phys. Chem.* 88: 4019.

Fox, T.G., and P.J. Flory. 1950. "Second-Order Transition Temperature and Related Properties of Polystyrene. I. Influence of Molecular Weight." *J. Appl. Phys.* 21: 581.

Fox, T.G., and P.J. Flory. 1951. "Further Studies on the Melt Viscosity of Polyisobutylene." *J. Phys. Chem.* 55: 221.

Fox, T.G., and P.J. Flory. 1954. "The Glass Transition and Related Properties of Polystyrene. Influence of Molecular Weight." *J. Polym. Sci.* 14: 315.

Frank, C. 1952. "Supercooling of Liquids." *Proc. Roy. Soc. A*. 215: 43.

Fredrickson, G.H. 1986. "Linear and Nonlinear Experiments for a Spin Model with Cooperative Dynamics." In *Dynamic Aspects of Structural Change in Liquids and Glasses*, C.A. Angell and M. Goldstein, eds. *Ann. N.Y. Acad. Sci.* 484: 185.

Fredrickson, G.H., and H.C. Andersen. 1984. "Kinetic Ising Model of the Glass Transition." *Phys. Rev. Lett.* 53: 1244.

Fredrickson, G.H., and H.C. Andersen. 1985. "Facilitated Kinetic Ising Models and the Glass Transition." *J. Chem. Phys.* 83: 5822.

Fredrickson, G.H., and S.A. Brawer. 1986. "Monte Carlo Investigation of a Kinetic Ising Model of the Glass Transition." *J. Chem. Phys.* 84: 3351.

Frenkel, D., and J.P. McTague. 1980. "Computer Simulations of Freezing and Supercooled Liquids." *Annu. Rev. Phys. Chem.* 31: 491.

Frick, B., B. Farago, and D. Richter. 1980. "Temperature Dependence of the Nonergodicity Parameter in Polybutadiene in the Neighborhood of the Glass Transition." *Phys. Rev. Lett.* 64: 2921.

Fuchs, M., W. Götze, S. Hildebrand, and A. Latz. 1992. "A Theory for the β-Relaxation Process near the Liquid-to-Glass Transition." *J. Phys. Cond. Matt.* 4: 7709.

Fujara, F., B. Geil, H. Sillescu, and G. Fleischer. 1992. "Translational and Rotational Diffusion in Supercooled Orthoterphenyl close to the Glass Transition." *Z. Phys. B*. 88: 195.

Fulcher, G.S. 1925. "Analysis of Recent Measurements of the Viscosity of Glasses." *J.*

Amer. Ceram. Soc. 8: 339.

Geiger, A., F.H. Stillinger, and A. Rahman. 1979. "Aspects of the Percolation Process for Hydrogen-Bond Networks in Water." *J. Chem. Phys.* 70: 4185.

Geiger, A., P. Mausbach, and J. Schnitker. 1986. "Computer Simulation Study of the Hydrogen-Bond Network in Metastable Water." In *Water and Aqueous Solutions*, G.W. Neilson and J.E. Enderby, eds., p. 15. Adam Hilger: Bristol.

Geszti, T. 1983. "Pre-vitrification by Viscosity Feedback." *J. Phys. C. Sol. State Phys.* 16: 5805.

Gibbs, J.H. 1956. "Nature of the Glass Transition in Polymers." *J. Chem. Phys.* 25: 185.

Gibbs, J.H., and E.A. DiMarzio. 1958. "Nature of the Glass Transition and the Glassy State." *J. Chem. Phys.* 28: 373.

Gillen, K.T., D.C. Douglass, and J.R. Hoch. 1972. "Self-Diffusion in Liquid Water to −31°C." *J. Chem. Phys.* 57: 5117.

Goldstein, M. 1965. "On the Temperature Dependence of Cooperative Relaxation Properties in Glass-Forming Liquids—Comments on a Paper by Adam and Gibbs." *J. Chem. Phys.* 43: 1852.

Goldstein, M. 1969. "Viscous Liquids and the Glass Transition: a Potential Energy Barrier Picture." *J. Chem. Phys.* 51: 3728.

Goldstein, M. 1975. "Validity of the Ehrenfest Equation for a System with More than One Ordering Parameter: Critique of a Paper by DiMarzio." *J. Appl. Phys.* 46: 4153.

Goldstein, M. 1976a. "Statistical Thermodynamics of Configurational Properties." In *The Glass Transition and the Nature of the Glassy State*, M. Goldstein and R. Simha, eds. *Ann. N.Y. Acad. Sci.* 279: 68.

Goldstein, M. 1976b. "Viscous Liquids and the Glass Transition. V. Sources of the Excess Specific Heat of the Liquid." *J. Chem. Phys.* 64: 4767.

Götze, W. 1985. "Properties of the Glass Transition Treated within the Mode-Coupling Theory." *Z. Phys. B. Cond. Matt.* 60: 195.

Götze, W. 1990. "Aspects of Structural Glass Transitions." In *Liquids, Freezing and Glass Transition*, J.-P. Hansen, D. Levesque, and J. Zinn-Justin, eds. chap. 5. Les Houches, Session LI. North-Holland: Amsterdam.

Götze, W., and L. Sjörgen. 1987. "The Glass Transition Singularity." *Z. Phys. B. Cond. Matt.* 65: 415.

Götze, W., and L. Sjörgen. 1991. "β Relaxation at the Glass Transition of Hard-Spherical Colloids." *Phys. Rev. A.* 43: 5442.

Götze, W., and L. Sjörgen. 1992. "Relaxation Processes in Supercooled Liquids." *Rep. Prog. Phys.* 55: 241.

Green, J.L., A.R. Lacey, and M.G. Sceats. 1986. "Determination of the Intrinsic Network Defect Density in Liquid Water." *Chem. Phys. Lett.* 130: 67.

Green, J.L., A.R. Lacey, and M.G. Sceats. 1987. "Collective Motions in H_2O/H_2O_2 Mixtures: Evidence for Defects and Network Reconstruction." *J. Chem. Phys.* 86: 1841.

Greer, A.L. 1995. "Metallic Glasses." *Science.* 267: 1947.

Grest, G.S., and M.H. Cohen. 1980. "Liquid-Glass Transition: Dependence of the Glass Transition on Heating and Cooling Rates." *Phys. Rev. B.* 21: 4113.

Grest, G.S., and M.H. Cohen. 1981. "Liquids, Glasses, and the Glass Transition: A Free-Volume Approach." *Adv. Chem. Phys.* 49: 455.

Grest, G.S., and S.R. Nagel. 1987. "Frequency-Dependent Specific Heat in a Simulation of the Glass Transition." *J. Phys. Chem.* 91: 4916.

Grimsditch, M. 1984. "Polymorphism in Amorphous SiO$_2$." *Phys. Rev. Lett.* 52: 2379.

Guggenheim, E.A. 1986. *Thermodynamics. An Advanced Treatment for Chemists and Physicists*. 8th ed. Chap. 3. North-Holland: Amsterdam.

Gujrati, P.D., and M. Goldstein. 1980. "Viscous Liquids and the Glass Transition. Non-configurational Contributions to the Excess Entropy of Disordered Phases." *J. Phys. Chem.* 84: 859.

Hallbrucker, A., and E. Mayer. 1987. "Calorimetric Study of the Vitrified Liquid Water to Cubic Ice Phase Transition." *J. Phys. Chem.* 91: 503.

Hallbrucker, A., E. Mayer, and G.P. Johari. 1989. "Glass-Liquid Transition and the Enthalpy of Devitrification of Annealed Vapor-Deposited Amorphous Solid Water. A Comparison with Hyperquenched Glassy Water." *J. Phys. Chem.* 93: 4986.

Hallett, J. 1963. "The Temperature Dependence of the Viscosity of Supercooled Water." *Proc. Phys. Soc.* 82: 1046.

Handa, Y.P., and D.D. Klug. 1988. "Heat Capacity and Glass Transition Behavior of Amorphous Ice." *J. Phys. Chem.* 92: 3323.

Handa, Y.P., O. Mishima, and E. Whalley. 1986. "High-Density Amorphous Ice. III. Thermal Properties." *J. Chem. Phys.* 84: 2766.

Hansen, J.-P., and I.R. McDonald. 1986. *Theory of Simple Liquids*. 2nd ed. Chap. 7. Academic Press: London.

Hansen, J.-P., and L. Verlet. 1969. "Phase Transitions of the Lennard-Jones System." *Phys. Rev.* 184: 152.

Hare, D.E., and C.M. Sorensen. 1986. "Densities of Supercooled H$_2$O and D$_2$O in 25 μm Glass Capillaries." *J. Chem. Phys.* 84: 5085.

Hare, D.E., and C.M. Sorensen. 1987. "The Density of Supercooled Water. II. Bulk Samples Cooled to the Homogeneous Nucleation Limit." *J. Chem. Phys.* 87: 4840.

Harris, J.G., and F.H. Stillinger. 1989. "Glass Transitions and Frustrated Tiling Models." *J. Phys. Chem.* 93: 6893.

Harris, J.G., and F.H. Stillinger. 1990. "Role of Frustration in the Thermal Properties of Tiling Models for Glasses." *Phys. Rev. B.* 41: 519.

Harrowell, P. 1993. "Visualizing the Collective Motions Responsible for the α and β Relaxations in a Model Glass." *Phys. Rev. E.* 48: 4359.

Hasted, J.B., and M. Shahidi. 1976. "The Low Frequency Dielectric Constant of Supercooled Water." *Nature.* 262: 777.

Haymet, A.D.J. 1983. "Orientational Freezing in Three Dimensions: Mean-Field Theory." *Phys. Rev. B.* 27: 1725.

Henderson, S.J., and R.J. Speedy. 1987. "Temperature of Maximum Density in Water at Negative Pressure." *J. Phys. Chem.* 91: 3062.

Herbst, C.A., R.L. Cook, and H.E. King, Jr. 1993. "High-Pressure Viscosity of Glycerol Measured by Centrifugal-Force Viscometry." *Nature.* 361: 518.

Hill, T.L. 1956. *Statistical Mechanics. Principles and Selected Applications*. Chap. 8. McGraw-Hill: New York.

Hillig, W.B., and D. Turnbull. 1956. "Theory of Crystal Growth in Undercooled Pure Liquids." *J. Chem. Phys.* 24: 914.

Hindman, J.C. 1974. "Relaxation Processes in Water: Viscosity, Self-Diffusion, and Spin-Lattice Relaxation. A Kinetic Model." *J. Chem. Phys.* 60: 4488.

Hindman, J.C., and A. Svirmickas. 1973. "Relaxation Processes in Water. Spin-Lattice

Relaxation of D_2O in Supercooled Water." *J. Phys. Chem.* 77: 2487.

Hindman, J.C., A. Svirmickas, and M. Wood. 1973. "Relaxation Processes in Water. A Study of the Proton Spin-Lattice Relaxation Time." *J. Chem. Phys.* 59: 1517.

Hiwatari, Y., and H. Miyagawa. 1990. "Molecular Dynamics Study of Binary Alloys near the Glass Transition." *J. Non-Cryst. Sol.* 117–118: 862.

Hoare, M. 1976. "Stability and Local Order in Simple Amorphous Packings." In *The Glass Transition and the Nature of the Glassy State*, M. Goldstein and R. Simha, eds. *Ann. N.Y. Acad. Sci.* 279: 186.

Hoare, M. 1978. "Packing Models and Structural Specificity." *J. Non-Cryst. Sol.* 31: 157.

Hodge, I.M., and C.A. Angell. 1978. "The Relative Permittivity of Supercooled Water." *J. Chem. Phys.* 68: 1363.

Honeycutt, J.D., and H.C. Andersen. 1984. "The Effect of Periodic Boundary Conditions on Homogeneous Nucleation Observed in Computer Simulations." *Chem. Phys. Lett.* 108: 535.

Honeycutt, J.D., and H.C. Andersen. 1986. "Small System Size Artifacts in the Molecular Dynamics Simulation of Homogeneous Crystal Nucleation in Supercooled Atomic Liquids." *J. Phys. Chem.* 90: 1585.

Hoover, W.G., and F.H. Ree. 1968. "Melting Transition and Communal Entropy for Hard Spheres." *J. Chem. Phys.* 49: 3609.

Hsu, C.S., and A. Rahman. 1979a. "Crystal Nucleation and Growth in Liquid Rubidium." *J. Chem. Phys.* 70: 5234.

Hsu, C.S., and A. Rahman. 1979b. "Interaction Potentials and Their Effect on Crystal Nucleation and Symmetry." *J. Chem. Phys.* 71: 4974.

Hudson, S., and H.C. Andersen. 1978. "The Glass Transition of Atomic Glasses." *J. Chem. Phys.* 69: 2323.

Itie, J.P., A. Polian, G. Calas, J. Petiau, A. Fontaine, and H. Tolentino. 1989. "Pressure-Induced Coordination Changes in Crystalline and Vitreous GeO_2." *Phys. Rev. Lett.* 63: 398.

Jenniskens, P., and D.F. Blake. 1994. "Structural Transitions in Amorphous Water Ice and Astrophysical Implications." *Science.* 265: 753.

Jeong, Y.H., S.R. Nagel, and S. Bhattacharya. 1986."Ultrasonic Investigation of the Glass Transition in Glycerol." *Phys. Rev. A.* 34: 602.

Jodrey, W.S., and E.M. Tory. 1981. "Computer Simulation of Isotropic, Homogeneous, Dense Random Packing of Equal Spheres." *Powder Technol.* 30: 111.

Jodrey, W.S., and E.M. Tory. 1985. "Computer Simulation of Close Random Packing of Equal Spheres." *Phys. Rev. A.* 32: 2347.

Johari, G.P. 1973. "Intrinsic Mobility of Molecular Glasses." *J. Chem. Phys.* 58: 1766.

Johari, G.P. 1976. "Glass Transition and Secondary Relaxations in Molecular Liquids and Crystals." In *The Glass Transition and the Nature of the Glassy State*, M. Goldstein and R. Simha, eds. *Ann. N.Y. Acad. Sci.* 279: 117.

Johari, G.P. 1995. "Phase Transition and Entropy of Amorphous Ices." *J. Chem. Phys.* 102: 6224.

Johari, G.P., A. Hallbrucker, and E. Mayer. 1987. "The Glass-Liquid Transition of Hyperquenched Water." *Nature.* 330: 552.

Johari, G.P., S. Ram, G. Astl, and E. Mayer. 1990. "Characterizing Amorphous and Microcrystalline Solids by Calorimetry." *J. Non-Cryst. Sol.* 116: 282.

Johari, G.P., G. Fleissner, A. Hallbrucker, and E. Mayer. 1994. "Thermodynamic Continuity Between Glassy and Normal Water." *J. Phys. Chem.* 98: 4719.

Jonsson, H., and H.C. Andersen. 1988. "Icosahedral Ordering in the Lennard-Jones Crystal and Glass." *Phys. Rev. Lett.* 60: 2295.

Jorgensen, W.L., J. Chandrasekhar, J. Madura, R.W. Impey, and M. Klein. 1983. *J. Chem. Phys.* 79: 926.

Kadanoff, L.P., and J. Swift. 1968. "Transport Coefficients Near the Liquid-Gas Critical Point." *Phys. Rev.* 166: 89.

Kanno, H., and C.A. Angell. 1979. "Water: Anomalous Compressibilities to 1.9 kbar and Correlation with Supercooling Limits." *J. Chem. Phys.* 70: 4008.

Kanno, H., and C.A. Angell. 1980. "Volumetric and Derived Thermal Characteristics of Liquid D_2O at Low Temperatures and High Pressures." *J. Chem. Phys.* 73: 1940.

Kauzmann, W. 1948. "The Nature of the Glassy State and the Behavior of Liquids at Low Temperatures." *Chem. Rev.* 43: 219.

Kim, B., and G.F. Mazenko. 1991. "Fluctuating Nonlinear Hydrodynamics, Dense Fluids, and the Glass Transition." *Adv. Chem. Phys.* 78:129.

Kirkpatrick, T.R. 1985. "Mode-Coupling Theory of the Glass Transition." *Phys. Rev. A.* 31: 939.

Kob, W., and H.C. Andersen. 1993a. "Relaxation Dynamics in a Lattice Gas: A Test of the Mode-Coupling Theory of the Ideal Glass Transition." *Phys. Rev. E.* 47: 3281.

Kob, W., and H.C. Andersen. 1993b. "Kinetic Lattice-Gas Model of Cage Effects in High-Density Liquids and Test of Mode-Coupling Theory of the Ideal Glass Transition." *Phys. Rev. E.* 48: 4364.

Kob, W., and H.C. Andersen. 1994. "Scaling Behavior in the β-Relaxation Regime of a Supercooled Lennard-Jones Mixture." *Phys. Rev. Lett.* 73: 1376

Kob, W., and H.C. Andersen. 1995. "Testing Mode-Coupling Theory for a Supercooled Binary Lennard-Jones Mixture: The van Hove Correlation Function." *Phys. Rev. E.* 51: 4626.

Kohlrausch, R. 1854. "Theorie des Elektrischen Rückstandes in der Leidener Flasche." *Ann. Phys. Chemie (Leipzig).* 91: 179.

Kowall, T., P. Mausbach, and A. Geiger. 1990. "Short-Wavelength Collective Dynamics in Metastable Water. A Molecular Dynamics Simulation Study." *Ber. Bunsenges. Phys. Chem.* 94: 279.

Kraus, G.F., and S.C. Greer. 1984. "Vapor Pressure of Supercooled H_2O and D_2O." *J. Phys. Chem.* 88: 4781.

Lang, E., and H.-D. Lüdemann. 1977. "Pressure and Temperature Dependence of the Longitudinal Proton Relaxation Times in Supercooled Water to $-87°C$ and 2500 bar." *J. Chem. Phys.* 67: 718.

Lang, E., and H.-D. Lüdemann. 1980. "Pressure and Temperature Dependence of the Longitudinal Deuterium Relaxation Times in Supercooled Heavy Water to 300 MPa and 188 K." *Ber. Bunsenges. Phys. Chem.* 84: 462.

Lang, E.W., and H.-D. Lüdemann. 1981. "High Pressure O-17 Longitudinal Relaxation Time Studies in Supercooled H_2O and D_2O." *Ber. Bunsenges. Phys. Chem.* 85: 603.

Laughlin, W.T., and D.R. Uhlmann. 1972. "Viscous Flow in Simple Organic Liquids." *J. Phys. Chem.* 76: 2317.

LaViolette, R.A., and F.H. Stillinger. 1985. "Multidimensional Geometric Aspects of the Solid-Liquid Transition in Simple Systems." *J. Chem. Phys.* 83: 4079.

LaViolette, R.A., and F.H. Stillinger. 1986. "Thermal Disruption of the Inherent Structure of Simple Liquids." *J. Chem. Phys.* 85: 6027.

Leutheusser, E. 1984. "Dynamical Model of the Liquid-Glass Transition." *Phys. Rev. A.* 29: 2765.

MacFarlane, D.R., and C.A. Angell. 1982. "An Emulsion Technique for the Study of Marginal Glass Formation in Molecular Liquids." *J. Phys. Chem.* 86: 1927

MacFarlane, D.R., R.K. Kadiyala, and C.A. Angell. 1983. "Homogeneous Nucleation and Growth of Ice from Solutions. TTT Curves, the Nucleation Rate, and the Stable Glass Criterion." *J. Chem. Phys.* 79: 3921.

Magazú, S., G. Maisano, D. Majolino, F. Mallamace, and P. Migliardo. 1989. "Relaxation Process in Deeply Supercooled Water by Mandelstam-Brillouin Scattering." *J. Phys. Chem.* 93: 942.

Mandell, M.J., J.P. McTague, and A. Rahman. 1976. "Crystal Nucleation in a Three-Dimensional Lennard-Jones System: A Molecular Dynamics Study." *J. Chem. Phys.* 64: 3699.

Mandell, M.J., J.P. McTague, and A. Rahman. 1977. "Crystal Nucleation in a Three-Dimensional Lennard-Jones System. II. Nucleation Kinetics for 256 and 500 Particles." *J. Chem. Phys.* 66: 3070.

Mayer, E. 1985a. "New Method for Vitrifying Water and Other Liquids by Rapid Cooling of their Aerosols." *J. Appl. Phys.* 58: 663.

Mayer, E. 1985b. "Infrared Spectrum of Vitrified Liquid Water. A Comparison with the Vapor Deposited Amorphous Form." *J. Phys. Chem.* 89: 3474.

Mayer, E. 1994. "Hyperquenched Glass Bulk Water: A Comparison with Other Amorphous Forms of Water, and with Vitreous but Freezable Water in a Hydrogel and on Hydrated Methemoglobin." In *Hydrogen Bond Networks*, M.-C. Bellissent-Funel and J.C. Dore, eds., p. 355. *NATO Advanced Study Institute Series C*, vol. 435. Kluwer Academic Publishers: Dordrecht.

Mayer, E., and P. Brüggeller. 1982. "Vitrification of Pure Liquid Water by High Pressure Jet Freezing." *Nature.* 298: 715.

McQuarrie, D.A. 1976. *Statistical Mechanics.* Chap. 13. Harper and Row: New York.

Mezei, F., W. Knaak, and B. Farago. 1987. "Neutron Spin-Echo Study of Dynamic Correlations near the Liquid-Glass Transition." *Phys. Rev. Lett.* 58: 571.

Mezei, M., and D.L. Beveridge. 1981. "Theoretical Studies of Hydrogen Bonding in Liquid Water and Dilute Aqueous Solutions." *J. Chem. Phys.* 74: 622.

Mezei, M., and R.J. Speedy. 1984. "Simulation Studies of the Dihedral Angle in Water." *J. Phys. Chem.* 88: 3180.

Michielsen, J.C.F., A. Bot, and J. van der Elsken. 1988. "Small-Angle X-Ray Scattering from Supercooled Water." *Phys. Rev. A.* 38: 6439.

Mishima, O. 1994. "Reversible First-Order Transition Between Two H_2O Amorphs at ca. 0.2 GPa and ca. 135 K." *J. Chem. Phys.* 100: 5910.

Mishima, O., L.D. Calvert, and E. Whalley. 1984. " 'Melting Ice' I at 77 K and 10 kbar: a New Method of Making Amorphous Solids." *Nature.* 310: 393.

Mishima, O., L.D. Calvert, and E. Whalley. 1985. "An Apparently First-Order Transition between Two Amorphous Phases of Ice Induced by Pressure." *Nature.* 314: 76.

Mishima, O., K. Takemura, and K. Aoki. 1991. "Visual Observation of the Amorphous-Amorphous Transition in H_2O Under Pressure." *Science.* 254: 406.

Miyagawa, H., and Y. Hiwatari. 1989. "Molecular-Dynamics Study of Binary Soft-

Sphere Glasses: Quench-Rate Effects and Aging Effects." *Phys. Rev. A.* 40: 6007.

Miyagawa, H., Y. Hiwatari, B. Bernu, and J.-P. Hansen. 1988. "Molecular Dynamics Study of Binary Soft-Sphere Mixtures: Jump Motions of Atoms in the Glassy State." *J. Chem. Phys.* 88: 3879.

Mohanty, U. 1988. "Configuration Entropy Dependence of Cooperative Relaxation Properties in Glass-Forming Liquids." *J. Chem. Phys.* 89: 3778.

Mohanty, U. 1990a. "Cooperative Relaxation in Glass-Forming Liquids." *Physica A.* 162: 362.

Mohanty, U. 1990b. "On the Temperature Dependence of Relaxation in Glass-Forming Liquids." *J. Chem. Phys.* 93: 8399.

Mohanty, U. 1991. "On the Nature of Supercooled and Glassy States of Matter." *Physica A.* 177: 345.

Mohanty, U. 1994. "Inhomogeneities and Relaxation in Supercooled Liquids." *J. Chem. Phys.* 100: 5905.

Mohanty, U. 1995. "Supercooled Liquids." *Adv. Chem. Phys.* 89: 89.

Mohanty, U., J. Oppenheim, and C.H. Taubes. 1994. "Low-Temperature Relaxation and Entropic Barriers in Supercooled Liquids." *Science.* 266: 425.

Mori, H. 1965a. "Transport, Collective Motion, and Brownian Motion." *Prog. Theor. Phys.* 33: 423.

Mori, H. 1965b. "A Continued-Fraction Representation of the Time-Correlation Functions." *Prog. Theor. Phys.* 34: 399.

Mountain, R.D. 1995. "Length Scales for Fragile Glass-Forming Liquids."*J. Chem. Phys.* 102: 5408.

Mountain, R.D., and A.C. Brown. 1984. "Molecular Dynamics Investigation of Homogeneous Nucleation for Inverse Power Potential Liquids and for a Modified Lennard-Jones Liquid." *J. Chem. Phys.* 80: 2730.

Mountain, R.D., and D. Thirumalai. 1987. "Molecular Dynamics Study of Glassy and Supercooled States of a Binary Mixture of Soft Spheres." *Phys. Rev. A.* 36: 3300.

Mountain, R.D., and D. Thirumalai. 1989. "Measures of Effective Ergodic Convergence in Liquids." *J. Phys. Chem.* 93: 6975.

Moynihan, C.T., P.B. Macedo, C.J. Montrose, P.K. Gupta, M.A. De Bolt, J.F. Dill, B.E. Dom, P.W. Drake, A.J. Easteal, R.P. Moeller, H. Sasabe, and J.A. Wilder. 1976. "Structural Relaxation in Vitreous Materials." In *The Glass Transition and the Nature of the Glassy State*, M. Goldstein and R. Simha, eds. *Ann. N.Y. Acad. Sci.* 279: 15.

Narten, A.H., C.G. Venkatesh, and S.A. Rice. 1976. "Diffractive Pattern and Structure of Amorphous Solid Water at 10 and 77° K." *J. Chem. Phys.* 64: 1106.

Nelson, D.R. 1983. "Order, Frustration, and Defects in Liquids and Glasses." *Phys. Rev. B.* 28: 5515.

Oguni, M., and C.A. Angell. 1980. "Heat Capacities of $H_2O + H_2O_2$, and $H_2O + N_2H_4$, Binary Solutions: Isolation of a Singular Component for c_p of Supercooled Water." *J. Chem. Phys.* 73: 1948.

O'Reilly, J.M. 1962. "The Effect of Pressure on Glass Temperature and Dielectric Relaxation Time of Polyvinyl Acetate." *J. Polym. Sci.* 57: 429.

Osipov, Y.A., B.V. Zheleznyi, and N.F. Bondarenko. 1977. "The Shear Viscosity of Water Supercooled to $-35°C$." *Russ. J. Phys. Chem.* 51: 748.

Palmer, R.G., D.L. Stein, E. Abrahams, and P.W. Anderson. 1984. "Models of Hierarchically Constrained Dynamics for Glassy Relaxation." *Phys. Rev. Lett.* 53: 958.

Pastore, G., B. Bernu, J.-P. Hansen, and Y. Hiwatari. 1988. "Soft-Sphere Model for the Glass Transition in Binary Alloys. II. Relaxation of the Incoherent Density-Density Correlation Functions." *Phys. Rev. A.* 38: 454.

Percus, J.K., and G.J. Yevick. 1958. "Analysis of Statistical Mechanics by Means of Collective Coordinates." *Phys. Rev.* 110: 1.

Petitet, J.P., R. Tufeu, and B. Le Neindre. 1983. "Determination of the Thermodynamic Properties of Water from Measurements of the Speed of Sound in the Temperature Range 251.15–293.15 K and in the Pressure Range 0.1–350 MPa." *Int. J. Thermophys.* 4: 35.

Poirier, J.P. 1985. "First-Order Phase Transitions between Amorphous Ices." *Nature.* 314: 12.

Ponyatovsky, E.G., and O.I. Barkalov. 1992. "Pressure-Induced Amorphous Phases." *Mater. Sci. Rep.* 8: 147.

Ponyatovsky, E.G., V.V. Sinand, and T.A. Pozdnyakova. 1994. "Second Critical Point and Low-Temperature Anomalies in the Physical Properties of Water." *JETP Lett.* 60: 360.

Poole, P.H., F. Sciortino, U. Essmann, and H.E. Stanley. 1992. "Phase Behavior of Supercooled Water." *Nature.* 360: 324.

Poole, P.H., F. Sciortino, U. Essmann, and H.E. Stanley. 1993a. "Spinodal of Liquid Water." *Phys. Rev. E.* 48: 3799.

Poole, P.H., U. Essmann, F. Sciortino, and H.E. Stanley. 1993b. "Phase Diagram for Amorphous Solid Water." *Phys. Rev. E.* 48: 4605.

Poole, P.H., F. Sciortino, T. Grande, H.E. Stanley, and C.A. Angell. 1994. "Effect of Hydrogen Bonds on the Thermodynamic Behavior of Liquid Water." *Phys. Rev. Lett.* 73: 1632.

Prielmeier, F.X., E.W. Lang, R.J. Speedy, and H.-D. Lüdemann. 1987. "Diffusion in Supercooled Water to 300 MPa." *Phys. Rev. Lett.* 59: 1128.

Prielmeier, F.X., E.W. Lang, R.J. Speedy, and H.-D. Lüdemann. 1988. "The Pressure Dependence of Self Diffusion in Supercooled Light and Heavy Water." *Ber. Bunsenges. Phys. Chem.* 92: 1111.

Privalko, V.P. 1980. "Excess Entropies and Related Quantities in Glass-Forming Liquids." *J. Phys. Chem.* 84: 3307.

Pruppacher, H.R. 1972. "Self-Diffusion Coefficient of Supercooled Water." *J. Chem. Phys.* 56: 101.

Pusey, P.N., and W. van Megen. 1987. "Observation of a Glass Transition in Suspensions of Spherical Colloidal Particles." *Phys. Rev. Lett.* 59: 2083.

Rasmussen, D.H., and A.P. MacKenzie. 1973. "Clustering in Supercooled Water." *J. Chem. Phys.* 59: 5003.

Rasmussen, D.H., A.P. MacKenzie, C.A. Angell, and J.C. Tucker. 1973. "Anomalous Heat Capacities of Supercooled Water and Heavy Water." *Science.* 181: 342.

Rössler, E. 1990. "Indications for a Change in Diffusion Mechanism in Supercooled Liquids." *Phys. Rev. Lett.* 65: 1595.

Rouch, J., C.C. Lai, and S.H. Chen. 1976. "Brillouin Scattering Studies of Normal and Supercooled Water." *J. Chem. Phys.* 65: 4016.

Ruocco, G., M. Sampoli, A. Torcini, and R. Vallauri. 1993. "Molecular Dynamics Results for Stretched Water." *J. Chem. Phys.* 99: 8095.

Sachdev, S. 1992. "Icosahedral Ordering in Supercooled Liquids and Metallic Glasses."

In *Bond-Orientational Order in Condensed Matter Systems*, K.P. Strandburg, ed., chap 6. Springer-Verlag: New York.

Sasai, M. 1990. "Instabilities of Hydrogen Bond Network in Liquid Water." *J. Chem. Phys.* 93: 7329.

Sasai, M. 1993. "The Random Graph Model of Hydrogen Bond Network." *Bull. Chem. Soc. Jpn.* 66: 3362.

Sastry, S. 1993. "Phase Behavior and Collective Dynamics of Liquid Water." Ph.D. thesis, Boston University, Boston, Mass.

Sastry, S., F. Sciortino, and H.E. Stanley. 1993. "Limits of Stability of the Liquid Phase in a Lattice Model with Water-Like Properties." *J. Chem. Phys.* 98: 9863.

Sato, H., K. Watanabe, J.M.H. Levelt Sengers, J.S. Gallagher, P.G. Hill, J. Straub, and W. Wagner. 1991. "Sixteen Thousand Evaluated Experimental Thermodynamic Property Data for Water and Steam." *J. Phys. Chem. Ref. Data.* 20: 1023.

Sceats, M.G., and S.A. Rice. 1980a. "The Enthalpy and Heat Capacity of Liquid Water and the Ice Polymorphs from a Random Network Model." *J. Chem. Phys.* 72: 3248.

Sceats, M.G., and S.A. Rice. 1980b. "The Entropy of Water from the Random Network Model." *J. Chem. Phys.* 72: 3260.

Sceats, M.G., and S.A. Rice. 1980c. "A Random Network Model Calculation of the Free Energy of Liquid Water." *J. Chem. Phys.* 72: 6183.

Sceats, M.G., and S.A. Rice. 1982. "Amorphous Solid Water and its Relationship to Liquid Water: A Random Network Model for Water." In *Water: a Comprehensive Treatise,* F. Franks, ed., vol. 7, chap. 2. Plenum: New York.

Sceats, M.G., M. Stavola, and S.A. Rice. 1979. "A Zeroth Order Random Network Model of Liquid Water." *J. Chem. Phys.* 70: 3927.

Scherer, G.W. 1986. *Relaxation in Glass and Composites.* Chap. 1. John Wiley and Sons: New York.

Schrödinger, E. 1989. *Statistical Thermodynamics.* Chap. III. Dover: New York.

Schufle, J.A. 1965. "The Specific Gravity of Supercooled Water in a Capillary." *Chem. Ind.* 16: 690.

Schufle, J.A., and M. Venugopalan. 1967. "Specific Volume of Liquid Water to $-40°C$." *J. Geophys. Res.* 27: 3271.

Sciortino, F., and S. Sastry. 1994. "Sound Propagation in Liquid Water: The Puzzle Continues." *J. Chem Phys.* 100: 3881.

Sciortino, F., P.H. Poole, H.E. Stanley, and S. Havlin. 1990a. "Lifetime of the Bond Network and Gel-Like Anomalies in Supercooled Water." *Phys. Rev. Lett.* 64: 1686.

Sciortino, F., A. Geiger, and H.E. Stanley. 1990b. "Isochoric Differential Scattering Functions in Liquid Water: The Fifth Neighbor as a Network Defect."*Phys. Rev. Lett.* 65: 3452.

Sciortino, F., A. Geiger, and H.E. Stanley. 1991. "Effects of Defects on Molecular Mobility in Liquid Water." *Nature.* 354: 218.

Sciortino, F., A. Geiger, and H.E. Stanley. 1992. "Network Defects and Molecular Mobility in Liquid Water." *J. Chem. Phys.* 96: 3857.

Sengers, J.V., and J.M.H. Levelt Sengers. 1978. "Critical Phenomena in Classical Fluids." In *Progress in Liquid Physics*, C.A. Croxton, ed., p. 103. Wiley: Chichester.

Sidebottom, D., R. Bergman, L. Börjesson, and L.M. Torell. 1992. "Observation of Scaling Behavior in the Liquid-Glass Transition Range from Dynamic Light Scattering in Poly (Propylene Glycol)." *Phys. Rev. Lett.* 68: 3587.

Sidebottom, D., R. Bergman, L. Börjesson, and L.M. Torell. 1993. "Two-Step Relaxation Decay in a Strong Glass Former." *Phys. Rev. Lett.* 71: 2260.

Signorini, G.F., J.-L. Barrat, and M.L. Klein. 1990. "Structural Relaxation and Dynamical Correlations in a Molten State near the Liquid-Glass Transition: A Molecular Dynamics Study." *J. Chem. Phys.* 92: 1294.

Sjörgen, L. 1980a. "Numerical Results on the Velocity Correlation Function in Liquid Argon and Rubidium." *J. Phys. C. Sol. State Phys.* 13: 705.

Sjörgen, L. 1980b. "Kinetic Theory of Current Fluctuations in Simple Classical Fluids." *Phys. Rev. A.* 22: 2866.

Sjörgen, L. 1991. "Dynamical Scaling Laws in Polymers near the Glass Transition." *J. Phys. Cond. Matt.* 3: 5023.

Sjörgen, L., and W. Götze. 1991. "α-Relaxation Near the Glass Transition." *J. Non-Cryst. Sol.* 131–133: 153.

Sjörgen, L., and A. Sjölander. 1979. "Kinetic Theory of Self-Motion in Monatomic Liquids." *J Phys. C. Sol. State Phys.* 12: 4369.

Smith, K.H., E. Shero, A. Chizmeshya, and G.H. Wolf. 1995. "The Equation of the State of Polyamorphic Germania Glass: A Two-Domain Description of the Viscoelastic Response." *J. Chem. Phys.* 102: 6851.

Sokolov, A.P., W. Steffen, and E. Rössler. 1995. "High-Temperature Dynamics of Glass-Forming Liquids." *Phys. Rev. E.* 52: 5105.

Speedy, R.J. 1982a. "Stability-Limit Conjecture. An Interpretation of the Properties of Water." *J. Phys. Chem.* 86: 982.

Speedy, R.J. 1982b. "Limiting Forms of the Thermodynamic Divergences at the Conjectured Stability Limits in Superheated and Supercooled Water." *J. Phys. Chem.* 86: 3002.

Speedy, R.J. 1992. "Evidence for a New Phase of Water: Water II." *J. Phys. Chem.* 96: 2332.

Speedy, R.J. 1994. "On the Reproducibility of Glasses." *J. Chem. Phys.* 100: 6684.

Speedy, R.J., and C.A. Angell. 1976. "Isothermal Compressibility of Supercooled Water and Evidence for a Thermodynamic Singularity at $-45°C$." *J. Chem. Phys.* 65: 851.

Speedy, R.J., and M. Mezei. 1985. "Pentagon-Pentagon Correlations in Water." *J. Phys. Chem.* 89: 171.

Speedy, R.J., J.D. Madura, and W.L. Jorgensen. 1987. "Network Topology in Simulated Water." *J. Phys. Chem.* 91: 909.

Speedy, R.J., P.G. Debenedetti, C. Huang, R.C. Smith, and B.D. Kay. 1995. "The Entropy of Glassy Water." In *Physical Chemistry of Aqueous Systems. Meeting the Needs of Industry. Proceedings of the 12th International Conference on the Properties of Water and Steam*, H.J. White, Jr., J.V. Sengers, D.B. Neumann, and J.C. Bellows, eds., p. 347. Begell House: New York.

Sperling, L.H. 1986. *Introduction to Physical Polymer Science.* Chap. 6. John Wiley and Sons: New York.

Stanley, H.E. 1979. "A Polychromatic Correlated-Site Percolation Problem with Possible Relevance to the Unusual Behavior of Supercooled H_2O and D_2O." *J. Phys. A. Math Gen.* 12: L329.

Stanley, H.E., and J. Teixeira. 1980. "Interpretation of the Unusual Behavior of H_2O and D_2O at Low Temperatures: Tests of a Percolation Model." *J. Chem. Phys.* 73: 3404.

Stanley, H.E., C.A. Angell, U. Essmann, M. Hemmati, P.H. Poole, and F. Sciortino.

1994. "Is There a Second Critical Point in Liquid Water?" *Physica A*. 206: 1.

Steinhardt, P.J., D.R. Nelson, and M. Ronchetti. 1981. "Icosahedral Bond Orientational Order in Supercooled Liquids." *Phys. Rev. Lett.* 47: 1297.

Steinhardt, P.J., D.R. Nelson, and M. Ronchetti. 1983. "Bond-Orientational Order in Liquids and Glasses." *Phys. Rev. B*. 28: 784.

Stillinger, F.H. 1980. "Water Revisited." *Science*. 209: 451.

Stillinger, F.H. 1985. "Role of Potential Energy Scaling in the Low-Temperature Relaxation Behavior of Amorphous Materials." *Phys. Rev. B*. 32: 3134.

Stillinger, F.H. 1988a. "Supercooled Liquids, Glass Transitions, and the Kauzmann Paradox." *J. Chem. Phys.* 88: 7818.

Stillinger, F.H. 1988b. "Relaxation and Flow Mechanisms in 'Fragile' Glass-Forming Liquids." *J. Chem. Phys.* 89: 6461.

Stillinger, F.H. 1995. "A Topographic View of Supercooled Liquids and Glass Formation." *Science*. 267: 1935.

Stillinger, F.H., and J.A. Hodgdon. 1994. "Translation-Rotation Paradox for Diffusion in Fragile Glass-Forming Liquids." *Phys. Rev. E*. 50: 2064.

Stillinger, F.H., and R.A. LaViolette. 1985. "Sensitivity of Liquid-State Inherent Structure to Details of Intermolecular Forces." *J. Chem. Phys.* 83: 6413.

Stillinger, F.H., and A. Rahman. 1974. "Improved Simulation of Liquid Water by Molecular Dynamics." *J. Chem. Phys.* 60: 1545.

Stillinger, F.H., and T.A. Weber. 1982. "Hidden Structure in Liquids." *Phys. Rev. A*. 25: 978.

Stillinger, F.H., and T.A. Weber. 1983a. "Dynamics of Structural Transitions in Liquids." *Phys. Rev. A*. 28: 2408.

Stillinger, F.H., and T.A. Weber. 1983b. "Inherent Structure in Water." *J. Phys. Chem.* 87: 2833.

Stillinger, F.H., and T.A. Weber. 1984a. "Packing Structures and Transitions in Liquids and Solids." *Science*. 225: 983.

Stillinger, F.H., and T.A. Weber. 1984b. "Inherent Pair Correlation in Simple Liquids." *J. Chem. Phys.* 80: 4434.

Stillinger, F.H., and T.A. Weber. 1985. "Inherent Structure Theory of Liquids in the Hard-Sphere Limit." *J. Chem. Phys.* 83: 4767.

Stillinger, F.H., and T.A. Weber. 1986. "Tiling, Prime Numbers, and the Glass Transition." In *Dynamic Aspects of Structural Change in Liquids and Glasses*, C.A. Angell and M. Goldstein, eds. *Ann. N.Y. Acad. Sci.* 484: 1.

Stixrude, L., and M.S.T. Bukowinski. 1990. "A Novel Topological Compression Mechanism in a Covalent Liquid." *Science*. 250: 541.

Swope, W.C., and H.C. Andersen. 1990. "10^6-Particle Molecular Dynamics Study of Homogeneous Nucleation of Crystals in a Supercooled Atomic Liquid." *Phys. Rev. B*. 41: 7042.

Taborek, P., R.N. Kleiman, and D.J. Bishop. 1986. "Power-Law Behavior in the Viscosity of Supercooled Liquids." *Phys. Rev. B*. 34: 1835.

Tamman, G., and G. Hesse. 1926. "Die Abhängigkeit der Viscosität von der Temperatur bei Unterkühlten Flüssigkeiten." *Z. Anorg. Allg. Chem.* 156: 245.

Tanemura, M., Y. Hiwatari, H. Matsuda, T. Ogawa, N. Ogita, and A. Ueda. 1977. *Prog. Theor. Phys.* 58: 1079.

Teixeira, J., and J. Leblond. 1978. "Brillouin Scattering from Supercooled Water." *J.*

Phys. (Paris). 39: L83.

Thiele, E. 1963. "Equation of State for Hard Spheres." *J. Chem. Phys.* 39: 474.

Thirumalai, D., and R.D. Mountain. 1987. "Relaxation of Anisotropic Correlations in (Two-Component) Supercooled Liquids." *J. Phys. C. Sol. State Phys.* 20: L399.

Thirumalai, D., R.D. Mountain, and T.R. Kirkpatrick. 1989. "Ergodic Behavior in Supercooled Liquids and Glasses." *Phys. Rev. A.* 39: 3563.

Trinh, E., and R.A. Apfel. 1978. "Method for the Measurement of the Sound Velocity in Metastable Liquids, with an Application to Water." *J. Acoust. Soc. Amer.* 63: 777.

Trinh, E., and R.E. Apfel. 1980. "Sound Velocity of Supercooled Water down to $-33°C$ Using Acoustic Levitation." *J. Chem. Phys.* 72: 6731.

Turnbull, D. 1969. "Under What Conditions Can a Glass Be Formed?" *Contemp. Phys.* 10: 473.

Turnbull, D., and M.H. Cohen. 1958. "Concerning Reconstructive Transformation and Formation of Glass." *J. Chem. Phys.* 29: 1049.

Turnbull, D., and M.H. Cohen. 1961. "Free-Volume Model of the Amorphous Phase: Glass Transition." *J. Chem. Phys.* 34: 120.

Turnbull, D., and M.H. Cohen. 1970. "On the Free-Volume Model of the Liquid-Glass Transition." *J. Chem. Phys.* 52: 3038.

Uhlmann, D.R. 1972. "A Kinetic Treatment of Glass Formation." *J. Non-Cryst. Sol.* 7: 337.

Ullo, J.J., and S. Yip. 1985. "Dynamical Transition in a Dense Fluid Approaching Structural Arrest." *Phys. Rev. Lett.* 54: 1509.

Ullo, J., and S. Yip. 1989. "Dynamical Correlations in Metastable Liquids." *Phys. Rev. A.* 39: 5877.

van Blaaderen, A., and P. Wiltzius. 1995. "Real-Space Structure of Colloidal Hard-Sphere Glasses." *Science.* 270: 1177.

van Megen, W., and P.N. Pusey. 1991. "Dynamic Light-Scattering Study of the Glass Transition in a Colloidal Suspension." *Phys. Rev. A.* 43: 5429.

van Megen, W., S.M. Underwood, and P.N. Pusey. 1991. "Nonergodicity Parameters of Colloidal Glasses." *Phys. Rev. Lett.* 67: 1586.

Verlet, L., and J.-J. Weis. 1972. "Equilibrium Theory of Simple Liquids." *Phys. Rev. A.* 5: 939.

Vogel, H. 1921. "Das Temperatur-abhängigkeitsgesetz der Viskosität von Flüssigkeiten." *Phys. Z.* 22: 645.

Weber, T.A., and F.H. Stillinger. 1987. "Tiling Model for Glass Formation with Incremental Domain-Size Kinetics." *Phys. Rev. B.* 36: 7043.

Weber, T.A., G.H. Fredrickson, and F.H. Stillinger. 1986. "Relaxation Behavior in a Tiling Model for Glasses." *Phys. Rev. B.* 34: 7641.

Weeks, J.D., D. Chandler, and H.C. Andersen. 1971. "Role of Repulsive Forces in Determining the Equilibrium Structure of Simple Liquids." *J. Chem. Phys.* 54: 5237.

Wertheim, M.S. 1963. "Exact Solution of the Percus-Yevick Integral Equation for Hard Spheres." *Phys. Rev. Lett.* 10: 321.

Wertheim, M.S. 1964. "Analytic Solution of the Percus-Yevick Equation." *J. Math. Phys.* 5: 643.

Whalley, E., O. Mishima, Y.P. Handa, and D.D. Klug. 1986. "Pressure Melting below the Glass Transition: A New Way of Making Amorphous Solids." In *Dynamic Aspects of Structural Change in Liquids and Glasses,* C.A. Angell and M. Goldstein, eds. *Ann.*

N.Y. Acad. Sci. 484: 81.

Whalley, E., D.D. Klug, and Y.P. Handa. 1989. "Entropy of Amorphous Ice." *Nature.* 342: 782.

Williams, E., and C.A. Angell. 1977. "Pressure Dependence of the Glass Transition Temperature in Ionic Liquids and Solutions. Evidence against Free Volume Theories." *J. Phys. Chem.* 81: 232.

Williams, G., and D.C. Watts. 1970. "Non-Symmetrical Dielectric Relaxation Behavior Arising from a Simple Empirical Decay Function." *Trans. Faraday Soc.* 66: 80.

Williams, M.L., R.F. Landel, and J.D. Ferry. 1955. "The Temperature Dependence of the Relaxation Mechanisms in Amorphous Polymers and Other Glass-Forming Liquids." *J. Amer. Chem. Soc.* 77: 3701.

Wood, W.W., and J.D. Jacobson. 1957. "Preliminary Results from a Recalculation of the Monte Carlo Equation of State for Hard Spheres." *J. Chem. Phys.* 27: 1207.

Woodcock, L.V. 1981. "Glass Transition in the Hard-Sphere Model and Kauzmann's Paradox." In *Structure and Mobility in Molecular and Atomic Glasses,* J.M. O'Reilly and M. Goldstein, eds. *Ann. N.Y. Acad. Sci.* 371: 274.

Woodcock, L.V., and C.A. Angell. 1981. "Diffusivity of the Hard-Sphere Model in the Region of Fluid Metastability." *Phys. Rev. Lett.* 47: 1129.

Xie, Y., K.F. Ludwig, Jr., G. Morales, D.E. Hare, and C.M. Sorensen. 1993. "Noncritical Behavior of Density Fluctuations in Supercooled Water." *Phys. Rev. Lett.* 71: 2050.

Zallen, R. 1983. *The Physics of Amorphous Solids.* Wiley-Interscience: New York.

Zarzycki, J. 1990. *Glasses and the Vitreous State.* Cambridge University Press: Cambridge.

Zheleznyi, B.V. 1969. "The Density of Supercooled Water." *Russ. J. Phys. Chem.* 43: 1311.

Ziman, J.M. 1979. *Models of Disorder. The Theoretical Physics of Homogeneously Disordered Systems.* Chap. 9. Cambridge University Press: Cambridge.

Zwanzig, R. 1960. "Ensemble Method in the Theory of Irreversibility." *J. Chem. Phys.* 33: 1338.

Note Added in Proof

Sastry, Debenedetti, Sciortino, and Stanley have presented a theoretical discussion of the singularity-free scenario for supercooled water in *Phys. Rev. E* 53, 6144, 1996.

Speedy, Debenedetti, Smith, Huang, and Kay have presented experimental evidence of continuity between supercooled and glassy water in *J. Chem. Phys.* 105, 240, 1996.

Kivelson, Kivelson, Zhao, Nussinov, and Tarjus have proposed a new theory of supercooled liquids based on the idea of an avoided phase transition, in *Physica A* 219, 27, 1995. See also Kivelson, Tarjus, Zhao, and Kivelson in *Phys. Rev. E* 53, 751, 1996.

Experimental observations not inconsistent with polyamorphism are presented by Ha, Cohen, Zhao, Lee, and Kivelson in *J. Phys. Chem.* 100, 1, 1996, and by Cohen, Ha, Zhao, Lee, Fischer, Strouse, and Kivelson in *J. Phys. Chem.* 100, 8518, 1996.

A powerful biased Monte Carlo simulation method has been applied by Frenkel and coworkers to study free energy barriers, critical nucleus size, and nucleation rates in moderately supercooled liquids. See ten Wolde, Ruiz-Montero, and Frenkel, *Phys. Rev. Lett.* 75, 2714, 1995, and *J. Chem. Phys.* 104, 9932, 1996.

5

Outlook

Twenty years ago, the study of metastable liquids was mostly that of superheated liquids. Skripov's *Metastable Liquids* is the best example of that viewpoint. Today, many of the most interesting questions and all of the most promising applications in the field involve supercooled liquids.

The conceptual boundaries between thermodynamics and kinetics are usually sharp. Not so with metastable liquids: the blurring of these boundaries is the subject's distinguishing feature. This is illustrated by the use of thermodynamic arguments in the calculation of nucleation rates; the relationship between absolute and attainable limits of stability; the existence of distinct relaxation mechanisms for stable and unstable systems; the relationship between configurational entropy and relaxation time in supercooled liquids; the agreement (or lack thereof) between thermodynamic and time-dependent estimates of the Kauzmann temperature; and the existence of plausible kinetic and thermodynamic interpretations of the glass transition.

Major progress in the fundamental understanding of metastable liquids has been made in the last twenty years. Density functional theory has allowed the rigorous calculation of free energy barriers to nucleation in nonideal fluids and mixtures. Important steps towards the formulation of fully kinetic theories of nucleation have been taken. The strong-fragile classification has emerged as a useful organizing principle for understanding many aspects of relaxation in supercooled liquids. Mode-coupling theory has provided a general mathematical description of relaxation dynamics that is quantitatively accurate for a large number of moderately supercooled fragile liquids. Computer simulations have become an indispensable research tool for testing theories (mode coupling, spinodal decomposition), searching for coherent behavior and length scales in supercooled liquids, and introducing constraints rigorously. The key unanswered questions that define the boundaries of current knowledge and chart directions for future research are listed below.

The rigorous thermodynamics of metastable systems is the thermodynamics of systems under constraints. Computer simulations are the ideal method for introducing constraints rigorously. Work in this area has just begun. Simulating superheated liquids allowing the formation of different-sized voids, and studying supercooled liquids while preventing the formation of ordered embryos of specified size, are examples of the type of basic study that needs to be done. Such studies will provide insight into the properties of liquids under different constraints, and they will address the basic questions of what are appropriate constraints, and when a given system is overconstrained. Eventually, this information could lead to rigorous analytical treatments of metastability.

Nucleation at very high supersaturations and the transition from nucleation to spinodal decomposition remain poorly understood. Valuable information has been gained from cluster analysis of lattice-based simulations (Section 3.3); it is important to extend this work off-lattice. An important question that such simulations should address is whether indeed the critical nucleus becomes fractal and ramified close to the classical spinodal. A novel biased Monte Carlo technique has been used by van Duijneveldt and Frenkel [*J. Chem. Phys.* 96: 4655 (1992)] to study nucleation in slightly supercooled liquids. Extending this idea to superheated and deeply supercooled liquids could provide insights on the transition between nucleation and spinodal decomposition, and more generally on nucleation far away from the binodal.

Density functional calculations involving only fluid phases assume a priori that the critical nucleus is spherical. It would be interesting to incorporate shape fluctuations in density functional calculations. Also of interest are density functional calculations on molecular fluids, which would permit a more detailed comparison with experiments than has been possible to date.

Identifying a length scale for cooperativity in supercooled liquids has proved elusive. This is an important gap in understanding because one of the most useful theories of relaxation in supercooled liquids, the Adam-Gibbs theory, is based on the idea of cooperativity but provides no means of calculating the characteristic length scale. Although recent simulation studies (Section 4.3.2) appear to have identified a growing length scale upon supercooling, this must be validated for other model systems and under different state conditions. Another question intimately related to cooperativity and relaxation is the poorly understood breakdown of the Stokes-Einstein equation in deeply supercooled liquids.

It is not possible to know a priori whether the relaxation behavior of a given fragile liquid will be predicted accurately by mode-coupling theory. Understanding why some of the predictions of this theory are accurate for substances with appreciable molecular complexity (to which the theory does not apply) and why some others fail is a challenging and interesting question. Here again, computer simulations of a series of model liquids with carefully controlled geometry should provide much needed insight.

The thermophysical properties, phase behavior, and stability limits of deeply supercooled water pose some of the most interesting open questions on the physics of liquids today. Experiments, simulations, and theory have all contributed significantly to what is known about supercooled water. Unfortunately, however, the difficulty of performing accurate measurements at deeper supercoolings and higher pressures than have been used to date, and even under tension, is formidable. Simulations and theory are therefore especially valuable ways of studying such highly metastable conditions. Phase transitions between amorphous phases of different density (polyamorphism) appear key to understanding the global phase behavior of water, silica, and several liquid metals. That such different substances can exhibit similar behavior (density anomalies, increase in the response functions upon cooling) in the supercooled regime is quite surprising. Understanding why this happens is a major theoretical challenge.

The ability of liquids to exist beyond their domain of absolute thermodynamic stability is exploited by nature in many ingenious ways. It influences, among other things, the climate, ozone depletion, and the economically important survival of fruit trees in winter. It offers opportunities for innovative engineering, especially in applying the kinetic inhibition of crystallization to such problems as the preservation and storage of cells and biochemicals, the transportation of natural gas from offshore reservoirs, and food processing and storage. Finally, it confronts the physical scientist with the challenging fundamental problems discussed in this book, problems that are not encountered in the study of stable liquids.

Appendix 1

Stability of Fluids: Thermodynamic and Mathematical Proofs

A.1.1 EQUIVALENCE OF EXTREMUM CONDITIONS

The second law of thermodynamics says that the entropy of a closed system increases during any spontaneous adiabatic process. Such a process will occur if a constraint is removed from a system enclosed by adiabatic boundaries, for example, by eliminating a semipermeable membrane or a rigid internal partition. Since an isolated system necessarily undergoes adiabatic interactions, the entropy of an isolated system at equilibrium is a maximum with respect to that of states resulting from all possible variations consistent with the given energy and any existing constraints. Consider (Gibbs, 1875–76, 1877–78, 1961) an equilibrium state a of an isolated system. If its entropy were not a maximum consistent with its energy, it would be possible to attain state b, having the same energy but higher entropy. But, from b, the system can always be brought by cooling to a state c, having the same entropy as a, but lower energy. Hence, if state b cannot be attained from a without removing a constraint, neither can state c be attained from a. At equilibrium, the entropy of an isolated system is a maximum consistent with the energy, and the energy is a minimum consistent with the entropy. These two conditions are identical. In symbols,

$$[\Delta S]_{U,V,N} \leq 0 \Leftrightarrow [\Delta U]_{S,V,N} \geq 0. \tag{A.1.1}$$

Let the closed system of interest be in contact with two large reservoirs. One of them, a thermal reservoir, performs no work, and interacts with the system through a rigid and impermeable boundary. In certain cases, we will imagine this boundary to be adiabatic. In others, it will be assumed to be nonadiabatic. In the former case, the thermal reservoir does not interact with the system; in the latter, the system has the thermal reservoir's temperature. The other reservoir is mechanical; it can undergo only adiabatic work interactions with the system of interest through an adiabatic and impermeable boundary that may be fixed or movable. In the former case, the mechanical reservoir does not interact with the system. In the latter case, the system has the mechanical reservoir's pressure.

Finally, let the entire composite system composed of the thermal reservoir, the mechanical reservoir, and the system of interest, be isolated.

Let the mechanical boundary be fixed and the thermal boundary be non-adiabatic. The system of interest has fixed mass, temperature, and volume (N, T, V). We consider variations away from a given equilibrium state. Then, for all possible variations, the entropy of the isolated composite decreases, and we must have

$$\Delta S + \Delta S^\tau \leq 0, \tag{A.1.2}$$

$$\Delta U + \Delta U^\tau = 0, \tag{A.1.3}$$

where the superscript τ denotes the thermal reservoir, and unsuperscripted quantities refer to the system of interest. Since the reservoir performs no work, its energy change is given by

$$\Delta U^\tau = T \, \Delta S^\tau, \tag{A.1.4}$$

where T is common to the thermal reservoir and to the system. Substituting (A.1.4) into (A.1.3), multiplying (A.1.2) by T, and subtracting, we obtain

$$\Delta U - T \Delta S \geq 0. \tag{A.1.5}$$

But $A = U - TS$. Therefore,

$$[\Delta A]_{T,V,N} \geq 0. \tag{A.1.6}$$

At equilibrium, the Helmholtz energy of a closed system is a minimum with respect to that of states resulting from variations that do not change its volume or temperature.

Let the mechanical boundary be movable and the thermal boundary be nonadiabatic. The system of interest has fixed mass, temperature, and pressure (N, T, P). We consider variations away from a given equilibrium state. Then, for all possible variations, the entropy of the isolated composite decreases, and we must have

$$\Delta S + \Delta S^\tau \leq 0, \tag{A.1.2}$$

$$\Delta U + \Delta U^\tau + \Delta U^\pi = 0, \tag{A.1.7}$$

$$\Delta V + \Delta V^\pi = 0, \tag{A.1.8}$$

where the superscript π denotes the mechanical reservoir. The energy change of the thermal reservoir is given by (A.1.4); that of the mechanical reservoir by

$$\Delta U^\pi = -P \Delta V^\pi = P \Delta V. \tag{A.1.9}$$

Substituting (A.1.4) and (A.1.9) in (A.1.7), multiplying (A.1.2) by T, and subtracting, we obtain

$$\Delta U - T \Delta S + P \Delta V \geq 0. \tag{A.1.10}$$

But $G = U - TS + PV$. Therefore,

$$[\Delta G]_{T,P,N} \geq 0. \tag{A.1.11}$$

At equilibrium, the Gibbs energy of a closed system is a minimum with respect to that of states resulting from variations that do not change its pressure or temperature.

Let the mechanical boundary be movable and the thermal boundary be adiabatic. Furthermore, let the entropy, volume, and mass of the entire composite be fixed [i.e., the composite is no longer isolated, as it was in the proofs leading to (A.1.6) and (A.1.11)]. We consider variations away from a given equilibrium state. Then, for all possible variations, the energy of the closed composite increases, and we must have

$$\Delta U + \Delta U^{\pi} \geq 0, \tag{A.1.12}$$

$$\Delta V + \Delta V^{\pi} = 0. \tag{A.1.8}$$

The energy change of the mechanical reservoir is given by (A.1.9). Substituting (A.1.9) into (A.1.12), we obtain

$$\Delta U + P \, \Delta V \geq 0. \tag{A.1.13}$$

But $H = U + PV$. Therefore,

$$[\Delta H]_{S,P,N} \geq 0. \tag{A.1.14}$$

At equilibrium, the enthalpy of a closed system is a minimum with respect to that of states resulting from variations that do not change its pressure or entropy.[1]

We conclude that the inequalities (A.1.1), (A.1.6), (A.1.11), and (A.1.14) are entirely equivalent. They all follow from the fact that the entropy of an isolated system at equilibrium is a maximum consistent with its energy. The particular constraints imposed on a given system dictate which thermodynamic potential is an extremum at equilibrium.

A.1.2 PROOF OF EQUATION (2.15)

Consider a nonreacting system of given entropy, volume, composition, and mass, at equilibrium. For the equilibrium state to be stable we require [see Equation (2.9)] that second-order variations of energy that conserve entropy, volume, mass, and composition be positive:

$$\left[\delta^2 U\right]_{S,V,N_1,\ldots,N_n} \geq 0, \tag{A.1.15}$$

---

[1] The system's entropy in this case is fixed because the total entropy is fixed, the thermal reservoir is inactive, and the mechanical reservoir undergoes only reversible adiabatic interactions.

where n denotes the number of components in the mixture. To compute the energy change in (A.1.15) we divide the system into two subsystems α and β, and we consider a generic fluctuation, starting from an equilibrium condition, that results in subsystems α and β having different intensive properties (Modell and Reid, 1983). The resulting second-order energy change is given by

$$\delta^2 U = \delta^2 U^\alpha + \delta^2 U^\beta = \sum_{i=1}^{n+2}\sum_{j=1}^{n+2} U_{ij}^\alpha \, \delta X_i^\alpha \, \delta X_j^\alpha + \sum_{i=1}^{n+2}\sum_{j=1}^{n+2} U_{ij}^\beta \, \delta X_i^\beta \, \delta X_j^\beta,$$

(A.1.16)

where

$$U_{ij} \equiv \frac{\partial^2 U}{\partial X_i \, \partial X_j}.$$

(A.1.17)

These partial derivatives are to be evaluated at the initial equilibrium condition. In (A.1.16) and (A.1.17), X_i ($i = 1, \ldots, n+2$) denotes any one of the variables $[S, V, N_1, \ldots, N_n]$. Because each X_i is a conserved quantity, the variations must satisfy the constraint

$$\delta X_i^\alpha = -\delta X_i^\beta.$$

(A.1.18)

Furthermore, subsystems α and β are identical at equilibrium, hence their intensive properties coincide at the outset. This means

$$N^\alpha \, U_{ij}^\alpha = N^\beta \, U_{ij}^\beta,$$

(A.1.19)

where N^α denotes the total number of moles in α. Using (A.1.18) and (A.1.19) in (A.1.16),

$$\delta^2 U = \delta^2 U^\alpha + \delta^2 U^\beta = \frac{N}{N^\beta} \sum_{i=1}^{n+2}\sum_{j=1}^{n+2} U_{ij}^\alpha \, \delta X_i^\alpha \, \delta X_j^\alpha,$$

(A.1.20)

where N denotes the total number of moles in the entire system. Since the prefactor multiplying the summations in (A.1.20) is positive, and we are interested in the sign of the energy variation, this prefactor can be dropped. So, too, can the superscript α, because the subsystem it denotes is completely arbitrary. Henceforth, therefore, we are interested in the sign of the generic quadratic form

$$Q = \sum_{i=1}^{m}\sum_{j=1}^{m} u_{ij} \, x_i \, x_j,$$

(A.1.21)

where $m = n + 2$, u_{ij} denotes U_{ij}, and x_i denotes δX_i. We now diagonalize the quadratic form. If u_{11} does not vanish, we can write

$$Q = u_{11}\left[x_1^2 + 2\frac{x_1}{u_{11}}\sum_{j=2}^{m} u_{1j} x_j\right] + \sum_{i=1}^{m}\sum_{j=2}^{m} u_{ij} x_i x_j$$

(A.1.22)

(Debenedetti, 1988), or, equivalently,

$$Q = u_{11}\left[\left(x_1 + \sum_{j=2}^{m}\frac{u_{1j}}{u_{11}}x_j\right)^2 - \left(\sum_{j=2}^{m}\frac{u_{1j}}{u_{11}}x_j\right)^2\right] + \sum_{i=2}^{m}\sum_{j=2}^{m}u_{ij}\,x_i\,x_j,$$

$$\text{(A.1.23)}$$

which, upon rearrangement, reads

$$Q = u_{11}\left[x_1 + \sum_{j=2}^{m}\frac{u_{1j}}{u_{11}}x_j\right]^2 + \sum_{j=2}^{m}\sum_{k=2}^{m}u'_{jk}\,x_j\,x_k,\qquad\text{(A.1.24)}$$

where

$$u'_{jk} = u_{jk} - \frac{u_{1j}\,u_{1k}}{u_{11}}.\qquad\text{(A.1.25)}$$

The double sum on the right-hand side of (A.1.24) is a quadratic form, formally identical to Q. We may thus apply the procedure leading from (A.1.21) to (A.1.24) to it, and, successively, to the remaining quadratic forms, to obtain finally

$$Q = u_{11}\left[x_1 + \sum_{j=2}^{m}\left(\frac{u_{1j}}{u_{11}}\right)x_j\right]^2 + u'_{22}\left[x_2 + \sum_{j=3}^{m}\left(\frac{u'_{2j}}{u'_{22}}\right)x_j\right]^2$$

$$+ u''_{33}\left[x_3 + \sum_{j=4}^{m}\left(\frac{u''_{3j}}{u''_{33}}\right)x_j\right]^2 + \cdots + u^{m-1}_{mm}x_m^2,\qquad\text{(A.1.26)}$$

where

$$u'_{ij} = u_{ij} - \frac{u_{1i}u_{1j}}{u_{11}},\qquad i,j \geq 2,\qquad\text{(A.1.27)}$$

$$u''_{ij} = u'_{ij} - \frac{u'_{2i}u'_{2j}}{u'_{22}},\qquad i,j \geq 3,\qquad\text{(A.1.28)}$$

$$u^{m-1}_{ij} = u^{m-2}_{ij} - \frac{u^{m-2}_{m-1,i}\,u^{m-2}_{m-1,j}}{u^{m-2}_{m-1,m-1}},\qquad i,j \geq m,\qquad\text{(A.1.29)}$$

and where the validity of (A.1.26) is only limited by the requirement

$$u^{i}_{i+1,i+1} > 0,\qquad 0 \leq i \leq m - 1\qquad\text{(A.1.30)}$$

with the notation $u_{11} = u^0_{11}$ implied. Since all the terms in brackets in the diagonalized quadratic form (A.1.26) are non-negative, the necessary and sufficient condition for Q to be non-negative is for each of the coefficients, u_{11}, etc., to be non-negative.

Reverting to the original notation ($u_{ij} \rightarrow U_{ij}$, $x_i \rightarrow \delta X_i$), we write the diagonalized quadratic form as

$$\delta^2 U \propto \sum_{j=1}^{n+2} y_{jj}^{(j-1)} \, \delta Z_j^2, \tag{A.1.31}$$

where

$$\delta Z_j = \begin{cases} \delta X_j + \displaystyle\sum_{i=j+1}^{n+2} y_{ji}^{(j)} \delta X_i, & j = 1, 2, \ldots, n+1, \\ \delta X_{n+2}, & j = n+2. \end{cases} \tag{A.1.32}$$

The proportionality sign in (A.1.31) accounts for the absence of the factor N/N^β [see (A.1.20)]. In (A.1.31) and (A.1.32) we have used Legendre transform notation [see Section 2.3.2, Equations (2.117)–(2.119)]. Also, the mathematical properties of Legendre transforms (Modell and Reid, 1983) have been invoked in rewriting (A.1.26) in the compact form (A.1.31).

In order for the diagonalized quadratic form to be positive for arbitrary variations, we must have

$$y_{jj}^{(j-1)} > 0, \qquad j = 1, 2, \ldots, n+1. \tag{A.1.33}$$

For a thermodynamic system, $y_{n+2,n+2}^{(n+1)}$ is identically zero, and is therefore not included in (A.1.33). To see this, consider a pure substance ($n = 1$), and order the independent variables in the energy representation as (S, V, N). Then, $y_{n+2,n+2}^{(n+1)} = (\partial \mu / \partial N)_{T,P}$, which is zero. We now return to (A.1.33). It follows from the properties of Legendre transforms that

$$y_{jj}^{(j-1)} = y_{jj}^{(j-2)} - \frac{\left(y_{j,j-1}^{(j-2)}\right)^2}{y_{j-1,j-1}^{(j-2)}}. \tag{A.1.34}$$

Because of (A.1.33), both $y_{jj}^{(j-1)}$ and $y_{j-1,j-1}^{(j-2)}$ must be positive for a stable or metastable system. The first term in the right-hand side is also positive because if the order of the independent variables X_i is altered such that $X_{j-1} \leftrightarrow X_j$, this second-order derivative would have been among those listed in (A.1.33). Hence, when, starting from a condition of stability, a system approaches a limit of stability, the condition

$$y_{n+1,n+1}^{(n)} > 0 \tag{A.1.35}$$

is always violated first.[2] This inequality is therefore the stability criterion. Finally, since

$$dy^{(n)} = -\sum_{i=1}^{n} X_i d\xi_i + \sum_{i=n+1}^{n+2} \xi_i dX_i, \tag{A.1.36}$$

---

[2]It is essential that stability be tested starting from a stable or metastable condition. The analysis is not applicable to unstable systems (Heidemann, 1975; Beegle et al., 1975).

we must have

$$y_{n+1,n+1}^{(n)} = \left(\frac{\partial^2 y^{(n)}}{\partial X_{n+1}^2} \right)_{\xi_1,\xi_2,\ldots,\xi_n,X_{n+2}} = \left(\frac{\partial \xi_{n+1}}{\partial X_{n+1}} \right)_{\xi_1,\xi_2,\ldots,\xi_n,X_{n+2}} > 0, \quad \text{(A.1.37)}$$

which completes the proof of (2.15).

A.1.3 STABILITY OF LIMITS OF STABILITY: CRITICALITY

In order for an equilibrium state to be stable, it must satisfy

$$[\Delta U]_{S,V,N} = \delta U + \frac{1}{2!} \delta^2 U + \frac{1}{3!} \delta^3 U + \cdots > 0. \quad \text{(A.1.38)}$$

The vanishing of the first-order term constitutes the equilibrium criterion; the positiveness of the second-order variation constitutes the stability criterion. At a limit of stability, the second-order variation vanishes. Thus, to test the stability of a limit of stability, we require that the lowest-order, nonvanishing energy variation be positive.

To illustrate how this can be applied to practical situations, consider a pure substance, and let its temperature, volume, and mass be fixed. Then the Helmholtz energy must be minimum at equilibrium. Among the variations with respect to which the Helmholtz energy must be a minimum, we investigate isothermal density fluctuations in subsystems α and β, such that

$$\delta V^\alpha = -\delta V^\beta \quad \text{(A.1.39)}$$

with the masses of each subsystem constant. Then,

$$\delta^2 A = \left[-\left(\frac{\partial P}{\partial V} \right)_{T,N} \delta V^2 \right]^\alpha + \left[-\left(\frac{\partial P}{\partial V} \right)_{T,N} \delta V^2 \right]^\beta,$$

$$\delta^3 A = \left[-\left(\frac{\partial^2 P}{\partial V^2} \right)_{T,N} \delta V^3 \right]^\alpha + \left[-\left(\frac{\partial^2 P}{\partial V^2} \right)_{T,N} \delta V^3 \right]^\beta,$$

$$\delta^4 A = \left[-\left(\frac{\partial^3 P}{\partial V^3} \right)_{T,N} \delta V^4 \right]^\alpha + \left[-\left(\frac{\partial^3 P}{\partial V^3} \right)_{T,N} \delta V^4 \right]^\beta. \quad \text{(A.1.40)}$$

Taking into account (A.1.39) and the fact that intensive properties are identical in α and β before the fluctuation,

$$\delta^2 A = -\left[\frac{1}{N^\alpha} + \frac{1}{N^\beta} \right] \left(\frac{\partial P}{\partial v} \right)_T (\delta V^\alpha)^2,$$

$$\delta^3 A = -\left[\frac{1}{(N^\alpha)^2} - \frac{1}{(N^\beta)^2}\right]\left(\frac{\partial^2 P}{\partial v^2}\right)_T (\delta V^\alpha)^3 ,$$

$$\delta^4 A = -\left[\frac{1}{(N^\alpha)^3} + \frac{1}{(N^\beta)^3}\right]\left(\frac{\partial^3 P}{\partial v^3}\right)_T (\delta V^\alpha)^4 . \qquad \text{(A.1.41)}$$

The second-order variation vanishes at a limit of stability. Since the sign of the volume fluctuation is arbitrary, the following conditions must be satisfied in order for the limit of stability to be stable:

$$\left(\frac{\partial P}{\partial v}\right)_T = 0,$$

$$\left(\frac{\partial^2 P}{\partial v^2}\right)_T = 0,$$

$$\left(\frac{\partial^3 P}{\partial v^3}\right)_T < 0. \qquad \text{(A.1.42)}$$

The above are clearly necessary conditions for the stability of a limit of stability of a single-component fluid. Carrying out the full expansions for $\delta^3 U$, $\delta^4 U$, etc., in (A.1.38), these are shown to be also sufficient conditions. In this way, one derives the general necessary and sufficient condition for the stability of a limit of stability of an n-component fluid mixture,

$$y^{(n)}_{n+1,n+1} = 0,$$

$$y^{(n)}_{n+1,n+1,n+1} = 0,$$

$$y^{(n)}_{n+1,n+1,n+1,n+1} > 0; \text{ but if zero, then}$$

$$y^{(n)}_{n+1,n+1,n+1,n+1,n+1} = 0,$$

$$y^{(n)}_{n+1,n+1,n+1,n+1,n+1,n+1} > 0; \text{ but if zero, then} \ldots \qquad \text{(A.1.43)}$$

Stable limits of stability are called critical states. The criticality criterion calls for the lowest-order even derivative of the nth Legendre transform with respect to the $(n+1)$st variable to be positive, and all lowest-order derivatives to vanish.

A.1.4 PROOF OF EQUATIONS (2.86) AND (2.87)

For a pure fluid, the stability criteria (A.1.33) read

$$y^{(o)}_{11} > 0, \qquad \text{(A.1.44)}$$

$$y^{(1)}_{22} > 0, \qquad \text{(A.1.45)}$$

where the zeroth-order transform is the energy. Ordering the independent variables (S, V, N), the above relations become

$$y_{11}^{(0)} = U_{SS} > 0, \tag{A.1.46}$$

$$y_{22}^{(1)} = y_{22}^{(0)} - \frac{\left(y_{12}^{(0)}\right)^2}{y_{11}^{(0)}} = U_{VV} - \frac{U_{SV}^2}{U_{SS}} > 0, \tag{A.1.47}$$

which are identical to (2.86) and (2.87). The first equality in (A.1.47) follows from the mathematical properties of Legendre transforms (Modell and Reid, 1983).

References

Beegle, B., M. Modell, and R.C. Reid. 1975. "Reply to 'The Criteria of Thermodynamic Stability.'" *Amer. Inst. Chem. Eng. J.* 21: 826.

Debenedetti, P.G. 1988. "Thermodynamic Stability of Single-Phase Fluids and Fluid Mixtures under the Influence of Gravity." *J. Chem. Phys.* 89: 6881.

Gibbs, J.W. 1875–76 and 1877–78. "On the Equilibrium of Heterogeneous Substances." *Trans. Conn. Acad.* III: 108 (Oct. 1875–May 1876) and III: 343 (May 1877–July 1878).

Gibbs, J.W. 1961. *The Scientific Papers of J. Willard Gibbs, Ph.D., LL.D. I. Thermodynamics*, pp. 55–57. Dover: New York.

Heidemann, R.A. 1975. "The Criteria of Thermodynamic Stability." *Amer. Inst. Chem. Eng. J.* 21: 824.

Modell, M., and R.C. Reid. 1983. *Thermodynamics and Its Applications.* 2nd. ed., Chaps. 5 and 9. Prentice-Hall: Englewood Cliffs, NJ.

Appendix 2

Thermodynamics of Fluid Interfaces

A.2.1 FUNDAMENTAL EQUATIONS

Following Gibbs (1875–76, 1877–78, 1961), we find it convenient to base the treatment of interfaces on an idealization whereby the real interface, which has finite width, is replaced by a geometric surface, as illustrated in Figure A.2.1.[1] This geometric surface is sensibly coincident with the real interface Σ, but its exact location is, for the time being, arbitrary. We refer to the geometric surface as the dividing surface S. Let N be a closed surface intersecting S, and including part of the homogeneous mass on each side of S. We require that the portion of N contained in the actual interface be normal to S. Let the mass contained in N be further divided into three parts by two surfaces, L' and L'', located at each side of S and lying entirely within the region where the fluid is homogeneous. We denote that part of the mass enclosed in N and limited by both L' and L'', that is to say, the part that contains the actual interface, by M, and the remaining homogeneous portions by M' and M'' (Gibbs, 1875–76, 1877–78, 1961).

For the equilibrium of the composite system (M, M', M'') it is necessary that variations that do not involve movement of the boundaries or changes in the total entropy satisfy

$$\delta U + \delta U' + \delta U'' \geq 0 \tag{A.2.1}$$

subject to the constraints

$$\delta S + \delta S' + \delta S'' = 0, \tag{A.2.2}$$

$$\delta N_i + \delta N_i' + \delta N_i'' = 0, \qquad i = 1, 2, \ldots, n, \tag{A.2.3}$$

where i denotes different components in a mixture. Therefore,

$$\left(T - T''\right)\delta S + \left(T' - T''\right)\delta S' + \sum_i \left(\mu_i - \mu_i''\right)\delta N_i$$

$$+ \sum_i \left(\mu_i' - \mu_i''\right)\delta N_i'' \geq 0 \tag{A.2.4}$$

[1] Alternative treatments that take into account the finite thickness of interfaces exist (e.g., the van der Waals–Cahn–Hilliard theory: see Section 3.2). Here we restrict our attention to the Gibbsian approach, which underlies classical nucleation theory.

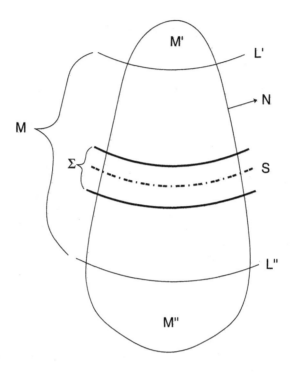

Figure A.2.1. Schematic of the Gibbs representation of the interface between two homogeneous fluid phases. Terms are defined in the text.

and, for equilibrium,

$$T = T' = T'', \tag{A.2.5}$$

$$\mu = \mu_i' = \mu_i'', \qquad i = 1, 2, \ldots, n. \tag{A.2.6}$$

At equilibrium, then, there is uniformity of temperature; furthermore, the chemical potential of every component present in both homogeneous phases is constant throughout the system.

Now, denote M^{iii} and M^{iv} the two portions of M such that M^{iii} is next to M' and M^{iv} next to M''. Imagine subsystem M^{iii} being filled by matter identical in intensive properties to M', and let M^{iv} be filled by matter having identical intensive properties to M''. M^{iii} and M^{iv}, in other words, are idealizations whose intensive properties are uniform and equal to those of the respective bulk phases all the way up to the dividing surface. Then, for variations at constant volume,

$$\delta U^{iii} = T' \delta S^{iii} + \sum_i \mu_i' \delta N_i^{iii}, \tag{A.2.7}$$

$$\delta U^{iv} = T'' \delta S^{iv} + \sum_i \mu_i'' \delta N_i^{iv}. \tag{A.2.8}$$

For reversible variations,

$$\delta U^{iii} = T \delta S^{iii} + \sum_i \mu_i \delta N_i^{iii}, \tag{A.2.9}$$

$$\delta U^{iv} = T \delta S^{iv} + \sum_i \mu_i \delta N_i^{iv}, \tag{A.2.10}$$

$$\delta U = T \delta S + \sum_i \mu_i \delta N_i, \tag{A.2.11}$$

where (A.2.11) pertains to M. Therefore,

$$\delta \left(U - U^{iii} - U^{iv} \right) = T \delta \left(S - S^{iii} - S^{iv} \right)$$
$$+ \sum_i \mu_i \delta \left(N_i - N_i^{iii} - N_i^{iv} \right). \tag{A.2.12}$$

Now, the quantity $\delta \left(X - X^{iii} - X^{iv} \right)$ where X is extensive, denotes the excess of X in M with respect to the amount that would exist if on each side of S the intensive properties had the same uniform value up to S that they have in the respective bulk phases. Therefore, defining

$$\delta X^{\sigma} = \delta \left(X - X^{iii} - X^{iv} \right), \tag{A.2.13}$$

we have, for reversible variations that do not alter the volume of any of the subsystems,

$$\delta U^{\sigma} = T \delta S^{\sigma} + \sum_i \mu_i \delta N_i^{\sigma}. \tag{A.2.14}$$

For variations in the form of S, we must add contributions due to changes in area and curvature,

$$\delta U^{\sigma} = T \delta S^{\sigma} + \sum_i \mu_i \delta N_i^{\sigma} + \sigma \delta F + K_1 \delta C_1 + K_2 \delta C_2, \tag{A.2.15}$$

or, equivalently,

$$\delta U^{\sigma} = T \delta S^{\sigma} + \sum_i \mu_i \delta N_i^{\sigma} + \sigma \delta F + \tfrac{1}{2} (K_1 + K_2) \delta (C_1 + C_2)$$
$$+ \tfrac{1}{2} (K_1 - K_2) \delta (C_1 - C_2), \tag{A.2.16}$$

where F is the area of S, and C_1 and C_2 are its principal curvatures. Now, it is always possible to choose the location of S so that $(K_1 + K_2)$ vanishes (Gibbs, 1875–76, 1877–78, 1961). The dividing surface is then referred to as the surface of tension, and σ as the superficial tension (Gibbs, 1975–76,

1877–78, 1961), or, in modern terminology, as the surface tension or interfacial tension. As for the last term, $K_1 = K_2$ when $C_1 = C_2$. This is the case when S is plane. As long as S can be considered to be composed of portions that are locally plane, then, the last term will be negligible. This condition is satisfied when the surface's radii of curvature are large compared to the actual thickness of the interface. Should this criterion be violated, continuum reasoning cannot be applied. For most cases of interest, then, we have, finally, the important equation

$$dU^\sigma = T\,dS^\sigma + \sum_i \mu_i\,dN_i^\sigma + \sigma\,dF. \qquad (\text{A.2.17})$$

This gives the changes of energy for reversible variations about an equilibrium state, when S is chosen so that $(K_1 + K_2)$ vanishes in the equilibrium state. Integrating (A.2.17) along the surface,

$$U^\sigma = TS^\sigma + \sigma F + \sum_i \mu_i N_i^\sigma. \qquad (\text{A.2.18})$$

Dividing by F,

$$u^\sigma = Ts^\sigma + \sigma + \sum_i \mu_i\Gamma_i, \qquad (\text{A.2.19})$$

where

$$u^\sigma = U^\sigma/F, \quad s^\sigma = S^\sigma/F, \quad \Gamma_i = N_i^\sigma/F. \qquad (\text{A.2.20})$$

From (A.2.17) and (A.2.18),

$$S^\sigma\,dT + F\,d\sigma + \sum_i N_i^\sigma\,d\mu_i = 0. \qquad (\text{A.2.21})$$

Dividing by F,

$$s^\sigma\,dT + d\sigma + \sum_i \Gamma_i\,d\mu_i = 0. \qquad (\text{A.2.22})$$

From (A.2.19) and (A.2.22),

$$du^\sigma = T\,ds^\sigma + \sum_i \mu_i d\Gamma_i. \qquad (\text{A.2.23})$$

Equations (A.2.17)–(A.2.19) and (A.2.21)–(A.2.23) are the fundamental equations of the thermodynamics of surfaces. Equations (A.2.21) and (A.2.22) are the surface analogues of the Gibbs-Duhem equation.

For mechanical equilibrium, that is to say, equilibrium with respect to the motion or deformation of S, we divide the composite into two portions, and we denote the energy, entropy, volume, etc., of each part, calculated on the assumption that each part is uniform all the way up to the dividing surface, by

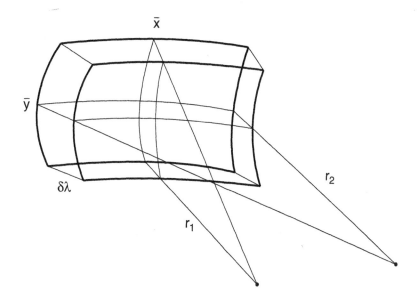

Figure A.2.2. Geometric construction used in the derivation of the condition of mechanical equilibrium of a curved interface between two homogeneous fluid phases.

U', S', etc., and U'', S'', etc. Then, for variations about an equilibrium state,

$$\delta U^\sigma + \delta U' + \delta U'' \geq 0, \tag{A.2.24}$$

$$\delta S^\sigma + \delta S' + \delta S'' = 0, \tag{A.2.25}$$

$$\delta N_i^\sigma + \delta N_i' + \delta N_i'' = 0, \tag{A.2.26}$$

$$\delta V' + \delta V'' = 0. \tag{A.2.27}$$

For mechanical equilibrium, we consider only those variations for which $\delta S^\sigma = \delta S' = \delta S'' = 0$ and $\delta N_i^\sigma = \delta N_i' = \delta N_i'' = 0$. Then,

$$\sigma\,\delta F - P'\delta V' - P''\,\delta V'' = 0. \tag{A.2.28}$$

Suppose the surface moves a distance $\delta\lambda$. Then (Figure A.2.2),

$$\frac{\overline{y}}{r_1 + \delta\lambda} = \frac{y}{r_1}, \tag{A.2.29}$$

$$\frac{\overline{x}}{r_2 + \delta\lambda} = \frac{x}{r_2}, \tag{A.2.30}$$

$$\begin{aligned}
\delta F = \overline{xy} - xy &= xy\left[\left(1 + \frac{\delta\lambda}{r_1}\right)\left(1 + \frac{\delta\lambda}{r_2}\right) - 1\right] \\
&= F\delta\lambda\left[C_1 + C_2 + O\left(\delta\lambda\right)\right], \tag{A.2.31}
\end{aligned}$$

$$\delta V' = F \, \delta \lambda, \tag{A.2.32}$$

$$\delta V'' = -F \, \delta \lambda. \tag{A.2.33}$$

Therefore

$$\sigma \, (C_1 + C_2) = P' - P'', \tag{A.2.34}$$

where P' relates to that part of M on which lie the centers of curvature. For a spherical surface, we recover Laplace's equation,

$$\Delta P = \frac{2\sigma}{r} \, . \tag{A.2.35}$$

A.2.2 WORK OF FORMATION OF AN EMBRYO OF A NEW PHASE

We consider a homogeneous fluid mass α in contact with a thermal reservoir τ and a work reservoir π, which impose on it a temperature T and a pressure P (see Appendix 1). For simplicity, we treat first the single-component case. Let an embryo of a second phase, β, be formed entirely within α. The entropy, volume, and mass of the composite (system plus reservoirs) is constant. In the final state, once β has been formed, we will use α and β to denote the bulk phases and their imagined continuation up to the dividing surface. Then, with subscripts 0 and f denoting initial and final conditions, the change in energy due to the formation of phase β is given by

$$\delta U = \delta U^\tau + \delta U^\pi + U_f^\alpha - U_0^\alpha + U^\beta + U^\sigma, \tag{A.2.36}$$

with

$$\delta U^\tau = T \, \delta S^\tau = -T \left(S^\beta + S^\sigma + S_f^\alpha - S_0^\alpha \right), \tag{A.2.37}$$

$$\delta U^\pi = -P \, \delta V^\pi = P \left(V^\beta + V_f^\alpha - V_0^\alpha \right), \tag{A.2.38}$$

$$U_f^\alpha = T \, S_f^\alpha - P \, V_f^\alpha + \mu^\alpha \, N_f^\alpha, \tag{A.2.39}$$

$$U_0^\alpha = T \, S_0^\alpha - P \, V_0^\alpha + \mu^\alpha \, N_0^\alpha, \tag{A.2.40}$$

$$U^\beta = T S^\beta - P^\beta V^\beta + \mu^\beta N^\beta, \tag{A.2.41}$$

$$U^\sigma = T S^\sigma + \sigma F + \mu^\sigma N^\sigma. \tag{A.2.42}$$

In (A.2.38) and (A.2.39) we have used the fact that μ^α does not change because T and P are constant in α. Furthermore, since both phases extend up to the dividing surface, $V^\sigma = 0$. Conservation of mass imposes the constraint

$$N_0^\alpha = N_f^\alpha + N^\beta + N^\sigma. \tag{A.2.43}$$

Substituting (A.2.37)–(A.2.43) in (A.2.36),

$$\begin{aligned} \delta U \;=\; & \left(P - P^\beta \right) V^\beta + \sigma F + \left[\mu^\beta \left(P^\beta \right) - \mu^\alpha \left(P \right) \right] N^\beta \\ & + \left[\mu^\sigma - \mu^\alpha \left(P \right) \right] N^\sigma. \end{aligned} \tag{A.2.44}$$

When (A.2.17) is adopted as the definition of surface energetics, the dividing surface is necessarily located so as to make $K_1 + K_2$ vanish in (A.2.16). The additional assumption is often made that $\mu^\sigma = \mu^\alpha$ in (A.2.42). This implies that the bulk phase α imposes its chemical potential on the interface σ though not, in general, upon the new phase β except at equilibrium. Equation (A.2.44) then simplifies to

$$\delta U = W_{\min} = \left(P - P^\beta\right) V^\beta + \sigma F + \left[\mu^\beta \left(P^\beta\right) - \mu^\alpha \left(P\right)\right] N^\beta. \quad \text{(A.2.45)}$$

Since the formation of the new phase occurred at constant entropy, constant volume, and in a closed system, δU is the minimum, reversible work needed to form an enclosed phase β within an existing phase α (e.g., a bubble in a liquid, a droplet in a vapor). This proves Equation (3.8). Note, however, that the rigorous result is (A.2.44).

Using identical arguments, the reversible work of forming a bubble or a droplet β inside phase a for an n-component mixture ($i = 1, \ldots, n$) is

$$\delta U = W_{\min} = \left(P - P^\beta\right) V^\beta + \sigma F + \sum_i \left[\mu_i^\beta \left(P^\beta\right) - \mu_i^\alpha \left(P\right)\right] N_i^\beta$$
$$+ \sum_i \left[\mu_i^\sigma - \mu_i^\alpha \left(P\right)\right] N_i^\sigma. \quad \text{(A.2.46)}$$

Contrary to the single-component case, this expression cannot be simplified so as to eliminate the surface excess terms because at most one of the N_i^σ terms can be made to vanish by appropriate location of the dividing surface. This fact is frequently overlooked. In deriving (A.2.46) it is assumed that the formation of the embryo β does not change the composition of the mother phase α. In other words, the embryo is small relative to the mother phase.

When the embryo is incompressible, for example, a droplet, we can write

$$\mu_i^\beta \left(P^\beta\right) = \mu_i^\beta \left(P\right) + \overline{v}_i^\beta \left(P^\beta - P\right), \quad \text{(A.2.47)}$$

where $\overline{v}_i^\beta$ is the partial molar volume of component i in phase β. Then the work of embryo formation simplifies to

$$\delta U = W_{\min} = \sigma F + \sum_i \left[\mu_i^\beta \left(P\right) - \mu_i^\alpha \left(P\right)\right] N_i^\beta$$
$$+ \sum_i \left[\mu_i^\sigma - \mu_i^\alpha \left(P\right)\right] N_i^\sigma. \quad \text{(A.2.48)}$$

The last summation is often incorrectly left out in the nucleation literature.

The composition of the critical nucleus follows from applying to (A.2.48) the conditions $(i = 1, \ldots, n)$

$$\frac{\partial W_{\min}}{\partial N_i^\beta} = 0, \tag{A.2.49}$$

$$\frac{\partial W_{\min}}{\partial N_i^\sigma} = 0. \tag{A.2.50}$$

(A.2.49) yields

$$\frac{\partial W_{\min}}{\partial N_j^\beta} = \sigma \frac{\partial F}{\partial N_j^\beta} + \Delta\mu_j + F \frac{\partial \sigma}{\partial N_j^\beta} + \sum_i N_i^\sigma \frac{\partial \mu_i^\sigma}{\partial N_j^\beta} + \sum_i N_i^\beta \frac{\partial \mu_i^\beta}{\partial N_j^\beta} \tag{A.2.51}$$

and (A.2.50) yields

$$\frac{\partial W_{\min}}{\partial N_j^\sigma} = \sigma \frac{\partial F}{\partial N_j^\sigma} + \Delta\tilde{\mu}_j + F \frac{\partial \sigma}{\partial N_j^\sigma} + \sum_i N_i^\sigma \frac{\partial \mu_i^\sigma}{\partial N_j^\sigma} + \sum_i N_i^\beta \frac{\partial \mu_i^\beta}{\partial N_j^\sigma}, \tag{A.2.52}$$

where

$$\Delta\mu_j = \mu_j^\beta (P) - \mu_j^\alpha (P), \tag{A.2.53}$$

$$\Delta\tilde{\mu}_j = \mu_j^\sigma - \mu_j^\alpha (P), \tag{A.2.54}$$

The sum of the third and fourth terms on the right-hand side in both (A.2.51) and (A.2.52) vanishes because of (A.2.21). The fifth terms cancel because of the Gibbs-Duhem equation. Finally,

$$F = 4\pi r^2 = 4\pi \left[\frac{3}{4\pi} \left(\sum_i \overline{v}_i N_i^\beta \right) \right]^{2/3}, \tag{A.2.55}$$

$$\frac{\partial F}{\partial N_j^\beta} = \frac{2\overline{v}_j^\beta}{r}, \tag{A.2.56}$$

$$\frac{\partial F}{\partial N_j^\sigma} = 0. \tag{A.2.57}$$

Using (A.2.56) in (A.2.51) and (A.2.57) in (A.2.52) yields the conditions of equilibrium:

$$\Delta\mu_j + \sigma \frac{\partial F}{\partial N_j^\beta} = 0 \Rightarrow \mu_j^\beta \left(P^\beta \right) = \mu_j^\alpha (P), \tag{A.2.58}$$

$$\Delta\tilde{\mu}_j = 0 \Rightarrow \mu_j^\sigma = \mu_j^\alpha (P). \tag{A.2.59}$$

As shown in Chapter 3, the critical nucleus is in unstable equilibrium with the mother phase. The equality of chemical potentials between the mother phase and the critical nucleus, (A.2.58), arises naturally from (A.2.48), which is rigorous for incompressible embryos.

Returning now to (A.2.45)–(A.2.47), if phase β is in equilibrium with α, the reversible work becomes

$$\delta U \equiv W_{\min} = \sigma F + \left(P - P^\beta\right) V^\beta. \tag{A.2.60}$$

If the new phase is spherical, we invoke (A.2.35), Laplace's equation, to obtain

$$W_{\min} = \frac{4\pi r^2 \sigma}{3} = \frac{2\pi r^3 \left(P^\beta - P\right)}{3} = \frac{16\pi\sigma^3}{3\left(P^\beta - P\right)^2} \tag{A.2.61}$$

(Gibbs, 1875–76, 1877–78, 1961). Equation (A.2.61) gives the reversible work of forming a spherical critical nucleus. It is completely general, and does not depend on any assumptions as to the nucleus's incompressibility.

A.2.3 SURFACE ENRICHMENT

In order to calculate the nucleation rate, it is necessary to determine the critical nucleus's size and composition. The nucleus, in classical theory, is assumed homogeneous. Because real nuclei are not uniform, the proper, thermodynamically consistent way of determining the nucleus size and composition within the classical theory is an important problem. It is often referred to as the surface enrichment problem in the nucleation literature (Wilemski, 1984, 1987).

We consider as an illustration the condensation of a binary vapor mixture. Treating the liquidlike embryos as incompressible, the thermodynamically consistent set of equations needed to calculate the composition, size, and work of formation of the critical nucleus is

$$\mu_1^\beta (P) + \left(P^\beta - P\right) \bar{v}_1^\beta = \mu_1^\alpha (P), \tag{A.2.62}$$

$$\mu_2^\beta (P) + \left(P^\beta - P\right) \bar{v}_2^\beta = \mu_2^\alpha (P), \tag{A.2.63}$$

$$r^* = \frac{2\sigma}{\left(P^\beta - P\right)}, \tag{A.2.64}$$

$$W_{\min}^* = \frac{4\pi}{3}\sigma \left(r^*\right)^2, \tag{A.2.65}$$

where the superscript * denotes the critical nucleus, β denotes the liquid, and α the vapor. (A.2.62) and (A.2.63) are used to solve for P^β and x_1^β (mole fraction of component 1). Then, with σ evaluated at the calculated composition, the size and work of formation of the critical nucleus are obtained from (A.2.64) and (A.2.65). The above equations are in one-to-one correspondence with (A.2.58), (A.2.35), and (A.2.61). The key point here is that the starting point, (A.2.48), is correct and thermodynamically consistent. If, as is usually done in the nucleation literature, one starts instead from the incorrect equation by truncating (A.2.48) through removal of the surface terms, erroneous conditions of equilibrium are obtained instead of (A.2.58) (Doyle, 1961; Reiss and Shugard, 1976).

These inconsistencies were noted by Wilemski (1984, 1987), who argued that the correct equations to be used in the calculation of the size and composition of the critical nucleus are (A.2.58), which indeed they are. Interestingly, Wilemski started with the incorrect equation, namely, the truncated version of (A.2.48), but he arrived at the correct result by partitioning the embryo into bulk and surface portions.

References

Doyle, G.J. 1961. "Self-Nucleation in the Sulfuric Acid–Water System." *J. Chem. Phys.* 35: 795.

Gibbs, J.W. 1875–76 and 1877–78. "On the Equilibrium of Heterogeneous Substances." *Trans. Conn. Acad.* III: 108 (Oct. 1875–May 1876) and III: 343 (May 1877–July 1878).

Gibbs, J.W. 1961. *The Scientific Papers of J. Willard Gibbs, Ph.D., LL.D. I. Thermodynamics*, pp. 219–58. Dover: New York.

Reiss, H., and M. Shugard. 1976. "On the Composition of Nuclei in Binary Systems." *J. Chem. Phys.* 65: 5280.

Wilemski, G. 1984. "Composition of the Critical Nucleus in Multicomponent Vapor Nucleation." *J. Chem. Phys.* 80: 1370.

Wilemski, G. 1987. "Revised Classical Binary Nucleation Theory for Aqueous Alcohol and Acetone Vapors." *J. Phys. Chem.* 91: 2492.

Appendix 3

Definitions of Microscopic and Statistical
Quantities

A.3.1 ENSEMBLES AND THERMODYNAMIC POTENTIALS

The relationship between molecular interactions and bulk properties is established by Boltzmann's entropy formula,

$$S(N, V, E) = k \ln \Omega. \qquad (A.3.1)$$

In this equation, S is the entropy, k is Boltzmann's constant (1.38×10^{-23} J K^{-1}), and Ω is the number of quantum states accessible to N particles with fixed energy E confined to a volume V. The independent variable set (N, V, E) corresponds to an isolated system.

A very large number of imaginary copies of the system under study, subject to the same macroscopic constraints chosen to specify the system's equilibrium thermodynamic properties, is called an ensemble (Gibbs, 1981). All members of an ensemble have the same bulk properties but, at any given instant, each may be in a different quantum state. An ensemble of isolated systems having a given number of particles, volume, and energy (N, V, E) is called microcanonical.

A basic postulate of statistical mechanics states that the value of a bulk property of a system at equilibrium can be calculated by either time or ensemble averaging, and the result will be the same. In the former case, the average is calculated over the values taken by the property of interest (e.g., pressure) in the given system at different instants. In the latter case, the average is instantaneous and is taken over the members of the ensemble, each quantum state being weighted by its probability of occurrence. A second basic postulate states that all quantum states in a microcanonical ensemble are equally probable.

For practical calculations, independent variables other than those of a microcanonical ensemble are often more convenient. Consider a closed system of fixed volume whose temperature is fixed by contact with a thermal reservoir (Appendix 1). The independent variables are then (N, V, T). These macroscopic quantities define the canonical ensemble. Let the composite (system + reservoir) be isolated. The probability of observing the system of interest in a given quantum state ν is proportional to the number of states available to the reservoir that are consistent with state ν,

$$p_\nu \propto \Omega\left(E - E_\nu\right). \tag{A.3.2}$$

Using (A.3.1) and the identity[1]

$$dS = T^{-1}dU + PT^{-1}\,dV - \mu T^{-1}\,dN, \tag{A.3.3}$$

we write

$$\ln\Omega\left(E - E_\nu\right) = \ln\Omega\left(E\right) - E_\nu\left(\frac{\partial \ln\Omega}{\partial E}\right)_{N,V}$$

$$= \ln\Omega\left(E\right) - \frac{E_\nu}{kT}, \tag{A.3.4}$$

where the reservoir is chosen to be much larger than the system, so that $E_\nu \ll E$, and the expansion can therefore be truncated at the linear terms. Substituting in (A.3.2),

$$p_\nu \propto \Omega\left(E\right)\exp\left(-E_\nu/kT\right) \equiv C\exp\left(-\beta E_\nu\right), \tag{A.3.5}$$

where $\beta = 1/kT$. To determine the normalization constant, we use the fact that $\sum_\nu p_\nu = 1$,

$$p_\nu = \frac{\exp\left(-\beta E_\nu\right)}{\displaystyle\sum_\nu \exp\left(-\beta E_\nu\right)}. \tag{A.3.6}$$

The denominator in (A.3.6) is called the canonical partition function, $Q(N, V, T)$. Its connection to thermodynamics follows from applying the generalization of Boltzmann's entropy formula to other ensembles (Chandler, 1987),

$$S = -k\sum_\nu p_\nu \ln p_\nu. \tag{A.3.7}$$

Substituting (A.3.6) in (A.3.7), we obtain

$$Q\left(N, V, T\right) = \sum_\nu \exp\left(-\beta E_\nu\right) = \exp\left[-\beta A\left(N, V, T\right)\right]. \tag{A.3.8}$$

Thus the canonical partition function is related to the Helmholtz energy. Thermodynamic quantities are obtained by differentiation of the partition function,

$$P = -\left(\frac{\partial A}{\partial V}\right)_{T,N} = kT\left(\frac{\partial \ln Q}{\partial V}\right)_{T,N}, \tag{A.3.9}$$

$$\mu = \left(\frac{\partial A}{\partial N}\right)_{T,V} = -kT\left(\frac{\partial \ln Q}{\partial N}\right)_{T,V}, \tag{A.3.10}$$

---

[1] We use the symbol U to denote the average energy, and E to denote the instantaneous energy. For a microcanonical system, both quantities are identical.

$$U = \left[\frac{\partial (A/T)}{\partial (1/T)} \right]_{N,V} = kT^2 \left[\frac{\partial \ln Q}{\partial T} \right]_{N,V}, \qquad (A.3.11)$$

$$S = - \left[\frac{\partial A}{\partial T} \right]_{N,V} = k \left[\ln Q + \left(\frac{\partial \ln Q}{\partial \ln T} \right)_{N,V} \right]. \qquad (A.3.12)$$

The thermodynamic potential corresponding to a canonical ensemble is the Helmholtz energy. Like all thermodynamic potentials, it is a Legendre transform of the energy (Chapter 2, Appendix 1):

$$A(T, V, N) = U(S, V, N) - TS. \qquad (A.3.13)$$

Applying the same procedure [i.e., Equations (A.3.2)–(A.3.7)], but extending it to the case where the system is open, isothermal, and has fixed volume,[2] we obtain

$$\Xi(\beta\mu, V, \beta) = \sum_N \exp(\beta\mu N) \sum_\nu \exp(-\beta E_\nu)$$

$$= \exp[\beta PV(\mu, V, T)]. \qquad (A.3.14)$$

The first summation is over all possible values of N, the number of particles in the system. The second summation is the canonical partition function corresponding to a particular number of particles. In the canonical ensemble, the reservoir imposes its temperature. In the grand canonical ensemble, both the temperature and the chemical potential are imposed, the former by thermal contact through nonadiabatic boundaries, and the latter by free exchange of particles. The partition function Ξ is called the grand partition function. It is appropriate for the description of open, isothermal systems in a fixed volume, for example, a fixed region in a fluid. The corresponding ensemble is called the grand canonical ensemble, and its thermodynamic potential is the grand potential, $-PV$, a Legendre transform of the energy,

$$-PV(T, \mu, V) = U(S, N, V) - TS - \mu N,$$

$$d(-PV) = -SdT - PdV - Nd\mu. \qquad (A.3.15)$$

Thermodynamic quantities are obtained by differentiating the partition function,

$$N = \left[\frac{\partial PV}{\partial \mu} \right]_{T,V} = kT \left[\frac{\partial \ln \Xi}{\partial \mu} \right]_{T,V} = \left[\frac{\partial \ln \Xi}{\partial \beta\mu} \right]_{\beta,V}, \qquad (A.3.16)$$

[2]In this case, the analogues of (A.3.2) and (A.3.3) are

$$p_\nu \propto \Omega(E - E_\nu, N - N_\nu),$$

$$\ln \Omega(E - E_\nu, N - N_\nu) = \ln \Omega(E, N) - \beta E_\nu + \beta\mu N_\nu.$$

$$U = \left(\frac{\partial PV}{\partial \ln \mu}\right)_{T,V} + T\left(\frac{\partial PV/T}{\partial \ln T}\right)_{\mu,V} = -\left[\frac{\partial \ln \Xi}{\partial \beta}\right]_{\beta\mu,V}$$

$$= kT\left[\left(\frac{\partial \ln \Xi}{\partial \ln \mu}\right)_{T,V} + \left(\frac{\partial \ln \Xi}{\partial \ln T}\right)_{V,\mu}\right], \tag{A.3.17}$$

$$S = \left[\frac{\partial PV}{\partial T}\right]_{\mu,V} = k\left[\ln \Xi + \left(\frac{\partial \ln \Xi}{\partial \ln T}\right)_{\mu,V}\right]. \tag{A.3.18}$$

If the pressure and temperature of a closed system are fixed, we imagine it interacting with the reservoir across an impermeable boundary that is nonadiabatic and can deform. As a result, the reservoir imposes its temperature and pressure. Following the same steps as before, we obtain

$$Y(N, P, T) = \sum_{\nu} \exp[-\beta(E_\nu + PV_\nu)]$$

$$= \exp[-\beta G(N, P, T)]. \tag{A.3.19}$$

The partition function Y is called the isothermal, isobaric partition function. The corresponding ensemble is called the isothermal, isobaric ensemble, and its thermodynamic potential is the Gibbs energy, also a Legendre transform of the energy:

$$G(T, P, N) = U(S, V, N) - TS - (-P)V = U - TS + PV. \tag{A.3.20}$$

Thermodynamic quantities are obtained by differentiation of the partition function,

$$V = \left(\frac{\partial G}{\partial P}\right)_{T,N} = -kT\left[\frac{\partial \ln Y}{\partial P}\right]_{T,N}, \tag{A.3.21}$$

$$H = \left(\frac{\partial G/T}{\partial 1/T}\right)_{P,N} = kT\left[\frac{\partial \ln Y}{\partial \ln T},\right]_{P,N}, \tag{A.3.22}$$

$$S = -\left(\frac{\partial G}{\partial T}\right)_{P,N} = k\left[\ln Y + \left(\frac{\partial \ln Y}{\partial \ln T}\right)_{P,N}\right]. \tag{A.3.23}$$

Boltzmann's entropy formula (A.3.1) and its generalization (A.3.7) provide the connection between the microscopic and the continuum domains. To each thermodynamic potential there corresponds an ensemble and a partition function. Additional ensembles are discussed by Hill (1956).

A.3.2 THE RADIAL DISTRIBUTION FUNCTION

The positions of particles in a fluid are correlated because of energetic and packing effects. Given that there is a particle at the origin of an arbitrary coordinate system, the other particles are not randomly distributed around it. The radial distribution (or pair correlation) function quantifies the correlations between pairs of particles. Consider a particle at position $r = 0$. The incremental number of particles located within a spherical shell of thickness dr, located a distance r away from the origin, is given by

$$dN(r) = \left[4\pi r^2 dr\right]\rho g(r). \tag{A.3.24}$$

The term in brackets is the volume of the spherical shell, and ρ is the bulk density. Their product yields the average number of particles in the spherical shell at the bulk density. The density in the spherical shell differs from the bulk density because of the particle at the origin. The radial distribution function $g(r)$ is the ratio of local to bulk density a distance r away from the origin, given that there is a particle at the origin. If there are no interparticle correlations (i.e., in an ideal gas), $g(r) = 1$. The radial distribution function vanishes at short separations, since two particles cannot overlap, and it tends to 1 for large enough r. This is because the influence of the particle located at the origin must vanish sufficiently far away from it.

Knowledge of the radial distribution function allows the calculation of thermodynamic properties, according to the relations

$$\frac{\beta U}{N} = \frac{3}{2} + 2\pi\beta\rho \int r^2 g(r)\,\phi(r)\,dr, \tag{A.3.25}$$

$$\frac{\beta P}{\rho} = 1 - \frac{2\pi\beta\rho}{3} \int r^3 g(r)\,\phi'(r)\,dr, \tag{A.3.26}$$

$$\rho kT K_T = 1 + 4\pi\rho \int r^2\left[g(r) - 1\right]dr, \tag{A.3.27}$$

$$\beta\mu = \ln \rho\Lambda^3 + 4\pi\beta\rho \int_0^1 d\xi \int r^2 \phi(r)g(r,\xi)\,dr \tag{A.3.28}$$

(McQuarrie, 1976). In (A.3.25), it is assumed that the total interparticle energy is the sum over all pair energies (pairwise additivity). The quantity $3/2$ is the energy per particle in a monatomic ideal gas, in units of kT. (A.3.26) also assumes pairwise additivity, and ϕ' denotes $d\phi/dr$. (A.3.27) is an identity, and it does not depend on the assumption of pairwise additivity. In (A.3.28), μ is the chemical potential, Λ is the de Broglie wavelength $h/(2\pi mkT)^{1/2}$, with h Planck's constant, and m the mass of a particle. The parameter ξ ($0 \leq \xi \leq 1$), called the coupling parameter (Kirkwood, 1935), is defined in such a way that

$$\phi(\{\mathbf{r}\}, \xi) = \sum_{2 \leq i < j \leq N} \phi_{ij} + \xi \sum_{j=2}^N \phi_{1j}, \tag{A.3.29}$$

where $\phi(\{\mathbf{r}\})$ denotes the total potential energy, which depends on the positions of every particle, and ϕ_{ij} is the contribution to the total potential energy due to the interaction between particles i and j. As ξ varies between 0 and 1, particle 1 is "switched on." When $\xi = 1$, the system is unchanged; when $\xi = 0$, particle 1 does not interact with the remaining $(N - 1)$ particles, hence the system contains only $(N - 1)$ particles. (A.3.28) assumes pairwise additivity.

Experimental information about $g(r)$ is obtained by x-ray scattering. The measured quantity is called the structure factor

$$S(k) = 1 + \rho \int d\mathbf{r} \, [g(r) - 1] \, e^{i\mathbf{k} \cdot \mathbf{r}}, \tag{A.3.30}$$

where

$$k = |\mathbf{k}| = (4\pi / \lambda) \sin (\theta / 2) . \tag{A.3.31}$$

In the above equation, $\mathbf{k}$ is the momentum transfer vector of scattered radiation, λ is the wavelength of the incident x rays, and θ is the scattering angle. The structure factor $S(k)$ is proportional to the intensity of radiation detected by an observer located at an angle θ with respect to the incident radiation.

It is often convenient to decompose interparticle correlations into a direct and an indirect part, by defining a direct correlation function $c(r)$

$$g(r_{12}) - 1 = c(r_{12}) + \rho \int c(r_{13}) \, [g(r_{23}) - 1] \, d\mathbf{r}_3, \tag{A.3.32}$$

where r_{ij} denotes the distance between particles i and j (Ornstein and Zernike, 1914). The above definition says that the influence of particle 1 on particle 2 consists of a direct part $c(r_{12})$, and an indirect part. The latter takes into account the influence of particle 1 on particle 2 through a third particle 3, summed over all such third particles. Taking the Fourier transforms of (A.3.32), we obtain

$$\hat{h}(\mathbf{k}) = \frac{\hat{c}(\mathbf{k})}{1 - \rho \hat{c}(\mathbf{k})}, \tag{A.3.33}$$

where the caret denotes a Fourier transform, and $h = g - 1$. Combining (A.3.33) and (A.3.27), we obtain

$$\rho k T K_T = S(0) = 1 + \rho \hat{h}(0) = \frac{1}{1 - \rho \hat{c}(0)}. \tag{A.3.34}$$

Analytical techniques for calculating the radial distribution function constitute an important branch of statistical mechanics known as integral equation theory. McQuarrie (1976) provides a good introductory discussion.

A.3.3 SPHERICALLY SYMMETRIC POTENTIALS

The Lennard-Jones potential is given by

$$\phi(r) = 4\varepsilon \left[\left(\frac{\sigma}{r} \right)^{12} - \left(\frac{\sigma}{r} \right)^{6} \right], \tag{A.3.35}$$

where ϕ is the potential energy corresponding to an interatomic separation r. The minimum value of ϕ is $-\varepsilon$. It occurs at $r/\sigma = 2^{1/6}$, where σ is the separation at which the potential is zero.

The soft sphere potential is given by

$$\phi(r) = 4\varepsilon \left(\frac{\sigma}{r}\right)^n. \tag{A.3.36}$$

When $n = 12$, this corresponds to the repulsive part of the Lennard-Jones potential.

The hard sphere potential is given by

$$\phi(r) = \begin{cases} \infty, & r \leq \sigma, \\ 0, & r > \sigma. \end{cases} \tag{A.3.37}$$

References

Chandler, D. 1987. *Introduction to Modern Statistical Mechanics*. Chap. 3. Oxford University Press: New York.

Gibbs, J.W. 1981. *Elementary Principles in Statistical Mechanics*. Chap. 4. Ox Bow Press: Woodbridge, N.J.

Hill, T.L. 1956. *Statistical Mechanics. Principles and Selected Applications*. Chap. 3. McGraw-Hill: New York.

Kirkwood, J.G. 1935. "Statistical Mechanics of Fluid Mixtures." *J. Chem. Phys.* 3: 300.

McQuarrie, D.A. 1976. *Statistical Mechanics*. Chap. 13. Harper and Row: New York.

Ornstein, L.S., and F. Zernike. 1914. "Accidental Deviations of Density and Opalescence at the Critical Point of a Single Substance." *Proc. Akad. Sci. (Amsterdam)*. 17: 793.

Index

About the Author

Pablo Debenedetti is Professor of Chemical Engineering at Princeton University.

CPSIA information can be obtained
at www.ICGtesting.com
Printed in the USA
BVHW040803140420
577480BV00018B/55